THE OXFORD ENGINEERING SCIENCE SERIES

THE OXFORD ENGINEERING SCIENCE SERIES

1 D. R. Rhodes: *Synthesis of planar antenna sources*
2 L. C. Woods: *Thermodynamics of fluid systems*
3 R. N. Franklin: *Plasma phenomena in gas discharges*
4 G. A. Bird: *Molecular gas dynamics*
5 H.-G. Unger: *Optical wave guides and fibres*
6 J. Heyman: *Equilibrium of shell structures*
7 K. H. Hunt: *Kinematic geometry of mechanisms*
8 D. S. Jones: *Methods in electromagnetic wave propagation*
9 W. D. Marsh: *Economics of electric utility power generation*
10 P. Hagedorn: *Non-linear oscillations*
11 R. Hill: *Mathematical theory of plasticity*
12 D. J. Dawe: *Matrix and finite element displacement analysis of structures*
13 N. W. Murray: *Introduction to the theory of thin-walled structures*
14 R. I. Tanner: *Engineering rheology*
15 M. F. Kanninen and C. H. Popelar: *Advanced fracture mechanics*
16 R. H. T. Bates and M. J. McDonnell: *Image restoration and reconstruction*
18 K. Huseyin: *Multiple-parameter stability theory and its applications*
19 R. N. Bracewell: *The Hartley transform*

Multiple Parameter Stability Theory and its Applications

Bifurcations, Catastrophes, Instabilities, . . .

KONCAY HUSEYIN
Department of Systems Design Engineering,
University of Waterloo, Ontario

CLARENDON PRESS · OXFORD
1986

Oxford University Press, Walton Street, Oxford OX2 6DP
Oxford New York Toronto
Delhi Bombay Calcutta Madras Karachi
Kuala Lumpur Singapore Hong Kong Tokyo
Nairobi Dar es Salaam Cape Town
Melbourne Auckland
and associated companies in
Beirut Berlin Ibadan Nicosia

Oxford is a trade mark of Oxford University Press

Published in the United States
by Oxford University Press, New York

British Library Cataloguing in Publication Data
Huseyin, Koncay
Multiple parameter stability theory and its applications: bifurcations, catastrophes, instabilities. —(The Oxford engineering science series)
1. System analysis 2. Nonlinear oscillations
I. Title
003 QA402
ISBN 0-19-856170-9

Library of Congress Cataloging in Publication Data
Huseyin, K. (Koncay), 1936–
Multiple parameter stability theory and its applications
(The Oxford engineering science series; 18)
Bibliography: p.
Includes index.
1. System analysis. 2. Stability. I. Title.
II. Series.
QS402.H85 1986 003 85-21600
ISBN 0-19-856170-9

Filmset and printed in Northern Ireland by The Universities Press (Belfast) Ltd.

To
my family

PREFACE

Real systems are usually under the influence of several parameters and are liable to exhibit various forms of instabilities as the parameters are varied. Such developments often entail drastic changes in the state of a system, and it is therefore of paramount importance for engineers and scientists to be able to understand, predict, and account for such phenomena in various contexts. The ability to perceive and analyse specific problems can be greatly enhanced by a good understanding of the fundamental concepts common to certain classes of systems, possible routes to instabilities, and associated behaviour patterns. A firm grasp of underlying methodologies is of course also essential.

With this motivation, and inspired by Koiter's work, a multiple-parameter non-linear theory of stability for discrete *potential* systems was first developed in a Ph.D. thesis (Huseyin 1967) which was later presented in a monograph (Huseyin 1975). This general theory, however, is restricted to potential (conservative) systems which lend themselves readily to classical variational principles and techniques. Many systems do not belong to this class, and require a different formulation. The main goal of this book is to present a parallel non-linear theory for such *non-potential* systems which can exhibit *dynamic* as well as *static instabilities*. This is achieved within the framework of an autonomous formulation. The emphasis is on the instabilities and incipient bifurcations associated with a certain *equilibrium path* (or *surface*) of the system. Dynamic instabilities are linked to the emergence of a family of periodic motions, while static instabilities involve equilibrium states only.

Chapter 1 is designed as an introduction to the basic notions and definitions of the stability theory, mathematical modelling of real systems, and certain classifications. A deliberate effort is made to clarify certain familiar terms which are often used interchangeably. For instance, there seems to be some confusion concerning the relationship (or differences) between 'conservative systems' and 'gradient systems'. Similarly, what is the connection between 'potential systems' and the above systems? What do we exactly mean by 'non-potential systems'? Which class of systems is catastrophe theory associated with? These and related questions are tackled in Section 1.6 from a point of view that is relevant to the developments of the subsequent chapters. The concepts of *static* and *dynamic* instability play a central role in the definitions given in this section, and the rest of the book is designed around these concepts.

A concise but refined and expanded general theory of stability for potential systems is presented in Chapter 2. A number of relatively more complex topics not covered in a previous book (Huseyin 1975) are treated here fully, and most of the earlier results are recoverable from the general theory presented in this chapter, as a special case. The connection to catastrophe theory, and the important questions concerning the imperfection sensitivity of certain critical points receive special attention.

The rest of the book is concerned with autonomous systems which embrace non-potential problems. Chapter 3 is devoted to the study of the static instabilities associated with such systems, and it starts with a brief description of the transformations to be used in the rest of the book. The formulation brings out the distinct features of this class of systems as compared to potential problems. The underlying methodology, however, remains basically the same as in Chapter 2, which revolves around a multiple-parameter perturbation technique. The technique yields *explicit* results concerning the post-critical behaviour of the systems considered. It is thus demonstrated analytically that many behaviour characteristics of potential systems are also exhibited by these more general systems—including the familiar catastrophe surfaces associated with gradient systems. The analysis is quantitative and yields the asymptotic equations of the equilibrium surfaces fully. While the analogies between the behaviour patterns are emphasized, certain important differences are also explored. It is demonstrated in Section 3.2, for example, that vanishing eigenvalues of the Jacobian do not necessarily mean static instability, as in the case of potential systems. Indeed a *general* critical point associated with a Jordan block of order two (with a two-fold zero eigenvalue) may appear to be at the threshold of a dynamic instability on certain equilibrium paths of a multiple-parameter system. This may have serious implications with regard to the stability properties of the equilibrium surface, and a detailed discussion of the phenomenon is presented in Sections 3.2.2 and 3.2.3. Imperfection sensitivities of *special* critical points are also discussed.

Chapter 4 is concerned with the dynamic instability phenomena, and starts with a discussion of the methodology. The multiple-parameter perturbation technique, which plays a central role in the treatment of the equilibrium configurations, is generalized and blended with the classical harmonic balancing method to produce a new approach for the analysis of the oscillatory instabilities and associated limit cycles. *The intrinsic harmonic balancing* technique created this way has a number of advantages. First of all, it eliminates the observed inconsistencies of the conventional harmonic balancing method, and provides for a reliable and conceptually simple perturbation method of analysis. Secondly, it comes as close as possible to unifying the methodology underlying both static and dynamic instabilities. A third advantage of the method is that it enables one to obtain *explicit formulas* concerning the behaviour of the system. Thus, the bifurcating paths, frequency-amplitude relationships, and asymptotic equations of the limit cycles are obtained in explicit terms for the case of Hopf bifurcation as well as many degenerate and generalized bifurcation cases.

Chapter 5 is devoted to specific problems, and the applicability of the general results is demonstrated on several problems selected from mechanics, electrical circuits, thermodynamics, bio-chemical processes, aerodynamics and urban systems.

The final chapter contains some observations about the seemingly erratic behaviour (*chaos*) of systems.

The characteristic features of this monograph may be summarized as follows:

1. It presents a *general* non-linear theory concerning the stability and instability behaviour of autonomous systems from an original engineering point of view which is relevant, relatively simple, and *applicable*.

2. Both the static and the dynamic instability phenomena are explored systematically. The methodology underlying this treatment comes as close as possible to *unifying* the analyses of these two distinct behaviour patterns. The emphasis is on analytical procedures which lead to construction of asymptotic solutions. The mathematical methods which are more suitable for proving theorems, and the associated jargon, are systematically avoided. Topological implications of the analytical results are discussed and illustrated. The analyses are essentially real.

3. Another distinctive feature of the theory treated in this book is that it does not only outline or describe general analytical approaches for the analyses of various phenomena but actually carries out the analyses in general terms. Thus, the results are *explicit* and include many useful formulae which can readily be applied to specific problems without necessarily repeating the entire analysis in each case. The applicability of the results is illustrated on several examples drawn from various disciplines. There are of course drawbacks associated with such a detailed analysis. The main problem is concerned with the notation, and a deliberate effort is made to define all symbols fully where they first appear in the text.

It is expected that these features of the book will appeal to a broad segment of the scientific community. Although it is envisaged as a monograph, the book is suitable for adoption as a text at graduate or senior undergraduate levels in a variety of departments—including mechanics, civil engineering, mechanical and aeronautical engineering, electrical engineering, systems engineering and engineering science, physics, chemistry and chemical engineering, applied mathematics, and other disciplines. The background required from a student is that of a

B.Sc. programme with emphasis on applied mechanics and mathematics.

Finally, I would like to express my gratitude to Professor W. H. Wittrick for his encouragement. I am indebted to Professor R. H. Plaut who has again performed a meticulous job in reading the manuscript and pointing out numerous corrections. Thanks are also due to Dr. V. Mandadi, Dr. A. S. Atadan and Mr. P. Yu whose contributions are reflected in this book. I am grateful to Miss Cathy Seitz and Mrs. Lise Permezel who patiently typed the manuscript. It is a pleasure to acknowledge the financial support of the National Science and Engineering Research Council of Canada, and Alexander von Humboldt-Stiftung of the Federal Republic of Germany.

Waterloo K.H.
January 1985

CONTENTS

1
INTRODUCTION

1.1 Introductory remarks

A variety of situations associated with natural as well as man-made systems and processes may be identified as instabilities, if the events are observed and evaluated properly. Indeed, some form of instability is not an uncommon event in the evolutionary processes of nature, and many engineering systems are liable to exhibit catastrophic instabilities under certain circumstances, unless such dangerous developments are prevented 'by design'. It is, therefore, of paramount importance to understand, predict, and account for instability phenomena in various contexts.

Generally, a *state* of a system can be described as the information characterizing it at a given time. On the other hand, real systems are usually under the influence of several independent parameters which may vary over certain specified sets, and the concept of instability is linked to the way the states evolve as the parameters vary slowly. Many systems exhibit smooth behaviour; that is, the states evolve slowly and smoothly in a certain region. As the parameters continue to vary smoothly, however, a critical stage may be reached, at which point the system exhibits a sudden jump from one state to another. This behaviour is often linked to instability of the original state, making it necessary for the system to seek and assume another stable state which may be qualitatively different from the original one. Thus, some slender structural systems buckle (and may collapse) and others flutter, turbulence occurs in a fast-moving fluid, water suddenly boils and ice melts, the aircraft at a high angle of attack starts oscillating, the density of population undergoes a saccadic leap, new patterns form as a self-organizing system evolves, rotating shafts and systems develop whirling divergence, pipelines flutter, electrical circuits exhibit self-excited oscillations, uncorrelated waves of ordinary light transform into coherent light in a laser, the light emission of a laser becomes chaotic (or turbulent), and giant stars collapse. These are examples of instabilities encountered in our environment.

Against such a broad spectrum and diversity, unifying analytical methods to study the phenomenon may not be easy to develop; nevertheless, there is a profound conceptual unity, and a thorough understanding of the basic concepts and principles underlying the instability phenomena in general is indispensable for effective analyses of specific problems. Such an overall view and insight can best be provided through a general theory which classifies the systems and delineates various instability phenomena systematically while establishing a framework and methodology for the analysis of specific systems—or classes of systems—at the same time.

There are a number of fundamental issues central to all instability phenomena. For instance, knowledge of *the stability boundary* and the values of parameters at which a sudden jump takes place is of vital concern to scientists and engineers. How can this boundary be determined and what are its basic properties? Does the abrupt transformation involve a fundamental change in the character of the original state? What are the properties of the resulting new state(s)? Since the analysis of post-instability behaviour is essential for a more complete understanding of the system, what would be the most appropriate approach for this purpose? These and similar problems can be tackled in a variety of ways. Clearly, the mathematical modelling of real systems may depend on the information actually sought as well as the reality itself. A linear formulation and analysis, for instance, may be capable of yielding useful information about the stability limit of certain systems, while the post-instability behaviour can only be examined via a non-linear analysis. More fundamentally, the character of interactions in a real system or process may be the most crucial factor—either by itself or in conjunction with the prescribed objectives—in determining the type of mathematical formulation and the most appropriate method of analysis. In this regard, it is well known, particularly in physical sciences, that many systems lend themselves to treatment through conventional variational principles associated with a potential function. The systems in this class are variously described as 'potential systems', 'gradient systems', or 'conservative systems'. In mechanics, for example, the systems which possess a potential energy—such as an elastic body subjected to gravitational dead loads—fall within the scope of this classification. On the other hand, an elastic body subjected to follower forces constitutes a non-potential system, since such forces generally cannot be derived from a well-defined potential energy. Follower forces—also called *circulatory* or *polygenic*—actually follow the configuration of the system according to a certain law (e.g., remain tangential to a trajectory or surface), and the work done by such forces during a displacement often depends on the manner in which the system attains a final state, and not just on the initial and final states as in the case of conservative forces (e.g., Bolotin 1963).

It follows from the above discussion that *potential systems* and *non-potential systems* may be identified as two broad classes of systems in order to facilitate an in-depth study of the salient features associated with each class via appropriate methods. Indeed, this classification has played a significant role in the development of bifurcation and stability theories in mechanics over the years, following the pioneering works of Euler

(1744) and Poincaré (1885). It is important to observe here that, although 'stability' is a dynamical concept by definition (e.g., Lyapunov's original definition), an essentially statical analysis based on the variational principles associated with a potential function can effectively elucidate equilibrium and stability properties of a potential system. A dynamical formulation and analysis is, of course, necessary if full information, including the transient oscillations related to instabilities, is required. In other words, different behaviour patterns may be explored within different mathematical frameworks.

In mechanics, the non-linear concepts of elastic stability associated with conservative (potential) systems have been developed substantially by Koiter (1945) within the context of statics. This classical work is largely based on the potential energy and its extremum properties. Thus, *the principle of virtual work* and *energy criterion* play a central role in obtaining information about the equilibrium states, stability, post-critical behaviour, and imperfection sensitivities. Following Koiter's stimulating work, a group of researchers was formed in the early 1960s at University College London by Henry Chilver (now Sir Henry, Vice-Chancellor of the Cranfield Institute of Technology, England). The work of this group concerning bifurcation and stability behaviour of conservative discrete systems has been extensive, and is described in a number of books and review articles which also contain lists of references. The books by Croll and Walker (1972), and Britvec (1973) are mostly concerned with a treatment of structural elements rather than a general theory. A review article by Roorda (1972) and a book edited by Supple (1973) outline some of the developments, and the books by Thompson and Hunt (1973) and Huseyin (1975) present most of the early work systematically. The latter also treats multiple-parameter systems and imperfection sensitivity of coincident critical points.

A general non-linear theory of stability associated with multiple-parameter conservative systems was first presented in a Ph.D. thesis (Huseyin 1967) which led to extensive developments of the basic concepts underlying the non-linear behaviour of a variety of conservative systems. An account of these developments can be found in the monograph mentioned above (Huseyin 1975). Essentially, this theory recognizes the inadequacy of *limit* and *bifurcation* points—associated with the topological properties of equilibrium paths—in describing the behaviour characteristics of a system under the influence of several independent parameters, and introduces new concepts, definitions, and classifications based on distinct forms of *an equilibrium surface*—a concept that arises in connection with multiple parameters. Catastrophe theory, developed in the context of Topology (Thom 1975), basically parallels this approach, and some of the elementary catastrophes can readily be linked to the *general* critical points introduced in 1967. The quantitative nature of the multiple-parameter stability theory and its advantages in engineering applications are discussed in a number of papers (Huseyin 1977, 1978*a*), and will also be explored in Chapter 2.

As in the case of elementary catastrophes, however, the non-linear theory presented in the monograph noted above (Huseyin 1975) is limited in scope to potential systems. Many systems are not in this category, and the work concerning non-potential systems has been largely confined to linear theories until recently (Ziegler 1956; Herrmann 1967; Leipholz 1970, 1977; Huseyin 1978*b*). If a system involves non-potential forces or, more generally, non-potential interactions, the convenience of basing the formulation and the analysis on a well-defined potential function—underlying the basic properties of the system—may be lost, and other appropriate models and methods are sought. As a matter of fact, non-potential systems can exhibit oscillatory instabilities, and a dynamical formulation becomes essential in such cases. This point must be kept in mind if the efforts are directed at establishing certain new variational principles.

A linear dynamical theory concerning multiple-parameter as well as one-parameter systems appeared in 1978 for the first time in book form (Huseyin 1978*b*). In this monograph, free vibrations and stability properties of conservative, pseudo-conservative, gyroscopic, and circulatory systems (identified physically as well as mathematically) are explored with particular emphasis on characteristic surfaces, stability boundaries, and the effect of internal and external damping on the response characteristics. Dynamic *flutter* instability and static *divergence* instability are central to the entire treatment. A more complete picture, however, can only be obtained through a non-linear formulation, and the primary objective of the present monograph is to fill the obvious gap by providing a general non-linear theory of stability capable of treating non-potential systems under the influence of several independent parameters. This will be done within the framework of an autonomous formulation, embracing both dynamical and statical behaviour. This volume is complementary to the earlier monographs (Huseyin 1975, 1978*b*).

1.2 Modelling of reality

As noted above, a real system may be represented by an appropriate mathematical model to describe, understand, and predict the behaviour of the system. The basis for constructing a model of a system is the fundamental physical laws that are known to prevail. It is important to realize, however, that no system can be modelled exactly, and the complexity of a model may have to be adjusted according to prescribed objectives as well as the tools available for analysis. In other words, a great deal of art and experience may be required in the construction of models. While an overly simplified model will not be capable of reflecting certain vital behaviour characteristics of the system, an overly complex model may obscure important results in a sea of irrelevant details if it can be analysed at all.

The concepts of *genericity* and *structural stability* crop up frequently in connection with our models. Consider, for example, differential equations, which arise quite often in a modelling process; the solutions of these equations describe the behaviour of a system, and in an attempt to classify the solutions, one is bound to be baffled as forbidding pathological cases emerge, one after another. This situation may be avoided by simply discarding exceptional cases, and considering the remaining 'generic' family of solutions. *Genericity*, however, depends on the mathematical universe, and what is generic in one context may be exceptional in a broader context. The concept of *structural stability* arises from a desire to reduce further the complicated generic cases by requiring that certain additional properties are satisfied. Generally speaking, the description of an object should not change if the object itself is varied slightly. A simple example, quoted frequently (e.g. Guckenheimer 1979), is concerned with the number of zeros of a function of a single variable; this number will not change if the function is varied slightly, providing the derivative of the function is non-zero at each zero of the function (Fig. 1.1). This *robustness* property may be interpreted as *structural stability*. A vector field $\boldsymbol{X}$ is structurally stable if it is robust in the above sense such that it has a neighbourhood consisting of vector fields topologically equivalent to $\boldsymbol{X}$. The robustness property is also reflected in the *phase space*.

The concept of structural stability, introduced by Andronov and Pontryagin (1937), plays a central role in the development of *universal unfoldings* associated with elementary catastrophes (Thom 1975). It is certainly important to understand how the behaviour of the system changes if its equations change, and it may seem that the structural stability must be an essential property of all models. Nevertheless, as noted earlier, the mathematical models developed in physical sciences

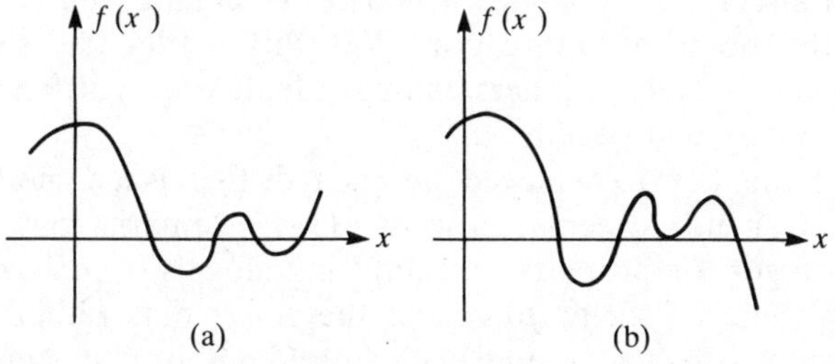

Fig. 1.1 Structural stability: (a) is structurally stable, (b) is structurally unstable.

and engineering for practical solution of problems have their own inherent idealizations, hardly ever representing reality with complete accuracy, and it appears rather unrealistic to require that all models should be structurally stable in any case. Clearly, much valuable information and insight into the behaviour of real systems can be gained from simplifications. Indeed, idealizations are often introduced into a model quite deliberately in an effort to enhance our ability to understand the actual behaviour that seems to be elusive. The literature is full of mathematical models defying the rules underlying the concepts of structural stability or universal unfoldings, yet many such models are considered satisfactory or even excellent when the predictions are compared with experimental results.

Some classical models are also structurally unstable. Consider, for example, the well-known oscillator and its linearized model, the linear oscillator. The differential equation of free oscillations is given by

$$\ddot{x} + \sin x = 0, \qquad (\cdot = \mathrm{d}/\mathrm{d}t), \tag{1.1}$$

or, letting $\dot{x} = y$,

$$\begin{aligned} \dot{x} &= y, \\ \dot{y} &= -\sin x. \end{aligned} \tag{1.2}$$

Similarly, the linear oscillator is described by

$$\begin{aligned} \dot{x} &= y, \\ \dot{y} &= -x. \end{aligned} \tag{1.3}$$

Both models, so central in a variety of disciplines, are structurally unstable, because arbitrarily small perturbations of the vector fields (right-hand sides of the equations) may alter the *phase portrait* qualitatively in the state space of x versus $\dot{x}$. Introduction of dissipation, for example, converts a *centre* into a *focus*.

Next, consider the point of bifurcation at which two distinct equilibrium paths intersect, and an exchange of stabilities occurs (e.g. Huseyin 1975); this model is structurally unstable because a small imperfection alters the topological character of equilibrium curves in the vicinity of the point of bifurcation. Yet this model has served with indisputable distinction in understanding, identifying, and analysing the behaviour of many real systems.

In view of the key role played by such distinguished models in the exploration of diverse systems, Howard's (1979) remarks that 'it appears to be almost a generic property of interesting models of real systems that they are not generic' seems to reflect the sentiments rather neatly. A specific criticism directed at universal unfoldings is that sometimes an irrelevant parameter is introduced, for the sake of structural stability, to construct a diagram which is no longer meaningful. A complete universal unfolding may be incomprehensible, impractical, and indeed irrelevant to engineers if the complexity increases beyond a limit. Conversely, an important parameter may have to be subsumed because the corresponding universal unfolding entails one parameter less than what is actually required to reveal a physically important phenomenon.

It is evident that the concepts of genericity, structural stability, and universal unfoldings have important roles to play in the construction of mathematical models. However, one has to exercise conventional wisdom in applying the principles underlying these concepts so that simplifications and idealizations that can be vital to our understanding of an otherwise incomprehensible real behaviour are not prematurely discarded in a zealous adherence to such principles. This view will be reflected in a number of developments in this volume.

1.3 Mathematical formulations and basic definitions

In this section, some of the ideas and concepts discussed rather loosely in the introduction will be made more precise, and additional definitions will be given.

To start with, it is natural to ask how a system is defined. A variety of definitions may be given, but here a system will be defined simply as a combination of interacting elements. Systems can be classified as natural or man-made, physical or conceptual, statical or dynamical, and in many other ways. The classifications are often introduced for convenience, and to provide insight into their broad range. *Physical systems* manifest themselves in physical terms, and dynamical systems evolve continuously with time. A system, whether natural or man-made, may be represented by a mathematical model as noted earlier, and at least initially a dynamical model has to be constructed since the concept of stability is essentially a dynamical one. Quite often, the basic laws underlying the evolutionary process take the form of a set of differential equations.

One may start by identifying certain observable and measurable quantities which define a *state*. The information characterizing a state of a mechanical system, for example, is often in the form of positions and velocities which are related to the energy of the system. However, the choice of state variables is not unique, and a different set of variables may be chosen as state variables, depending on the system. The selected variables must be independent and must describe the effect of the past history on the future response. Having identified the state variables, one establishes next the basic laws that describe the evolution. In the state-variable approach, this process leads to first order ordinary differential equations if the system is *discrete* (i.e. *lumped system*) such that it can be described by a finite number of state variables. On the other hand, *continuous systems* (i.e. *distributed systems*) do not have a finite set of state variables, and have to be modelled by partial differential equations. However, there are many ways of discretizing a continuous system, and this can be achieved either at the outset (modelling stage) or during the analysis through an appropriate method—such as application of the centre manifold theorem, Lyapunov–Schmidt reduction, Galerkin, or Rayleigh–Ritz procedures. The attention in this book will be focused on systems governed by ordinary differential equations.

Thus, suppose some system is described by a finite set of n variables x^i $(i = 1, 2, \ldots, n)$, where $x^i \equiv x^i(t)$ represent the state of the system at time t, and the state evolves via the vector differential equation

$$\dot{\boldsymbol{x}} = \boldsymbol{X}(\boldsymbol{x}, t), \qquad (\cdot = \mathrm{d}/\mathrm{d}t), \tag{1.4}$$

which can also be written as

$$\dot{x}^i = X_i(x^1, x^2, \ldots, x^n, t). \tag{1.5}$$

Here, the X_i are smooth functions (having continuous derivatives of all orders) of the variables x^i and t. Since time t appears explicitly in the equations, the system is said to be *non-autonomous*. In many applications, however, this is not the case, and eqn (1.4) takes the form

$$\dot{\boldsymbol{x}} = \boldsymbol{X}(\boldsymbol{x}) \tag{1.6}$$

which describes an *autonomous system*. The emphasis in this book is on autonomous problems, and $\boldsymbol{X}(\boldsymbol{x})$ will normally be real unless stated otherwise.

In order to illustrate how these equations actually arise, consider the mechanical system shown in Fig. 1.2, where v_1 and v_2 are velocities of the masses m_1 and m_2, respectively. Selecting y_1, v_1, y_2, v_2 as state variables, assuming that all elements are linear, observing that $\dot{y}_1 = v_1$

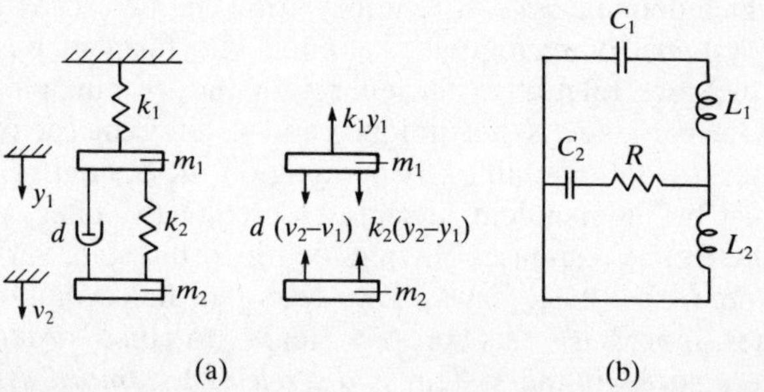

Fig. 1.2 (a) A mechanical system and (b) its electrical analogue.

and $\dot{y}_2 = v_2$, and applying Newton's second law to each mass leads to

$$\begin{aligned} \dot{y}_1 &= v_1, \\ \dot{v}_1 &= \frac{1}{m_1}[d(v_2 - v_1) + k_2(y_2 - y_1) - k_1 y_1], \\ \dot{y}_2 &= v_2, \\ \dot{v}_2 &= \frac{1}{m_2}[-d(v_2 - v_1) - k_2(y_2 - y_1)]. \end{aligned} \tag{1.7}$$

If the masses m_1 and m_2, the spring constants k_1 and k_2, and the coefficient of viscous damping d are known, these linear equations can be solved, provided the initial conditions $y_1(0)$, $v_1(0)$, $y_2(0)$, and $v_2(0)$ are given. Also, note that an alternative set of state variables for this system would be

$$x^1 = y_1, \quad x^2 = v_1, \quad x^3 = y_2 - y_1, \quad \text{and} \quad x^4 = v_2 - v_1,$$

resulting in a linear set of equations of the form $\dot{\boldsymbol{x}} = \boldsymbol{A}\boldsymbol{x}$ where $\boldsymbol{A}$ is a constant matrix in terms of element values.

Suppose now that the dashpot and the springs are non-linear elements such that corresponding forces are characterized by non-linear functions $P_d = P_d(x^4)$, $P_1 = P_1(x^1)$, and $P_2 = P_2(x^3)$, replacing the forces $d(v_2 - v_1)$, $k_1 y_1$, and $k_2(y_2 - y_1)$ on the free-body diagrams, respectively. Newton's second law then yields

$$\begin{aligned} \dot{x}^1 &= x^2, \\ \dot{x}^2 &= \frac{1}{m_1}[P_d(x^4) + P_2(x^3) - P_1(x^1)], \\ \dot{x}^3 &= x^4, \\ \dot{x}^4 &= \frac{1}{m_2}[-P_d(x^4) - P_2(x^3)]. \end{aligned} \tag{1.8}$$

Clearly, for a given set of force functions eqn (1.8) takes the form of eqn (1.6), representing a non-linear autonomous system.

It is interesting to note that the electrical analogue (Fig. 1.2*b*) of the mechanical system considered above can be formulated in exactly the same way by observing the analogy between the two systems. The analogy here is such that the mass m, spring constant k, coefficient of friction d, and displacement y correspond to inductance L, reciprocal of capacitance $1/C$, resistance R, and charge q, respectively.

It is also noted that a set of two second-order equations would also describe the motion of the system considered, and such higher order equations may emerge quite often in the formulation of many systems. Lagrangian formulation of a dynamical system, for example, leads to second-order differential equations—an approach used widely in mechanics. It is well known, however, that such higher order equations can readily be transformed into a set of first-order equations by introducing new variables, and the procedure followed above can be viewed as a demonstration of this transformation. The effect of structural parameters, such as masses, and dissipation and stiffness coefficients, are incorporated along with state variables in the all-embracing vector function $\boldsymbol{X}(\boldsymbol{x})$ of the first-order system (eqn 1.6), and it seems, therefore, that this formulation provides a suitable vehicle for a unified general treatment of autonomous systems.

Let the vector $\boldsymbol{x}=\boldsymbol{x}(t)$, describing the state of the system (eqn 1.6) at time t, be represented by a point in *state space* R^n, spanned by $\boldsymbol{x}$, which is assumed to be a *Euclidean space*. The path traced by the point $\boldsymbol{x}(t)$ in R^n is referred to as a *trajectory* or a *motion* of the system (1.6). If $\boldsymbol{x}(t)$ remains in one position for all t such that $\boldsymbol{x}(t)=\boldsymbol{x}_e$ and

$$\dot{\boldsymbol{x}}_e = 0, \tag{1.9}$$

then $\boldsymbol{x}_e$ represents a *stationary point* or an *equilibrium state*. It follows that an equilibrium state $\boldsymbol{x}_e$ satisfies

$$\boldsymbol{X}(\boldsymbol{x}_e)=\boldsymbol{0} \qquad \text{for all } t. \tag{1.10}$$

It may happen that $\boldsymbol{x}(t)$ traces a closed *orbit* in a *periodic* way, such that $\boldsymbol{x}(t)\neq\boldsymbol{x}(0)$ for $0<t<T$ and $\boldsymbol{x}(T)=\boldsymbol{x}(0)$, where $T>0$ is the period; then, the orbit represents a *periodic motion*. In general, the system (1.6) may exhibit other types of solutions (trajectories). It will be assumed that the system $\dot{\boldsymbol{x}}=\boldsymbol{X}(\boldsymbol{x})$ has a unique *solution* for a given initial state $\boldsymbol{x}_0$ at a given initial time t_0. Such a solution (or trajectory) is sometimes written as $\boldsymbol{x}(t;\boldsymbol{x}_0,t_0)$ to draw attention to the initial conditions. However, it is much cleaner to express a solution simply by $\boldsymbol{x}(t)$ if no confusion can arise, and in recent years this alternative notation has gained acceptance with the further assumption that $\boldsymbol{x}(t)$ denotes the point in R^n reached at time t by starting from $\boldsymbol{x}$ at time $t=0$. In other words $\boldsymbol{x}(t)$ is the solution of eqn (1.6) satisfying the initial condition $\boldsymbol{x}(0)=\boldsymbol{x}$. If there is a possibility of confusion, particular solutions will be denoted by another symbol; e.g. $\boldsymbol{u}(t)=\boldsymbol{x}(t;\boldsymbol{x}_0,t_0)$.

As noted above, an autonomous system described by eqn (1.6) may exhibit different types of solutions. *Transients* decay to zero as t tends to infinity, and attention is normally focused on *steady-states* that remain after the transients disappear. In other words, the *salient features* of the system are those that persist for a long time (*asymptotic behaviour*). The most prominent steady-states are the *equilibria* and *periodic motions* defined above. In fact, it was implicitly assumed for a long time that these were the only alternatives for asymptotic behaviour of solutions for a deterministic system. This assumption is valid for two-dimensional systems, but it has become clear following Lorenz's discovery (1963) that even simple looking higher order systems can exhibit a complicated '*chaotic*' motion which appears to be random or unpredictable. This rather puzzling behaviour may be associated with *strange attractors* and will be discussed in Chapter 6 very briefly.

An important class of periodic motions consists of *limit cycles*, which are *isolated* periodic orbits. For example, both the linear oscillator ($\dot{x}=-y$, $\dot{y}=x$) and the system

$$\begin{aligned}\dot{x}&=-y+ax(1-x^2-y^2),\\ \dot{y}&=x+ay(1-x^2-y^2),\end{aligned}\tag{1.11}$$

exhibit undistinguishable oscillatory motions. However, the latter system has only one periodic solution (a limit cycle) while the periodic orbits of the linear oscillator cover the entire state plane, and the system operates on one of these orbits depending on the initial conditions. On the other hand, every solution of the non-linear system (1.11), except $x=0$, $y=0$, eventually approaches the limit cycle

$$\begin{aligned}x&=\cos t,\\ y&=\sin t.\end{aligned}\tag{1.12}$$

This can readily be verified by introducing the polar coordinates $x=r\cos\theta$, $y=r\sin\theta$ into eqns (1.11), and observing that the solutions approach $r=1$ as $t\to\infty$. Indeed, in polar coordinates, eqns (1.11) take the form

$$\begin{aligned}\dot{r}\cos\theta-r\dot{\theta}\sin\theta&=-r\sin\theta+ar\cos\theta(1-r^2),\\ \dot{r}\sin\theta+r\dot{\theta}\cos\theta&=r\cos\theta+ar\sin\theta(1-r^2).\end{aligned}\tag{1.13}$$

Multiplying the first equation by $\cos\theta$ and the second equation by $\sin\theta$, and adding, leads to

$$\dot{r}=ar(1-r^2).\tag{1.14}$$

Similarly, multiplying the first equation by $\sin\theta$ and the second equation by $\cos\theta$, and subtracting, yields

$$\dot{\theta}=1.\tag{1.15}$$

The solution can be obtained by integration as

$$\begin{aligned}r&=1\Big/\left\{1+\left(\frac{1}{r_0^2}-1\right)e^{-2at}\right\}^{1/2},\\ \theta&=t+\theta_0,\end{aligned}\tag{1.16}$$

where $r=r_0$ at $t=0$, and $r=1$ represents the limit cycle. This solution takes the form of eqns (1.12) if $\theta_0=0$ at $t=0$, and all solutions approach eqns (1.12) or $r_0=1$ as $t\to\infty$ (Fig. 1.3).

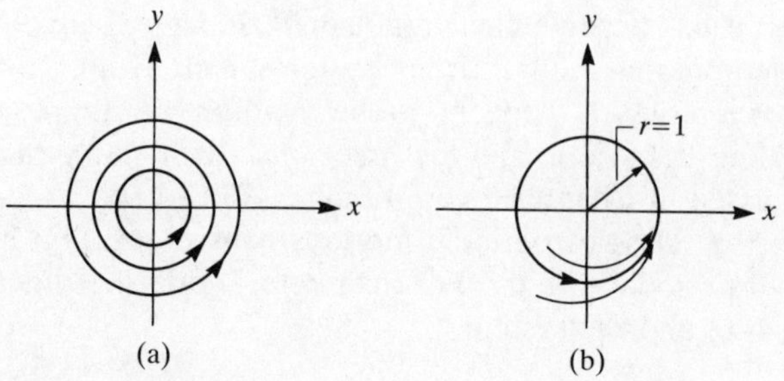

Fig. 1.3 (a) Periodic orbits of a linear oscillator, (b) limit cycle.

In this book emphasis is placed on exploring limit cycles and equilibrium solutions of autonomous systems and the way in which they change as certain parameters are varied. Finally, it is noted that while the family of periodic motions associated with a linear oscillator is structurally unstable, the behaviour with regard to a limit cycle is structurally stable.

1.4 Stability in autonomous systems

The concept of stability may be as elusive as it is important, and it is difficult to give a universal definition that is satisfactory in all applications. As a matter of fact, a variety of definitions have been laid down in different contexts. All definitions, however, are based on the response characteristics of a system subjected to certain perturbations. It is often vitally important to know how sensitive a given state of equilibrium (or a trajectory) is to perturbations, and a state is said to be *stable* if the motions of a slightly perturbed system remain in the vicinity of the state. Different definitions emerge with regard to the nature and magnitude of both the perturbations and deviations from the given state (Huseyin 1978*b*). Lyapunov's definition (1892) of stability will be fundamental in the treatment of autonomous systems in this book.

Stability of an equilibrium state. Without loss of generality, suppose $\boldsymbol{x} = 0$ is an equilibrium state of the autonomous system (eqn 1.6). A simple transformation of the type $\boldsymbol{x} = \boldsymbol{x}_e + \bar{\boldsymbol{x}}$ will achieve this for any given state of equilibrium $\boldsymbol{x}_e$. In other words, it is assumed that the stability of the origin is under study, and $\boldsymbol{X}(\mathbf{0}) = \mathbf{0}$. The origin $\boldsymbol{x} = \mathbf{0}$ is *stable* in the sense of Lyapunov if it is possible to find a positive $\delta(\varepsilon)$ for any given $\varepsilon > 0$ such that the motion following the initial disturbance

$$\|\boldsymbol{x}^0\| < \delta, \qquad \text{where } \boldsymbol{x}^0 = \boldsymbol{x}(0) \tag{1.17}$$

satisfies

$$\|\boldsymbol{x}(t)\| < \varepsilon \tag{1.18}$$

for all $t \geq 0$. Otherwise the state $\boldsymbol{x} = \mathbf{0}$ is *unstable*.

The state $\boldsymbol{x} = \mathbf{0}$ is said to be *asymptotically stable* if it is stable and, in addition,

$$\lim_{t \to \infty} \|\boldsymbol{x}(t)\| = 0. \tag{1.19}$$

These definitions imply that the equilibrium state represented by the origin of R^n is stable if the variable point $x(t)$, describing the trajectories of the system in R^n, remains within a hypersphere of arbitrarily small radius ε, provided the initial disturbance is such that the point $\boldsymbol{x}^0$ is within a hypersphere of sufficiently small radius δ. If the point $\boldsymbol{x}(t)$ also tends to the origin as $t \to \infty$, then the origin is asymptotically stable. On the other hand, the origin is unstable if the trajectory of the point $\boldsymbol{x}(t)$ crosses the hypersphere of radius ε. A two-dimensional representation of these phenomena is depicted in Fig. 1.4.

The equilibrium state $\boldsymbol{x} = \mathbf{0}$ is said to be *globally asymptotically stable* if it is stable and, in addition, every motion converges to the origin as $t \to \infty$. This property is also referred to as *asymptotic stability in the large*.

Stability of a motion. A given motion $\boldsymbol{x}(t)$ is said to be stable if for any given $\varepsilon > 0$ there exists a positive $\delta(\varepsilon)$ such that the motion $\boldsymbol{x}^*(t)$, following any initial disturbance for which

$$\|\boldsymbol{x} - \boldsymbol{x}^*\| < \delta, \tag{1.20}$$

satisfies

$$\|\boldsymbol{x}(t) - \boldsymbol{x}^*(t)\| < \varepsilon \tag{1.21}$$

for all $t \geq 0$. It is understood that the original motion $\boldsymbol{x}(t)$ and the perturbed motion $\boldsymbol{x}^*(t)$ satisfy the initial conditions $\boldsymbol{x}(0) = \boldsymbol{x}$ and $\boldsymbol{x}^*(0) = \boldsymbol{x}^*$, respectively.

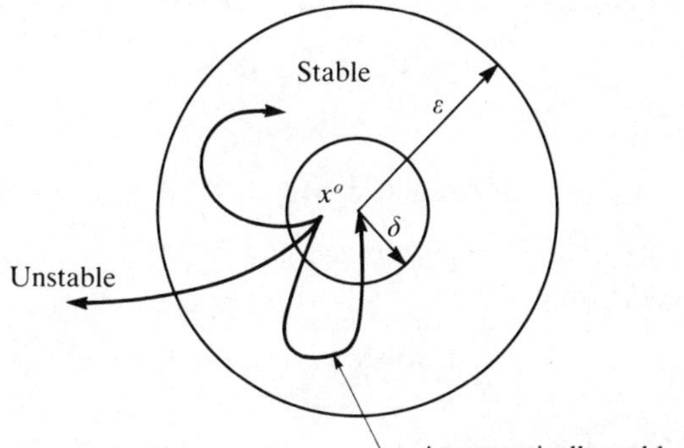

Fig. 1.4 Two-dimensional representation of a stable, unstable, and asymptotically stable origin of R^n.

The motion $\boldsymbol{x}(t)$ is *asymptotically stable* if it is stable and, in addition,

$$\lim_{t\to\infty} \|\boldsymbol{x}(t)-\boldsymbol{x}^*(t)\|=0, \tag{1.22}$$

in analogy with the case of equilibrium states.

In other words, a trajectory $\boldsymbol{x}(t)$ is stable if all the other trajectories which start in a sufficiently close neighbourhood of $\boldsymbol{x}(t)$ remain in a prescribed neighbourhood for later times.

It is interesting that the problem of the stability of a motion can always be reduced to the stability of the origin. Thus, in order to study the stability properties of the particular solution $\boldsymbol{x}=\boldsymbol{u}(t)$, introduce

$$\boldsymbol{x}=\boldsymbol{u}(t)+\boldsymbol{y} \tag{1.23}$$

into eqn (1.6) to obtain

$$\dot{\boldsymbol{u}}+\dot{\boldsymbol{y}}=\boldsymbol{X}(\boldsymbol{u}+\boldsymbol{y}). \tag{1.24}$$

Since $\boldsymbol{u}$ is a function of t, one may define

$$\boldsymbol{Y}(\boldsymbol{y}, t)=\boldsymbol{X}(\boldsymbol{u}+\boldsymbol{y})-\boldsymbol{X}(\boldsymbol{u}). \tag{1.25}$$

Observing now that

$$\dot{\boldsymbol{u}}=\boldsymbol{X}(\boldsymbol{u}) \tag{1.26}$$

leads to

$$\dot{\boldsymbol{y}}=\boldsymbol{Y}(\boldsymbol{y}, t) \tag{1.27}$$

where $\boldsymbol{y}=\boldsymbol{0}$ is the particular solution $\boldsymbol{x}=\boldsymbol{u}(t)$ and $\boldsymbol{Y}(\boldsymbol{0}, t)=\boldsymbol{0}$ for all t. Note, however, that the resulting system is in general non-autonomous.

If the motion $\boldsymbol{x}(t)$ is periodic, describing a closed orbit $\boldsymbol{c}$ (e.g. a limit cycle), another definition of stability may be helpful in many applications. Indeed, if interest is focused on the form of the orbit rather than its time dependence, one may define *orbital stability*. Thus, a closed orbit $\boldsymbol{c}$ is said to be *orbitally stable* if for any $\varepsilon>0$ there exists a positive $\delta>0$ such that

$$\|\boldsymbol{x}-\boldsymbol{x}^*\|<\delta \tag{1.28}$$

implies that

$$\text{distance}\,[\boldsymbol{x}^*(t), \boldsymbol{c}]<\varepsilon, \tag{1.29}$$

for $t\geq 0$. The orbit $\boldsymbol{c}$ is *asymptotically orbitally stable* if it is orbitally stable and in addition

$$\lim_{t\to\infty} \textit{distance}\,[\boldsymbol{x}^*(t), \boldsymbol{c}]=0. \tag{1.30}$$

In other words, if all motions originating from a point very close to $\boldsymbol{c}$ remain in the vicinity of $\boldsymbol{c}$, the orbit $\boldsymbol{c}$ is *orbitally stable*. In this definition, the points $\boldsymbol{x}(t)$ and $\boldsymbol{x}^*(t)$ in R^n do not have to remain close all the time as required by the definition of stability. Clearly, the latter implies orbital stability, but the converse is not generally true. An interesting example demonstrating this distinction is given by the system $\dot{\theta} = r$ and $\dot{r} = 0$. Here, the circular orbits are orbitally stable but not stable (Haken 1978) since the angular velocity $\dot{\theta}$ increases with r.

If the solutions of eqn (1.6) can be obtained in closed form, the stability definitions may be applied directly. It is often desired, however, to obtain information about the stability properties of a system without actually solving the differential equations. This is particularly so when solving a set of differential equations poses serious difficulties. *Lyapunov's direct (second) method* is primarily aimed at circumventing these difficulties, and is developed on the model of Lagrange's stability theorem (1788) for mechanical systems with a potential energy. According to this theorem, *an equilibrium state is stable if the total potential energy has a complete relative minimum at that state* (Huseyin 1975). This is a sufficient condition for stability, and the underlying idea has been generalized by Lyapunov.

Let $V(\boldsymbol{x})$ be a real scalar function of the vector $\boldsymbol{x}$, and assume that it is continuous, with its first partial derivatives in a certain region G, containing the origin. $V(\boldsymbol{x})$ is positive definite if $V(\mathbf{0}) = 0$ and $V(\boldsymbol{x}) > 0$ outside the origin (within G). The function $V(\boldsymbol{x})$ is said to be a *Lyapunov function* if it is positive definite and $\dot{V}(\boldsymbol{x})$ is negative semi-definite in G. In other words, the origin of R^n is an isolated minimum of the Lyapunov function $V(\boldsymbol{x})$. It is noted that the time derivative $\dot{V}(\boldsymbol{x})$ may be expressed as

$$\dot{V} = \frac{\partial V}{\partial x^i}\dot{x}^i = \langle \dot{\boldsymbol{x}}, \operatorname{grad} V \rangle = \langle \boldsymbol{X}, \operatorname{grad} V \rangle \tag{1.31}$$

where $\langle\ \rangle$ denotes scalar product of vectors.

Lyapunov has formulated the following two theorems:

i) *Stability theorem*. The null solution of the autonomous system (eqn 1.6), i.e. the origin or the trivial solution, is stable if there exists a Lyapunov function $V(\boldsymbol{x})$ in a region G about the origin.

ii) *Asymptotic stability theorem*. The origin is asymptotically stable if, in addition, $\dot{V}(\boldsymbol{x})$ is negative definite in G.

In order to prove these theorems, let the positive definite function $V(\boldsymbol{x})$ describe a cup-shaped surface in $V(\boldsymbol{x})$-$\boldsymbol{x}$ space (Fig. 1.5) such that its cross sections with $V(\boldsymbol{x}) = k$ (=constant) planes form a set of concentric curves, enclosing each other in the state space. In view of the basic definition of stability, let k be selected such that the curve $V(\boldsymbol{x}) = k$ is enclosed by the sphere of radius ε tangentially; one can then find a δ such that the sphere of radius δ is just enclosed by $V(\boldsymbol{x}) = k$ tangentially. It follows that any path with the initial state $\boldsymbol{x}^0$ within the sphere of radius δ cannot reach $V(\boldsymbol{x}) = k$ because $V(\boldsymbol{x}) < k$ and $\dot{V}(\boldsymbol{x}) \leq 0$. This implies

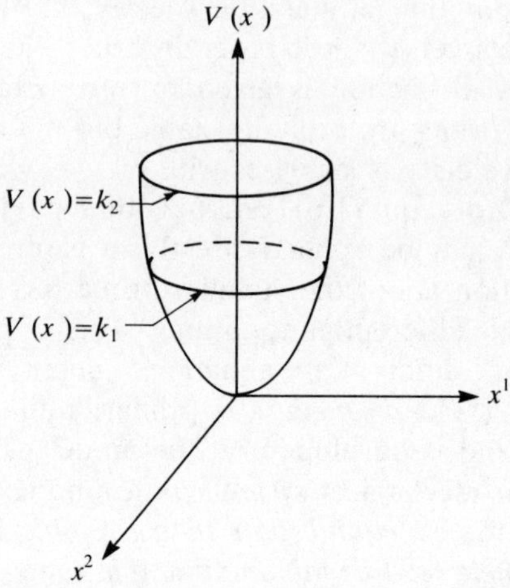

Fig. 1.5 Illustration for proof of Lyapunov's stability theorem and asymptotic stability theorem.

that the path remains within the sphere of radius ε. Furthermore, in the case of $\dot{V}(\boldsymbol{x})<0$, $V(\boldsymbol{x})$ decreases along the path which has to tend to the origin as $V(\boldsymbol{x})\to 0$, since $V(\boldsymbol{x})$ can only be zero at the origin ($V(\mathbf{0})=0$).

Lyapunov's theorems described above provide only sufficient conditions for stability, and if one cannot find a Lyapunov function, it does not follow necessarily that the system is unstable. There exists a number of instability theorems (La Salle and Lefschetz 1961, Willems 1970, Leipholz 1970); here only one of these theorems will be given:

iii) *Instability theorem*. The origin of system (1.6) is unstable if there exists a function $V(\boldsymbol{x})$ with continuous partial derivatives in G and $V(\mathbf{0})=0$, such that $\dot{V}(\boldsymbol{x})$ is positive (negative) definite and $V(\boldsymbol{x})$ can assume positive (negative) values arbitrarily near the origin.

To prove this theorem, a similar line of argument as in the case of the stability theorems can be adopted to demonstrate that there is at least one solution, starting arbitrarily near the origin, which is unstable.

Practical criteria of stability and/or instability may be based on the eigenvalues of the Jacobian matrix of the so-called *linearized system*. The Jacobian associated with $\dot{\boldsymbol{x}}=\boldsymbol{X}(\boldsymbol{x})$ is defined as the matrix

$$\boldsymbol{A}(\boldsymbol{x})=\left[\frac{\partial \boldsymbol{X}(\boldsymbol{x})}{\partial \boldsymbol{x}}\right]=\begin{bmatrix} \partial X_1/\partial x^1 & \partial X_1/\partial x^2 & \dots & \partial X_1/\partial x^n \\ \partial X_2/\partial x^1 & \partial X_2/\partial x^2 & \dots & \partial X_2/\partial x^n \\ \vdots & \vdots & & \vdots \\ \partial X_n/\partial x^1 & \partial X_n/\partial x^2 & \dots & \partial X_n/\partial x^n \end{bmatrix} \tag{1.32}$$

which becomes a constant $n\times n$ matrix when evaluated at a given state.

Thus, evaluation at the origin, satisfying $\boldsymbol{X}(\boldsymbol{0}) = \boldsymbol{0}$, yields

$$\boldsymbol{A} = \left[\frac{\partial \boldsymbol{X}(\boldsymbol{x})}{\partial \boldsymbol{x}}\right]_{\boldsymbol{x}=\boldsymbol{0}}. \tag{1.33}$$

The equation of the first approximation about the origin is

$$\dot{\boldsymbol{x}} = \boldsymbol{A}\boldsymbol{x} \tag{1.34}$$

which is also said to describe the *linearized system*.

The behaviour of linear systems characterized by equations of the form (1.34) has been explored rather fully (e.g. Huseyin 1978*b*), and only some results will be summarized here.

It has been shown via several different approaches, for example, that:

(i) The origin of system (1.34) is asymptotically stable if and only if $\boldsymbol{A}$ is a stable matrix—that is, all eigenvalues of the matrix $\boldsymbol{A}$ have negative real parts.

(ii) The origin of system (1.34) is stable if and only if $\boldsymbol{A}$ has no eigenvalues with a positive real part, and the eigenvalues with zero real parts correspond to Jordan blocks of order one—i.e. *multiplicity* and *index* of eigenvalues are equal so that there is no reduction in the dimension of the eigenvector space (Huseyin 1978*b*).

(iii) The origin of system (1.34) is unstable if $\boldsymbol{A}$ has an eigenvalue with a positive real part or the multiplicity of an eigenvalue with zero real part exceeds its index, resulting in a reduction in the dimension of the eigenvector space.

The stability of the origin of the system (1.34) can be studied by making use of quadratic forms. Consider the quadratic form

$$V(\boldsymbol{x}) = \langle \boldsymbol{x}, \boldsymbol{Q}\boldsymbol{x} \rangle \tag{1.35}$$

where $\boldsymbol{Q}$ is a symmetric constant matrix. It follows that

$$\begin{aligned} \dot{V}(\boldsymbol{x}) &= \langle \dot{\boldsymbol{x}}, \boldsymbol{Q}\boldsymbol{x} \rangle + \langle \boldsymbol{x}, \boldsymbol{Q}\dot{\boldsymbol{x}} \rangle \\ &= \langle \boldsymbol{A}\boldsymbol{x}, \boldsymbol{Q}\boldsymbol{x} \rangle + \langle \boldsymbol{x}, \boldsymbol{Q}\boldsymbol{A}\boldsymbol{x} \rangle \\ &= \langle \boldsymbol{x}, (\boldsymbol{A}'\boldsymbol{Q} + \boldsymbol{Q}\boldsymbol{A})\boldsymbol{x} \rangle \end{aligned} \tag{1.36}$$

where $\boldsymbol{A}'$ is the transpose of $\boldsymbol{A}$ and $[\boldsymbol{A}'\boldsymbol{Q} + \boldsymbol{Q}\boldsymbol{A}]$ is necessarily symmetric. If a positive definite matrix $\boldsymbol{Q}$ can be found such that

$$\boldsymbol{A}'\boldsymbol{Q} + \boldsymbol{Q}\boldsymbol{A} = \boldsymbol{P} \tag{1.37}$$

is negative definite, the system is asymptotically stable. It can be shown that if $\boldsymbol{A}$ is a stable matrix, a unique positive definite $\boldsymbol{Q}$ exists for any given negative definite $\boldsymbol{P}$. The solution of eqn (1.37) for a given $\boldsymbol{P} < 0$ is expressed as

$$\boldsymbol{Q} = -\int_0^\infty e^{\boldsymbol{A}'t}\boldsymbol{P}e^{\boldsymbol{A}t}\,dt. \tag{1.38}$$

where

$$\mathrm{e}^{\boldsymbol{A}t} = \boldsymbol{I} + \boldsymbol{A}t + \ldots + \frac{\boldsymbol{A}^n t^n}{n!} + \ldots,$$

and $\boldsymbol{I}$ is the identity matrix.

If $\boldsymbol{A}$ is not a stable matrix, and the solution of eqn (1.37), with $\boldsymbol{P} < 0$, yields a matrix $\boldsymbol{Q}$ which is indefinite or negative, then the origin of the system (1.34) is unstable.

Another useful criterion of stability for the system (1.34) is established on the basis of the characteristic polynomial

$$f(\lambda) = \lambda^n + a_1\lambda^{n-1} + a_2\lambda^{n-2} + \ldots + a_n = 0 \tag{1.39}$$

which follows from the characteristic equation

$$|(\boldsymbol{A} - \lambda\boldsymbol{I})| = 0. \tag{1.40}$$

The polynomial (1.39) is called a *Hurwitz polynomial* if all its roots have negative real parts. According to the Routh–Hurwitz criterion, a necessary and sufficient condition for the polynomial to be a Hurwitz polynomial is that

$$\begin{aligned}
\Delta_1 &= a_1 > 0,\\
\Delta_2 &= \begin{vmatrix} a_1 & 1 \\ a_3 & a_2 \end{vmatrix} > 0,\\
\Delta_3 &= \begin{vmatrix} a_1 & 1 & 0 \\ a_3 & a_2 & a_1 \\ a_5 & a_4 & a_3 \end{vmatrix} > 0,\\
&\vdots\\
\Delta_n &= a_n\Delta_{n-1} > 0.
\end{aligned} \tag{1.41}$$

The proof of this important theorem can be found in many books (e.g. Huseyin 1978*b*). Clearly, if $f(\lambda)$ is a Hurwitz polynomial satisfying eqns (1.41), the origin of system (1.34) is asymptotically stable.

The trajectories in the state space depend on the eigenvalues, and a geometrical representation can be very illustrative with regard to the behaviour of the system. Although the visualization of trajectories in n dimensions is extremely complex, for the case of $n = 2$ such a *phase portrait* takes a relatively simple form, and is of great value.

The solution of eqn (1.34) may be expressed as

$$\boldsymbol{x} = \mathrm{e}^{\boldsymbol{A}t}\boldsymbol{x}^0, \qquad \boldsymbol{x}^0 = \boldsymbol{x}(0). \tag{1.42}$$

Let the eigenvalues of $\boldsymbol{A}$ be λ_1 and λ_2; several interesting situations may be identified:

(1) Complex conjugate eigenvalues ($\gamma \pm \mathrm{i}\omega$, where γ and ω are real and not zero). In this case the origin of system (1.34) is a *stable focus* if $\gamma < 0$, and an *unstable focus* if $\gamma > 0$ (Fig. 1.6).

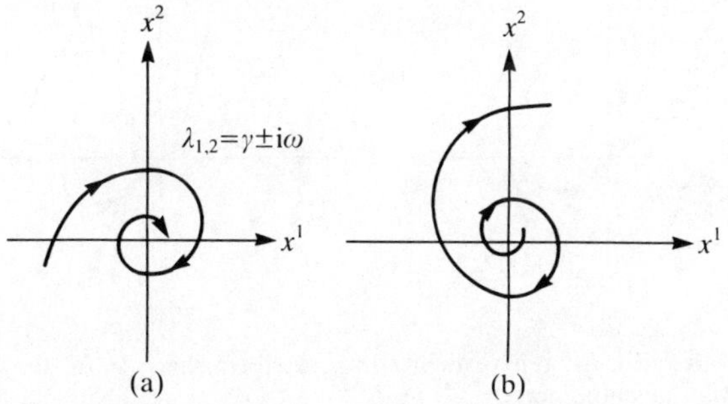

Fig. 1.6 Phase portrait using complex conjugate eigenvalues, showing (a) stable focus ($\gamma<0$) and (b) unstable focus ($\gamma>0$).

(2) Complex conjugate imaginary eigenvalues ($\pm i\omega$; i.e. $\gamma=0$). The origin in this case is called a *centre* (Fig. 1.7), which is stable.

(3) Real distinct non-zero eigenvalues. The origin in this case is a *stable node* if $\lambda_1<0$, $\lambda_2<0$; an *unstable node* if $\lambda_1>0$, $\lambda_2>0$ (Fig. 1.8); and a *saddle point* if $\lambda_1>0$, $\lambda_2<0$ (or vice versa), which is unstable (Fig. 1.9).

(4) $\lambda_1=0$, λ_2 is real. In this case, the origin is stable (unstable) if the non-zero eigenvalue is negative (positive). Also, note that $|\mathbf{A}|=0$ and the origin is not the only equilibrium state (Fig. 1.10).

(5) Coincident eigenvalues with distinct eigenvectors. In this case, **A** is of *simple* structure and it can be diagonalized. The origin is a *degenerate stable node* (*degenerate unstable node*) if $\lambda_1=\lambda_2<0$ ($\lambda_1=\lambda_2>0$), as is shown in Fig. 1.11.

(6) Coincident eigenvalues with coincident eigenvectors. Here the matrix **A** does not have a simple structure (it is *defective*) and cannot be diagonalized. The origin is again a *degenerate stable node* (*degenerate unstable node*) if $\lambda_1=\lambda_2<0$ ($\lambda_1=\lambda_2>0$), as illustrated in Fig. 1.12.

(7) $\lambda_1=\lambda_2=0$. The origin is unstable (Fig. 1.13).

The most interesting question here is concerned with the possible (and intuitively expected) connection between the behaviour characteristics of the non-linear system (eqn 1.6) and those of the linearized system (eqn 1.34) outlined above. As far as the trajectories are concerned, it turns out

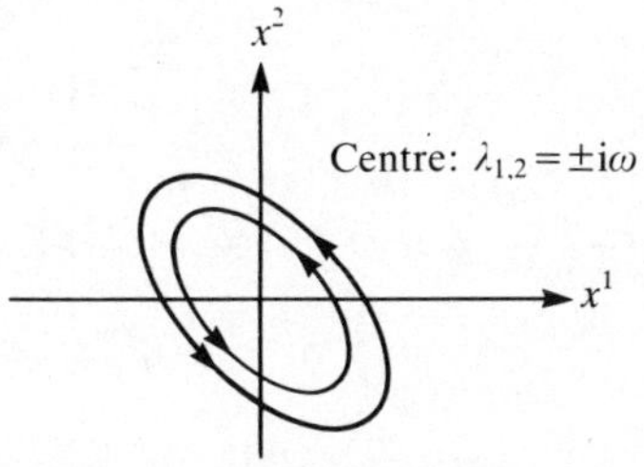

Fig. 1.7 Phase portrait using complex conjugate imaginary eigenvalues.

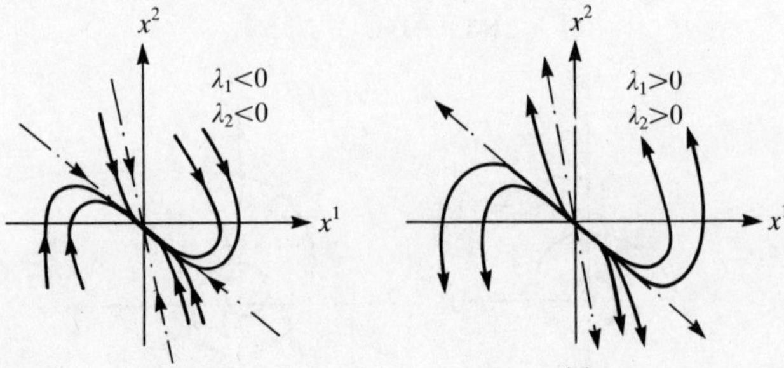

(a) (b)

Fig. 1.8 Phase portrait using real distinct non-zero eigenvalues. In (a) the origin is a stable node and in (b) an unstable node.

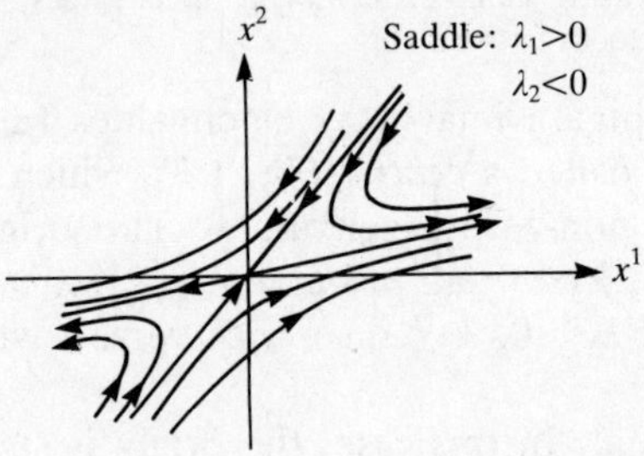

Fig. 1.9 Phase portrait using real distinct non-zero eigenvalues. The origin is a saddle point.

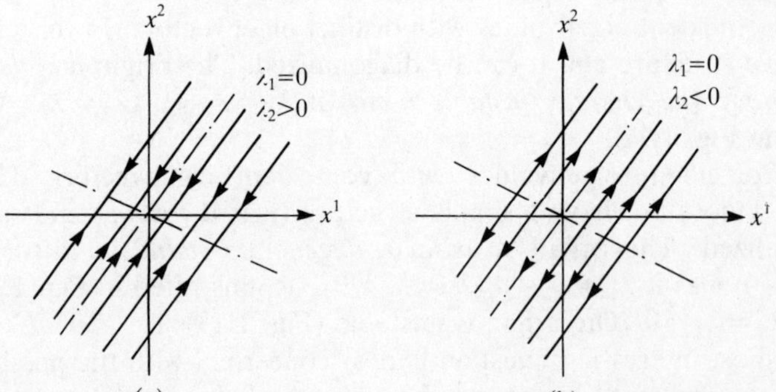

(a) (b)

Fig. 1.10 Phase portrait using $\lambda_1 = 0$, λ_2 is real. In (a) the origin is stable and in (b) it is unstable.

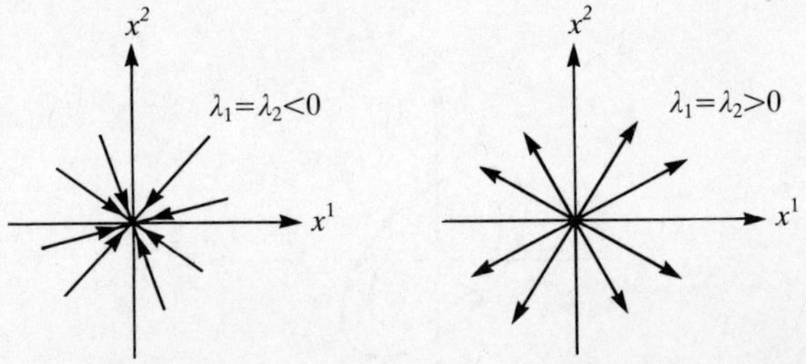

(a) (b)

Fig. 1.11 Phase portrait using coincident eigenvalue with distinct eigenvectors. In (a) the origin is a degenerate stable node and in (b) it is a degenerate unstable node.

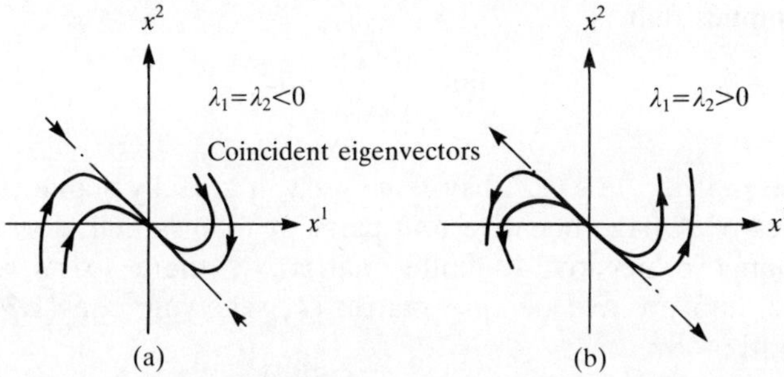

Fig. 1.12 Phase portrait using coincident eigenvalues with coincident eigenvectors. In (a) the origin is a degenerate stable node and in (b) it is a degenerate unstable node.

that the non-linear system exhibits a qualitatively similar behaviour in the vicinity of the origin if the Jacobian matrix $\boldsymbol{A}$ has no eigenvalues with zero real parts. On the other hand, the non-linear terms may transform the steady harmonic oscillations associated with a centre to growing or decaying oscillations. Similarly, the stability properties of the non-linear system (1.6) coincides with those of the linearized system (1.34) if $\boldsymbol{A}$ has no pure imaginary eigenvalues; in this case, the origin is *hyperbolic*. More generally, an equilibrium state $\boldsymbol{x}_e$, satisfying $\boldsymbol{X}(\boldsymbol{x}_e)=0$, is said to have a *hyperbolic structure* if the Jacobian $\boldsymbol{A}(\boldsymbol{x}_e)$ at this point has no pure imaginary eigenvalues. If all eigenvalues of $\boldsymbol{A}(\boldsymbol{x}_e)$ have negative real parts, the equilibrium state $\boldsymbol{x}_e$ is a *sink*, and all nearby solutions approach $\boldsymbol{x}_e$ exponentially.

Hyperbolicity is a generic property, and represents a *strong* behaviour. In order to show the link between the non-linear system (1.6) and its linearized model (1.34) in the hyperbolic case, suppose $\dot{\boldsymbol{x}}=\boldsymbol{X}(\boldsymbol{x})$ is expressed as

$$\dot{\boldsymbol{x}}=\boldsymbol{A}\boldsymbol{x}+\boldsymbol{h}(\boldsymbol{x}), \qquad \boldsymbol{h}(\boldsymbol{0})=0, \tag{1.43}$$

where the function $\boldsymbol{h}(\boldsymbol{x})$ has a Taylor's series expansion in the neighbourhood of the origin, beginning with terms of at least second order in the

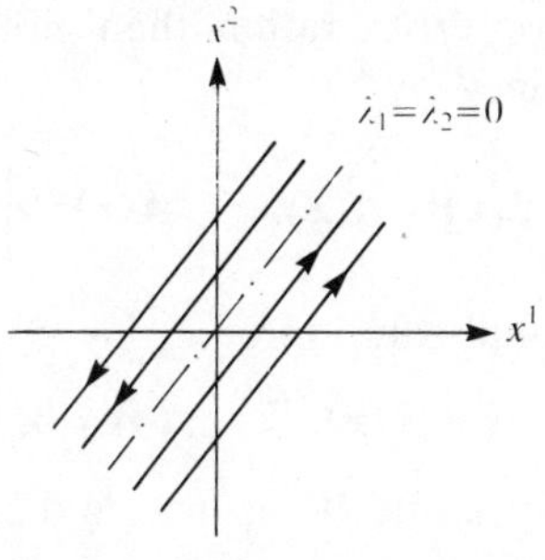

Fig. 1.13 Phase portrait using $\lambda_1=\lambda_2=0$. The origin is unstable.

x^i. This implies that

$$\lim_{\|\boldsymbol{x}\|\to 0} \frac{\|\boldsymbol{h}(\boldsymbol{x})\|}{\|\boldsymbol{x}\|} = 0 \tag{1.44}$$

Suppose that the linearized system is asymptotically stable, so that all eigenvalues of $\boldsymbol{A}$ have negative real parts. It follows that, corresponding to a symmetric negative definite matrix $\boldsymbol{P}$, there exists a unique, symmetric, and positive definite matrix $\boldsymbol{Q}$, satisfying eqn (1.37). Select the quadratic form

$$V(\boldsymbol{x}) = \langle \boldsymbol{x}, \boldsymbol{Qx} \rangle \tag{1.45}$$

as a Lyapunov function; then, $\dot{V}(\boldsymbol{x})$ along the trajectories of eqn (1.43) takes the form

$$\begin{aligned} \dot{V}(\boldsymbol{x}) &= \langle \dot{\boldsymbol{x}}, \boldsymbol{Qx} \rangle + \langle \boldsymbol{x}, \boldsymbol{Q}\dot{\boldsymbol{x}} \rangle \\ &= \langle (\boldsymbol{Ax}+\boldsymbol{h}), \boldsymbol{Qx} \rangle + \langle \boldsymbol{x}, \boldsymbol{Q}(\boldsymbol{Ax}+\boldsymbol{h}) \rangle \\ &= \langle \boldsymbol{x}, (\boldsymbol{A}'\boldsymbol{Q} + \boldsymbol{AQ})\boldsymbol{x} \rangle + 2\langle \boldsymbol{h}, \boldsymbol{Qx} \rangle \\ &= \langle \boldsymbol{x}, \boldsymbol{Px} \rangle + 2\langle \boldsymbol{h}, \boldsymbol{Qx} \rangle . \end{aligned} \tag{1.46}$$

Since $\langle \boldsymbol{h}, \boldsymbol{Qx} \rangle$ consists of higher order terms compared to $\langle \boldsymbol{x}, \boldsymbol{Px} \rangle$, in the vicinity of $\boldsymbol{x} = 0$,

$$\dot{V}(\boldsymbol{x}) < 0, \tag{1.47}$$

and the non-linear system (1.43) is asymptotically stable, as its linearized model.

If the matrix $\boldsymbol{A}$ has a positive real part, similar arguments show that the non-linear system (1.47) is unstable.

If the origin is not hyperbolic, and some of the eigenvalues of $\boldsymbol{A}$ are pure imaginary while the remainder have negative real parts, the origin is a *critical equilibrium state.* Linearizations in this case do not lead to definite conclusions, and the stability of the non-linear system can only be established by involving the higher order terms in the analysis (La Salle and Lefschetz 1961). Also, the criteria based on the linearized model are concerned with local properties rather than global behaviour. It can be shown that if $\boldsymbol{D}(\boldsymbol{x})$, defined by

$$\boldsymbol{D}(\boldsymbol{x}) = \boldsymbol{A}'(\boldsymbol{x}) + \boldsymbol{A}(\boldsymbol{x}), \qquad \boldsymbol{A}(\boldsymbol{x}) = \left[\frac{\partial \boldsymbol{X}(\boldsymbol{x})}{\partial \boldsymbol{x}}\right], \tag{1.48}$$

is negative definite the null solution of

$$\dot{\boldsymbol{x}} = \boldsymbol{X}(\boldsymbol{x}), \qquad \boldsymbol{X}(\boldsymbol{0}) = \boldsymbol{0}, \tag{1.49}$$

is globally asymptotically stable (Krasowski's theorem) provided

$$\langle \boldsymbol{X}(\boldsymbol{x}), \boldsymbol{X}(\boldsymbol{x}) \rangle \to \infty \qquad \text{as } \|\boldsymbol{x}\| \to \infty.$$

Consider the Lyapunov function

$$V(\boldsymbol{x}) = \langle \boldsymbol{X}(\boldsymbol{x}), \boldsymbol{X}(\boldsymbol{x}) \rangle \tag{1.50}$$

which is positive definite since $\boldsymbol{X}(\boldsymbol{x}) \neq \boldsymbol{0}$ for $\boldsymbol{x} \neq \boldsymbol{0}$ and by virtue of negative definiteness of eqn (1.48). Then,

$$\begin{aligned}\dot{V}(\boldsymbol{x}) &= \langle \dot{\boldsymbol{X}}(\boldsymbol{x}), \boldsymbol{X}(\boldsymbol{x}) \rangle + \langle \boldsymbol{X}(\boldsymbol{x}), \dot{\boldsymbol{X}}(\boldsymbol{x}) \rangle \\ &= \langle \boldsymbol{A}(\boldsymbol{x})\dot{\boldsymbol{x}}, \boldsymbol{X}(\boldsymbol{x}) \rangle + \langle \boldsymbol{X}(\boldsymbol{x}), \boldsymbol{A}(\boldsymbol{x})\dot{\boldsymbol{x}} \rangle \\ &= \langle \boldsymbol{A}(\boldsymbol{x})\boldsymbol{X}(\boldsymbol{x}), \boldsymbol{X}(\boldsymbol{x}) \rangle + \langle \boldsymbol{X}(\boldsymbol{x}), \boldsymbol{A}(\boldsymbol{x})\boldsymbol{X}(\boldsymbol{x}) \rangle \\ &= \langle \boldsymbol{X}(\boldsymbol{x}), \{\boldsymbol{A}'(\boldsymbol{x}) + \boldsymbol{A}(\boldsymbol{x})\}\boldsymbol{X}(\boldsymbol{x}) \rangle = \langle \boldsymbol{X}(\boldsymbol{x}), \boldsymbol{D}(\boldsymbol{x})\boldsymbol{X}(\boldsymbol{x}) \rangle < 0,\end{aligned} \tag{1.51}$$

showing that $\boldsymbol{x} = \boldsymbol{0}$ is a globally asymptotically stable equilibrium state. Note that in a sufficiently close neighbourhood of $\boldsymbol{x} = \boldsymbol{0}$, the stability condition may be reduced to

$$(\boldsymbol{A}' + \boldsymbol{A}) < 0, \qquad \boldsymbol{A} = \left[\frac{\partial \boldsymbol{X}}{\partial \boldsymbol{x}}\right]_{\boldsymbol{x}=\boldsymbol{0}}. \tag{1.52}$$

This result can be obtained independently, on the basis of eqn (1.43). Thus, consider the Lyapunov function

$$V(\boldsymbol{x}) = \langle \boldsymbol{Ax}, \boldsymbol{Ax} \rangle = \langle \boldsymbol{x}, \boldsymbol{A}'\boldsymbol{Ax} \rangle \tag{1.53}$$

under the assumption that $\boldsymbol{A}$ is non-singular (i.e. $|\boldsymbol{A}| \neq 0$), implying that $\boldsymbol{A}'\boldsymbol{A} > 0$. Then,

$$\begin{aligned}\dot{V}(\boldsymbol{x}) &= \langle \dot{\boldsymbol{x}}, \boldsymbol{A}'\boldsymbol{Ax} \rangle + \langle \boldsymbol{x}, \boldsymbol{A}'\boldsymbol{A}\dot{\boldsymbol{x}} \rangle \\ &= \langle (\boldsymbol{Ax} + \boldsymbol{h}), \boldsymbol{A}'\boldsymbol{Ax} \rangle + \langle \boldsymbol{x}, \boldsymbol{A}'\boldsymbol{A}(\boldsymbol{Ax} + \boldsymbol{h}) \rangle \\ &= \langle \boldsymbol{Ax}, (\boldsymbol{A}' + \boldsymbol{A})\boldsymbol{Ax} \rangle + 2\langle \boldsymbol{x}, \boldsymbol{A}'\boldsymbol{Ah} \rangle.\end{aligned} \tag{1.54}$$

Since $\langle \boldsymbol{x}, \boldsymbol{A}'\boldsymbol{Ah} \rangle$ involves higher order terms compared to the first inner product in eqn (1.54), $\dot{V} < 0$ if eqn (1.52) holds—i.e. if $(\boldsymbol{A} + \boldsymbol{A}')$ is negative definite the origin is asymptotically stable.

1.5 Autonomous systems with parameters

As emphasized in the introductory remarks, systems are often under the influence of several independent parameters, η^{α} ($\alpha = 1, 2, \ldots, m$), and interest is then focused on the behaviour of the system as these parameters vary. In elastic stability problems, for example, external loads, imperfections, certain magnitudes and moduli have an effect on the behaviour of the system and when these effects are parametrized and incorporated in the model, the resulting system becomes a multiple-parameter system. Consider, for instance, the system (1.8); the force functions P_d, P_1 and P_2 may all depend on some parameters when expressed explicitly, and if these are denoted by η^{α}, the system (eqn 1.8) takes the general form of $\dot{\boldsymbol{x}} = \boldsymbol{X}(\boldsymbol{x}, \boldsymbol{\eta})$ where $\boldsymbol{\eta}$ is the parameter vector.

Systems of the form

$$\dot{\boldsymbol{x}} = \boldsymbol{X}(\boldsymbol{x}, \boldsymbol{\eta}); \qquad \boldsymbol{x} \in R^n, \quad \boldsymbol{\eta} \in R^m \tag{1.55}$$

will be considered in the following chapters, and it will always be assumed that the X_i are smooth functions of x^i $(i = 1, 2, \ldots, n)$ and η^α $(\alpha = 1, 2, \ldots, m)$. The $(n+m)$-dimensional *state-parameter* space, R^{n+m}, spanned by x^i and η^j, is assumed to be a Euclidean space. The n-dimensional *state space* R^n and m-dimensional *parameter-space* R^m are sub-spaces of the *state-parameter* space. It is assumed that there is a one-to-one correspondence between a set of variables $(\boldsymbol{x}, \boldsymbol{\eta})$ and the points of R^{n+m}.

For a given parameter vector $\boldsymbol{\eta}$, the system (eqn 1.55) is in the form of eqn (1.6), and all the definitions and theorems given for eqn (1.6) are valid. The points of R^{n+m} representing the equilibrium states of eqn (1.55) satisfy

$$\boldsymbol{X}(\boldsymbol{x}, \boldsymbol{\eta}) = \mathbf{0} \tag{1.56}$$

which generally defines an m-dimensional *equilibrium surface* (*manifold*) in R^{n+m}. This surface may have a complicated shape, and will be explored in detail. It is also noted that the equilibrium surface may consist of stable, unstable, and critical states.

At this stage, the instability mechanism has to be clarified. Consider a stable equilibrium state corresponding to a given parameter vector $\boldsymbol{\eta}_0$; this state is represented by a point in state-parameter space and has to be on the equilibrium surface. If $\boldsymbol{\eta}$ is varied slowly in some prescribed way (e.g., consider a ray of the form $\eta^\alpha = \eta_0^\alpha + l^\alpha \sigma$, where l^α are certain constants and σ a scalar variable), the equilibrium point moves on an equilibrium path (sub-manifold of the equilibrium surface), and the eigenvalues of the Jacobian

$$[X_{ij}(\boldsymbol{x}, \boldsymbol{\eta})] = \left[\frac{\partial X_i(\boldsymbol{x}, \boldsymbol{\eta})}{\partial x^j}\right], \tag{1.57}$$

evaluated on this path, change accordingly. Mainly two types of instability may occur: *static instability* (*divergence*) or *dynamic instability* (*flutter*). The former phenomenon takes place when at least one eigenvalue (λ) becomes positive (real) after passing through zero at $\sigma = \sigma_c$ where the Jacobian becomes singular. Generically, a real eigenvalue (negative for $\sigma < \sigma_c$) crosses the origin of the complex λ-plane at $\sigma = \sigma_c$ and becomes positive for $\sigma > \sigma_c$. However, the Jacobian also becomes singular when a conjugate pair of eigenvalues coincides at the origin of the λ-plane and proceeds in opposite directions on the real axis. Clearly, an equilibrium state is *divergent unstable* when at least one eigenvalue is positive (real). On the other hand, *dynamic instability* or *flutter* is exhibited, for instance, when a pair of complex conjugate eigenvalues crosses the imaginary axis of λ-plane, resulting in a conjugate

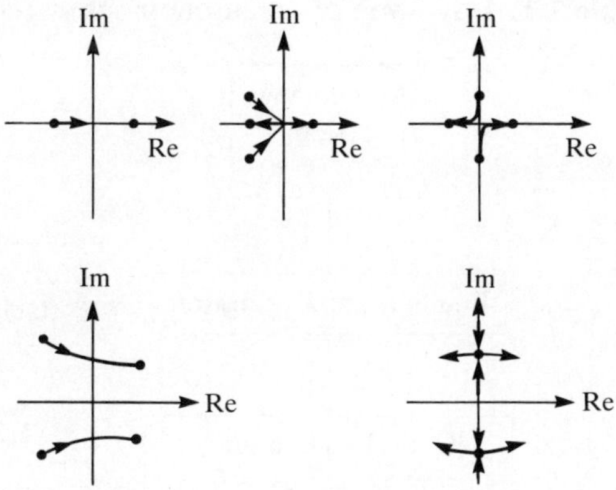

Fig. 1.14 Instabilities as eigenvalues pass through the complex λ-plane.

pair with positive real part. Flutter may also occur when two imaginary eigenvalues approach each other and coalesce at $\sigma = \sigma_c$, eventually leaving the imaginary axis in opposite directions, and producing an eigenvalue with a positive real part which leads to self-excited oscillations (Fig. 1.14). This situation is not generic, and occurs in the models of non-potential elastic stability problems which ignore damping completely. Flutter, however, is again *characterized by a complex eigenvalue with positive real part as in the first case*.

At a divergence (critical) point, bifurcation of equilibrium paths and exchange of stabilities may occur. More generally, the equilibrium surface exhibits topologically distinct forms in the vicinity of a divergence point, depending on other factors related to the latter, and dynamic jumps may also occur. However, the salient features of the behaviour (for $t \to \infty$) are normally concerned with equilibria and statics (Tables 1.1 and 1.2). On the other hand, a flutter (critical) point represents the birth of a distinctly different behaviour, namely periodic motions. Indeed, as a pair of complex conjugate eigenvalues crosses the imaginary axis, a family of limit cycles may branch off the critical equilibrium state (e.g. Hopf bifurcation), and the system assumes a dynamic mode (limit cycle).

Consider again the multiple-parameter system: the critical points in the state-parameter space (R^{n+m})—whether divergence or flutter—are said to form a *critical zone*, and the points of the parameter space (R^m) corresponding to the *critical zone* associated with an initial loss of stability constitute the *stability boundary* in R^m. Clearly, the stability boundary may consist of both a *divergence boundary* and *flutter boundary* (Huseyin 1978*b*), and a parameter ray emanating from an initial point may first intersect either the divergence or the flutter boundary depending on its direction.

Table 1.1. Equilibria of an autonomous system

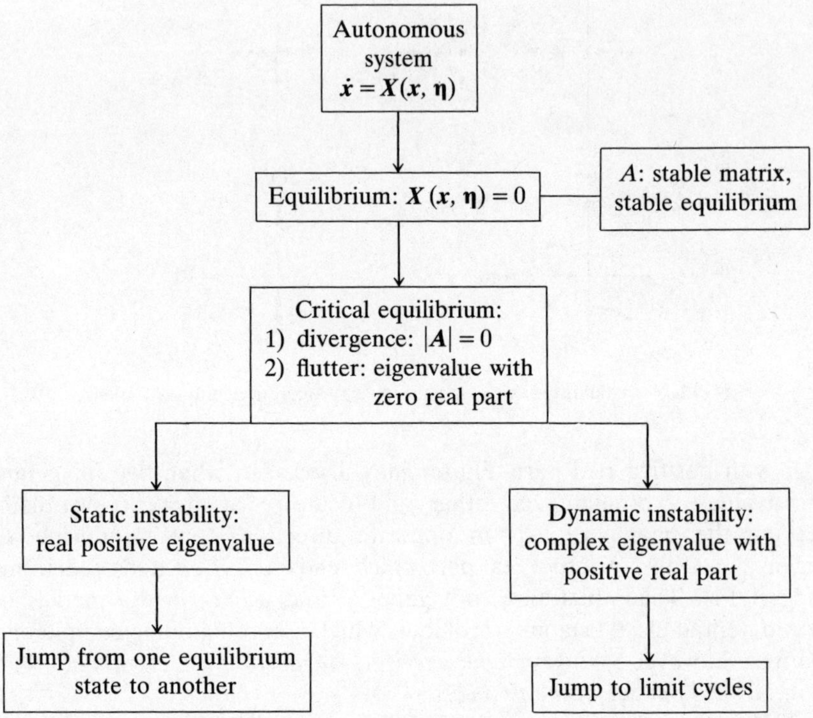

1.6 Classification of autonomous systems

An autonomous system described by eqns (1.6) or (1.55) is capable of embracing several classes of systems with distinct features and behaviour characteristics. Identifying important sub-classes of autonomous systems may prove useful in recognizing certain basic properties associated with each class as well as setting the objectives and directions of analysis. Indeed, the classification adopted for the treatment of linear autonomous systems (Huseyin 1978*b*) has been instrumental in exposing many interesting features of various autonomous systems in a linear context. Here, however, emphasis is placed on a more fundamental classification, whereby the concepts of *static* and *dynamic instabilities* play a predominant role. The following three classes of systems will help identify the underlying principles.

1) *Gradient systems*. The simplest class of autonomous systems that come to mind is so-called *gradient systems* associated with elementary catastrophe theory. Essentially, a gradient system is a special autonomous

Table 1.2.

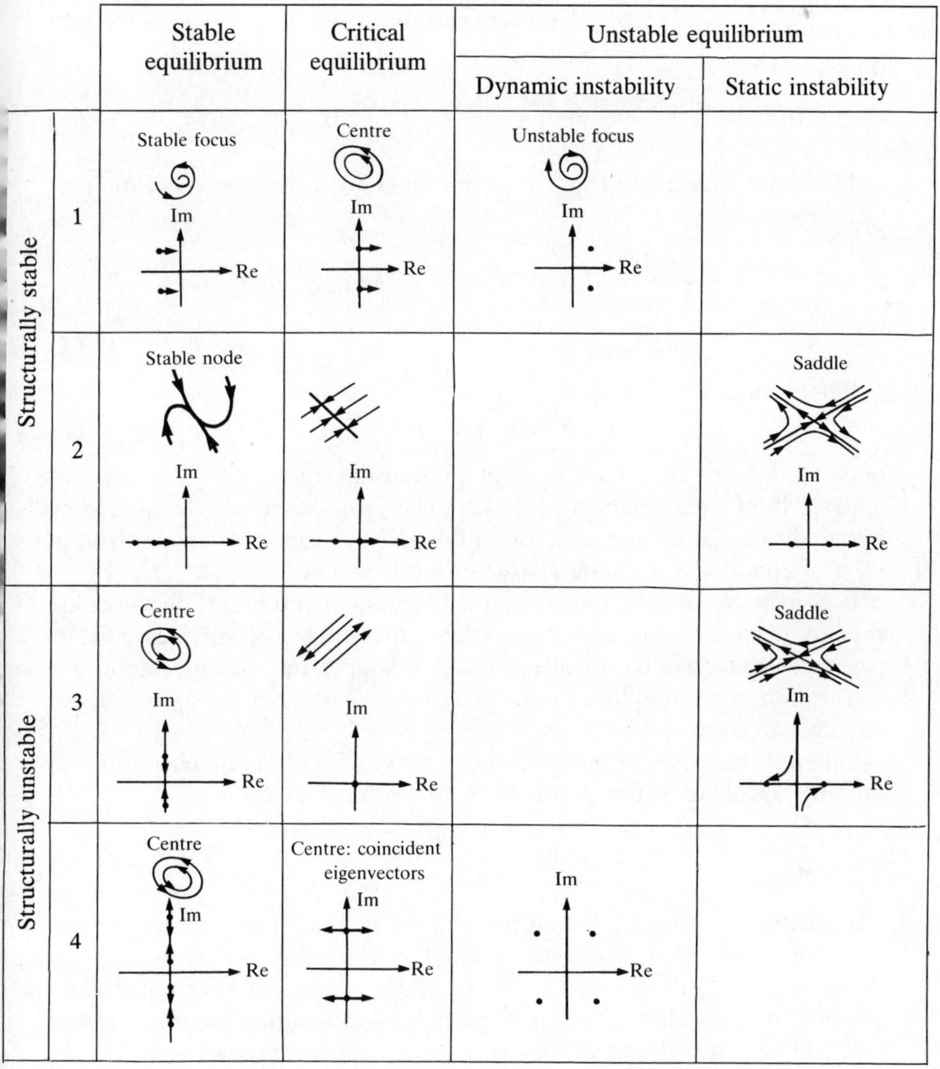
Stable equilibrium
Critical equilibrium
Unstable equilibrium
Dynamic instability
Static instability
Structurally stable
Structurally unstable
1
2
3
4
Stable focus
Centre
Unstable focus
Stable node
Saddle
Centre
Saddle
Centre
Centre: coincident eigenvectors
Im
Re

system (eqn 1.55), with a vector field of the form $\boldsymbol{X}(\boldsymbol{x}, \boldsymbol{\eta}) = -\mathrm{grad}_{\boldsymbol{x}} F(\boldsymbol{x}, \boldsymbol{\eta})$, and is described by

$$\dot{\boldsymbol{x}} = -\mathrm{grad}_{\boldsymbol{x}} F(\boldsymbol{x}, \boldsymbol{\eta}), \tag{1.58}$$

where the potential $F(\boldsymbol{x}, \boldsymbol{\eta})$ is a smooth function of $\boldsymbol{x}$ and $\boldsymbol{\eta}$, and

$$\mathrm{grad}_{\boldsymbol{x}} F = \left(\frac{\partial F}{\partial x^1}, \frac{\partial F}{\partial x^2}, \ldots, \frac{\partial F}{\partial x^n}\right). \tag{1.59}$$

The equilibria of the gradient system (eqn 1.58) are then given by

$$\text{grad}_{\boldsymbol{x}} F(\boldsymbol{x}, \boldsymbol{\eta}) = \boldsymbol{0}, \tag{1.60}$$

or

$$\frac{\partial F}{\partial x^1} = \frac{\partial F}{\partial x^2} = \ldots = \frac{\partial F}{\partial x^n} = 0. \tag{1.61}$$

The time derivative of the potential function (with η held constant) is given by

$$\begin{aligned}\dot{F} &= \frac{\partial F}{\partial x^i}\frac{\mathrm{d}x^i}{\mathrm{d}t} = \langle \text{grad}_{\boldsymbol{x}} F, \dot{\boldsymbol{x}} \rangle = -\langle \text{grad}_{\boldsymbol{x}} F, \text{grad}_{\boldsymbol{x}} F \rangle \\ &= -\|\text{grad}_{\boldsymbol{x}} F\|^2. \end{aligned}\tag{1.62}$$

It follows that

$$\dot{F} < 0 \quad \text{for} \quad F \neq 0, \tag{1.63}$$

and $\dot{F} = 0$ if and only if $\boldsymbol{x} = \boldsymbol{x}_e$ is an equilibrium state.

Also, if an equilibrium state $\boldsymbol{x} = \boldsymbol{x}_e$ (where the function F is stationary) is actually a relative minimum of F, then F is a Lyapunov function and in view of eqn (1.63) the state is asymptotically stable.

It is also noted that the direction $(-\text{grad}_{\boldsymbol{x}} F)$ represents the greatest rate of decrease, and the trajectories cross the level surfaces $F(\boldsymbol{x}) =$ constant orthogonally. Steady states of a gradient system, therefore, consists of the equilibria, and as $t \to \infty$ all trajectories approach an equilibrium state.

Suppose the origin of the gradient system (1.58) is an equilibrium state; then the Jacobian at this point takes the form

$$\mathbf{A}(\mathbf{0}) = -\left[\frac{\partial^2 F}{\partial x^i\, \partial x^j}\right]_{\boldsymbol{x}=\mathbf{0}}. \tag{1.64}$$

The Jacobian matrix is, therefore, symmetric and all its eigenvalues are real. The associated eigenvectors are linearly independent and can be assumed to be orthogonal; i.e. the Jacobian matrix (1.64) can always be diagonalized by means of an orthogonal transformation (Huseyin 1978*b*).

It follows that *flutter instability cannot occur, and the only type of instability exhibited by a gradient system is static instability* (*divergence*).

2) *Hamiltonian systems*. In classical mechanics, free vibrations of a conservative, elastic, and discrete system may be formulated via Lagrange's equations, which yield (e.g. Huseyin 1978*b*)

$$\boldsymbol{M}\ddot{\boldsymbol{q}} + \boldsymbol{k}(\boldsymbol{q}) = \boldsymbol{0} \tag{1.65}$$

where the mass matrix $\boldsymbol{M}$ is symmetric positive definite and the restoring forces represented by the vector $\boldsymbol{k}(\boldsymbol{q})$ have non-linear elements expressed as smooth functions of the generalized co-ordinates q^i $(i = 1, 2, \ldots, n)$. Assume, without loss of generality, that $\boldsymbol{M} = \boldsymbol{I}$ (the identity matrix) and $\boldsymbol{k}(\boldsymbol{q}) = \text{grad}\, V(\boldsymbol{q})$, where $V(\boldsymbol{q})$ is the potential energy of the system. Then the equations of motion (1.65) can be expressed as

$$\ddot{\boldsymbol{q}} + \text{grad}\, V(\boldsymbol{q}) = \boldsymbol{0}. \tag{1.66}$$

The kinetic energy of the system is assumed to be given by

$$T = \tfrac{1}{2}\langle \dot{\boldsymbol{q}}, \dot{\boldsymbol{q}} \rangle, \tag{1.67}$$

and the total energy $H = T + V$ is called a Hamiltonian:

$$H(\boldsymbol{q}, \dot{\boldsymbol{q}}) = \tfrac{1}{2}\langle \dot{\boldsymbol{q}}, \dot{\boldsymbol{q}} \rangle + V(\boldsymbol{q}). \tag{1.68}$$

An alternative way of describing the free vibrations of a conservative system is to use Hamilton's equations which are equivalent to eqn (1.66). To demonstrate this, let $\dot{\boldsymbol{q}} = \boldsymbol{p}$ so that $\ddot{\boldsymbol{q}} = \dot{\boldsymbol{p}}$; then eqn (1.66) can be expressed as

$$\dot{\boldsymbol{q}} = \boldsymbol{p}, \qquad \dot{\boldsymbol{p}} = -\text{grad}\, V(\boldsymbol{q}). \tag{1.69}$$

On the other hand, it follows from $H(q, p)$ in eqn (1.68) that

$$\frac{\partial H}{\partial \boldsymbol{q}} = \text{grad}\, V(\boldsymbol{q}) \quad \text{and} \quad \frac{\partial H}{\partial \boldsymbol{p}} = \boldsymbol{p}, \tag{1.70}$$

which, in conjunction with eqn (1.69), lead to

$$\dot{\boldsymbol{q}} = \frac{\partial H}{\partial \boldsymbol{p}} \quad \text{and} \quad \dot{\boldsymbol{p}} = -\frac{\partial H}{\partial \boldsymbol{q}}, \tag{1.71}$$

the well-known Hamilton's equations in the *state-space* spanned by $\boldsymbol{q}$ and $\boldsymbol{p}$.

Hamilton's equations can further be combined into

$$\dot{\boldsymbol{x}} = \boldsymbol{L}\, \text{grad}_{\boldsymbol{x}} H(\boldsymbol{x}), \tag{1.72}$$

where the vector field is the *symplectic gradient* of the Hamiltonian $H(\boldsymbol{x})$,

$$\boldsymbol{L} = \begin{bmatrix} \boldsymbol{0} & \boldsymbol{I} \\ -\boldsymbol{I} & \boldsymbol{0} \end{bmatrix}, \qquad \boldsymbol{x} = (\boldsymbol{q}, \boldsymbol{p}), \tag{1.73}$$

and $\boldsymbol{I}$ denotes the $n \times n$ unit matrix. Eqn (1.72) follows from

$$\begin{bmatrix} \dot{\boldsymbol{q}} \\ \dot{\boldsymbol{p}} \end{bmatrix} = \begin{bmatrix} \boldsymbol{0} & \boldsymbol{I} \\ -\boldsymbol{I} & \boldsymbol{0} \end{bmatrix} \begin{bmatrix} \text{grad}_{\boldsymbol{q}} H \\ \text{grad}_{\boldsymbol{p}} H \end{bmatrix}, \tag{1.74}$$

which is identical to eqn (1.71).

The skew-symmetric matrix $\boldsymbol{L}$ and eqn (1.72) indicate that Hamiltonian systems have a *symplectic structure*, distinguishing this class of systems from gradient systems. The equilibrium states satisfy

$$\text{grad}_{\boldsymbol{x}} H(\boldsymbol{x}) = \boldsymbol{0}, \tag{1.75}$$

or equivalently

$$\frac{\partial H}{\partial \boldsymbol{p}} = \mathbf{0} \quad \text{and} \quad \frac{\partial H}{\partial \boldsymbol{q}} = \mathbf{0}, \tag{1.76}$$

which imply

$$\operatorname{grad} V(\boldsymbol{q}) = \mathbf{0} \quad \text{or} \quad \frac{\partial V}{\partial \boldsymbol{q}} = \mathbf{0}. \tag{1.77}$$

Clearly, if the system is under the influence of a parameter vector $\boldsymbol{\eta}$, the potential energy V and the Hamiltonian H contain $\boldsymbol{\eta}$ as well as $\boldsymbol{x}$, and eqn (1.75) takes the form of eqn (1.60).

The time derivative of the potential function $H(\boldsymbol{x})$ vanishes, because

$$\begin{aligned} \dot{H} &= \frac{\partial H}{\partial \boldsymbol{x}} \frac{\mathrm{d}\boldsymbol{x}}{\mathrm{d}t} = \langle \operatorname{grad}_{\boldsymbol{x}} H, \dot{\boldsymbol{x}} \rangle \\ &= \langle \operatorname{grad}_{\boldsymbol{x}} H, \boldsymbol{L} \operatorname{grad}_{\boldsymbol{x}} H \rangle = 0 \end{aligned} \tag{1.78}$$

for all $\boldsymbol{x}$ and t, $\boldsymbol{L}$ being a skew-symmetric matrix.

It follows that the potential function H is constant on the solutions of eqn (1.72), in contrast with the potential function (F) of gradient systems which is a decreasing function of t everywhere except at equilibrium states. This is the well-known *law of conservation of energy* associated with mechanical systems. If the potential energy $V(\boldsymbol{q})$ is positive definite, eqn (1.68) shows that H is also positive definite, and H is therefore a Lyapunov function, indicating that the corresponding state is *stable*. Thus, *Lagrange's theorem* is established: *An equilibrium state is stable if the potential energy has a complete relative minimum at that point*. Obviously, *asymptotic stability is ruled out in a Hamiltonian system*.

Suppose the origin of system (1.72) is an equilibrium state; the Jacobian at this point is

$$\boldsymbol{A}(\mathbf{0}) = \boldsymbol{L}\boldsymbol{H} = \begin{bmatrix} \mathbf{0} & \boldsymbol{I} \\ -\boldsymbol{I} & \mathbf{0} \end{bmatrix} \begin{bmatrix} \boldsymbol{V} & \mathbf{0} \\ \mathbf{0} & \boldsymbol{I} \end{bmatrix} = \begin{bmatrix} \mathbf{0} & \boldsymbol{I} \\ -\boldsymbol{V} & \mathbf{0} \end{bmatrix} \tag{1.79}$$

where

$$\boldsymbol{H} = [\partial^2 H / \partial x^i \, \partial x^j]_{x^i=0} \quad \text{and} \quad \boldsymbol{V} = [\partial^2 V / \partial q^i \, \partial q^j]_{q^i=0}.$$

The eigenvalues of $A(0)$ can, then, be determined from the characteristic determinant

$$\begin{vmatrix} -\lambda \boldsymbol{I} & \boldsymbol{I} \\ -\boldsymbol{V} & -\lambda \boldsymbol{I} \end{vmatrix} = 0, \tag{1.80}$$

which is the same as

$$|\boldsymbol{V} + \lambda^2 I| = 0. \tag{1.81}$$

Since the matrix $\boldsymbol{V}$ is symmetric, its eigenvalues are always real, and the roots λ of eqn (1.81) are, therefore, either real or imaginary. This implies that *dynamic instability (flutter) cannot occur, and the only type of instability exhibited by an equilibrium state of a Hamiltonian system is static instability (divergence)*. Note, however, that unlike gradient systems, the equilibria are not the only steady states in a Hamiltonian system, and periodic motions can occur. Indeed, an equilibrium state where the eigenvalues of the Jacobian are all imaginary is a *centre*. As remarked earlier, centres are structurally unstable, and the addition of damping to the system transforms a *centre* into a *focus*. This type of system will be considered next.

3) *Damped Hamiltonian systems*. Suppose a positive definite damping term is added to eqn (1.65), and assume again that the mass matrix is transformed to the identity matrix $\boldsymbol{I}$; then the equations of motion are governed by

$$\ddot{\boldsymbol{q}} + \boldsymbol{D}\dot{\boldsymbol{q}} + \text{grad}\, V(\boldsymbol{q}) = \boldsymbol{0}, \tag{1.82}$$

where $\boldsymbol{D} = \boldsymbol{D}' > 0$.

Alternatively, the system (1.82) can be expressed as

$$\begin{aligned} \dot{\boldsymbol{q}} &= \boldsymbol{p} \\ \dot{\boldsymbol{p}} &= -\boldsymbol{D}\boldsymbol{p} - \text{grad}\, V(\boldsymbol{q}) \end{aligned} \tag{1.83}$$

in analogy with eqn (1.69). This is a damped Hamiltonian system, and its equilibrium states satisfy

$$\frac{\partial H}{\partial \boldsymbol{q}} = \frac{\partial H}{\partial \boldsymbol{p}} = \boldsymbol{0} \quad \text{or} \quad \text{grad}\, V(\boldsymbol{q}) = \boldsymbol{0}. \tag{1.84}$$

In other words, the equilibria of a Hamiltonian system remain unaffected when damping is added to the system.

The time derivative of the Hamiltonian $H(\boldsymbol{q}, \boldsymbol{p})$ on the solutions of eqns (1.83) now takes the form

$$\begin{aligned} \dot{H} &= \frac{\partial H}{\partial \boldsymbol{q}} \dot{\boldsymbol{q}} + \frac{\partial H}{\partial \boldsymbol{p}} \dot{\boldsymbol{p}} \\ &= \langle \text{grad}\, V(\boldsymbol{q}), \boldsymbol{p} \rangle + \langle \boldsymbol{p}, (-\boldsymbol{D}\boldsymbol{p} - \text{grad}\, V(\boldsymbol{q})) \rangle \\ &= -\langle \boldsymbol{p}, \boldsymbol{D}\boldsymbol{p} \rangle. \end{aligned} \tag{1.85}$$

Clearly $\dot{H} < 0$ since $\boldsymbol{D} > 0$, and if an equilibrium state of the undamped Hamiltonian system ($\boldsymbol{D} = 0$) is stable ($V(\boldsymbol{q})$ minimum), H is a Lyapunov function with $\dot{H} < 0$, indicating stability.

Suppose again that the origin of system (1.83) is an equilibrium state; the Jacobian of eqn (1.83) at this point is

$$\boldsymbol{A}(\mathbf{0}) = \begin{bmatrix} \mathbf{0} & \boldsymbol{I} \\ -\boldsymbol{V} & -\boldsymbol{D} \end{bmatrix}, \tag{1.86}$$

and the eigenvalues of $\boldsymbol{A}(\mathbf{0})$ can be determined from

$$\begin{vmatrix} -\lambda \boldsymbol{I} & \boldsymbol{I} \\ -\boldsymbol{V} & -\boldsymbol{D} - \lambda \boldsymbol{I} \end{vmatrix} = 0, \tag{1.87}$$

which is equivalent to

$$|\lambda^2 \boldsymbol{I} + \lambda \boldsymbol{D} + \boldsymbol{V}| = 0. \tag{1.88}$$

It is first noted that when the Jacobian (1.86) becomes singular,

$$\begin{vmatrix} \mathbf{0} & \boldsymbol{I} \\ -\boldsymbol{V} & -\boldsymbol{D} \end{vmatrix} = 0, \tag{1.89}$$

Schur's formula implies that $|\boldsymbol{V}| = 0$ and, therefore, the Jacobian (1.79) of the corresponding Hamiltonian system also becomes singular since

$$\begin{vmatrix} \mathbf{0} & \boldsymbol{I} \\ -\boldsymbol{V} & \mathbf{0} \end{vmatrix} = |\boldsymbol{V}| = 0. \tag{1.90}$$

It is then inferred that if the damped Hamiltonian system is at a divergence point, so is the corresponding Hamiltonian system. Furthermore, it can be shown that although the eigenvalues of eqn (1.88) may be complex, a complex eigenvalue can never have a positive real part (Huseyin 1978*b*, pp. 82–83), and if the real part is positive, the imaginary part is zero, implying that an eigenvalue with a positive real part is actually real. The obvious conclusion is that *a damped* ($\boldsymbol{D} > 0$) *Hamiltonian system cannot exhibit flutter instability*.

The background for a major classification of autonomous systems has now been established. It is observed that the three classes of systems considered above—that is, gradient, Hamiltonian, and damped Hamiltonian systems—have certain common features with regard to their behaviour characteristics in general and stability properties in particular. These remarkable features can be summarized as follows:

1) All systems in the three classes considered are described by a smooth potential function such that at the equilibria of these systems the gradient of the potential function vanishes (grad $F = \mathbf{0}$, or grad $H = \mathbf{0}$). Furthermore, there may exist an additional function with this property (grad $V(\boldsymbol{q}) = \mathbf{0}$).

2) Away from the equilibria the potential function decreases or remains constant as t increases.

3) Dynamic instability (flutter) is ruled out, and a stable equilibrium

state can become unstable only through divergence (static instability) as the parameters of the system are varied slowly.

In addition to the classes of systems identified above, there may be other autonomous systems possessing these properties, and the entire group will be described as *potential systems*. A general autonomous system of the form (1.55) which is not in this category will be called a *non-potential system*. In general, both static and dynamic instabilities can be exhibited by *non-potential systems* (Table 1.3).

A statical analysis based on a potential function reveals the salient features of a potential system—such as the equilibria and instabilities. In particular, if the system is a gradient system or damped Hamiltonian system the transients die away rather quickly, and excluding dynamics from the model is not necessarily a disadvantage. Indeed, a model containing *statics* only has its own advantages—e.g. conceptual simplicity and enhanced geometric intuition. In the case of Hamiltonian systems, a dynamical analysis is, of course, necessary to examine the oscillatory behaviour. Nevertheless, a statical analysis is again capable of yielding useful information concerning the equilibria and instabilities.

Table 1.3. Classification of potential and non-potential systems

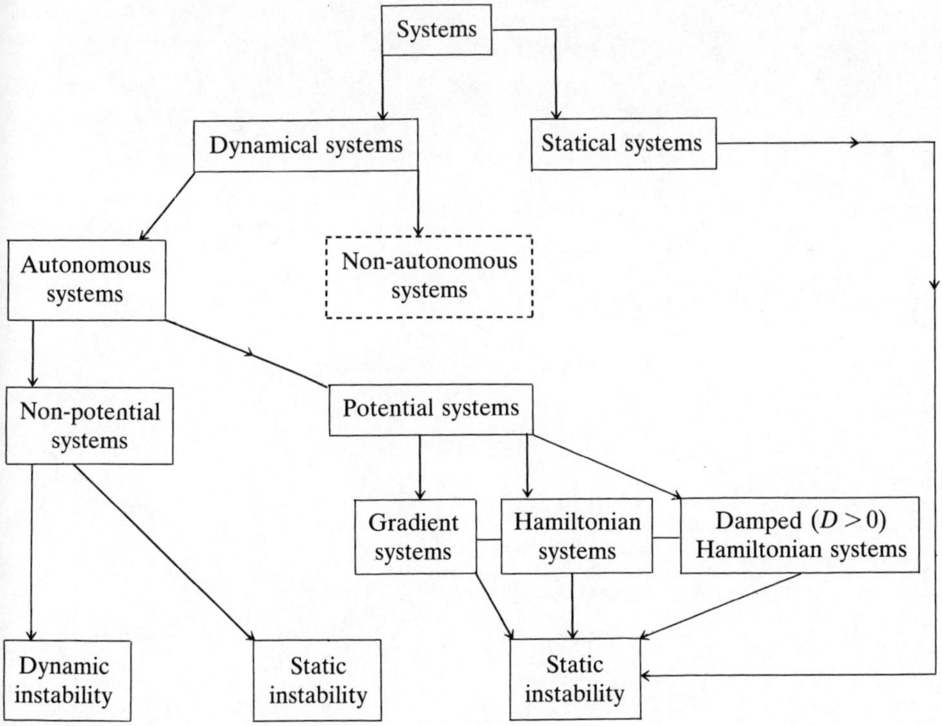

The next chapter is concerned with a concise general theory of multiple-parameter potential systems. The treatment represents a generalization of the theory presented in an earlier monograph (Huseyin 1975), and most of the results obtained there can be derived as a special case from the theory in the next chapter.

2

POTENTIAL SYSTEMS

2.1 Potential function and stability of equilibrium states

Consider a finite-dimensional system characterized by a smooth potential function

$$V = V(\boldsymbol{q}, \boldsymbol{\eta}); \qquad \boldsymbol{q} \in R^n, \qquad \boldsymbol{\eta} \in R^m, \tag{2.1}$$

where the generalized coordinates q^i $(i = 1, 2, \ldots, n)$ describe the behaviour of the system and may, therefore, be called *behaviour variables* as well. The η^α $(\alpha = 1, 2, \ldots, m)$ represent m independent parameters influencing the behaviour of the system, and they are also referred to as *control parameters*. The $(n + m)$-dimensional *configuration space* R^{n+m}, spanned by q^i and η^α, is assumed to be a Euclidean space, and the statical behaviour of the system is described in this space. It is understood that if dynamics were to be introduced in the formulation here, the configuration space would have to be expanded into a $(2n + m)$ dimensional *state-parameter* space, depending on the nature of the dynamics (Section 1.6).

The equilibrium states of the potential system described by the potential function (2.1) are defined by

$$\operatorname{grad}_{\boldsymbol{q}} V(\boldsymbol{q}, \boldsymbol{\eta}) = 0, \tag{2.2}$$

which can also be written as

$$\frac{\partial V}{\partial q^i} = 0, \qquad (i = 1, 2, \ldots, n). \tag{2.3}$$

Normally, these equations define an m-dimensional *equilibrium surface* (*manifold*) in R^{n+m} which may have a complicated shape (Huseyin 1975). A general non-linear analysis leading to global features of the equilibrium surface is almost impossible, but the local characteristics in the neighbourhood of appropriately chosen equilibrium states may provide valuable information about the behaviour of the system, and can be explored in general terms. *Critical equilibrium states*, for example, are such pertinent states, and the properties of the equilibrium surface in the vicinity of critical points may indeed be most revealing.

It was shown in Section 1.6 that an equilibrium state c is stable if the potential function $V(\boldsymbol{q}, \boldsymbol{\eta})$ has a complete relative minimum at that point. A sufficient condition for $V(\boldsymbol{q}, \boldsymbol{\eta})$ to have a relative minimum is that the second variation $\delta^2 V$ is positive definite. This implies that the Hessian matrix evaluated at the equilibrium state c is positive definite, i.e.

$$[V_{ij}] \triangleq \left[\frac{\partial^2 V(\boldsymbol{q}, \boldsymbol{\eta})}{\partial q^i \, \partial q^j}\right]_c > 0. \tag{2.4}$$

If the Hessian matrix is negative definite, negative semi-definite or indefinite, it is impossible for V to have a minimum, and the system is unstable. On the other hand, if the Hessian matrix at c is positive semi-definite, no conclusion can be reached concerning the stability of the equilibrium state before exploring the higher order variations of V. Such a state of equilibrium is called *primary critical*, and it is clear that the determinant of the Hessian matrix vanishes at such a point, i.e.

$$|V_{ij}| = 0. \tag{2.5}$$

Here, any equilibrium point satisfying eqn (2.5) will be called *critical*, and it follows that if a critical point is not primary, then it is unstable. Primary critical points, where an initially stable system may lose its stability, are obviously of great practical significance. As noted above, the second variation of V fails to provide sufficient information concerning the extremum properties of V at such points; it is evident, however, that the third variation of V must vanish if V is to have a minimum. If this is the case, then the positive definiteness of the fourth variation emerges as a sufficient condition for stability, ensuring a relative minimum for V. Similarly, a vanishing fifth variation followed by a positive-definite sixth variation become the necessary and sufficient conditions for stability, respectively, if the fourth variation turns out to be positive semi-definite. This pattern may continue until the maximum–minimum properties of V can be established.

Since the stability of a primary critical point determines the stability of post-critical behaviour (Koiter 1945, Huseyin 1975), it is extremely important to have explicit necessary and/or sufficient conditions for stability. Such conditions have, in fact, been derived (Huseyin 1972*a*, 1975) on the basis of a transformed potential function. Consider an arbitrary equilibrium state c: $(\boldsymbol{q}_c, \boldsymbol{\eta}_c)$ on the equilibrium surface and introduce the orthogonal transformation

$$\boldsymbol{q} = \boldsymbol{q}_c + \boldsymbol{T}\boldsymbol{u}, \qquad \boldsymbol{T}\boldsymbol{T}' = \boldsymbol{I}, \tag{2.6}$$

and

$$\boldsymbol{\eta} = \boldsymbol{\eta}_c + \boldsymbol{\mu}, \tag{2.7}$$

to obtain

$$H(\boldsymbol{u}, \boldsymbol{\mu}) \equiv V(\boldsymbol{q}_c + \boldsymbol{T}\boldsymbol{u}, \boldsymbol{\eta}_c + \boldsymbol{\mu}) \tag{2.8}$$

where the Hessian matrix

$$[H_{ij}] \triangleq \left[\frac{\partial^2 H}{\partial u^i\, \partial u^j}\right]_c = \begin{bmatrix} h_1 & & & 0 \\ & h_2 & & \\ & & \ddots & \\ 0 & & & h_n \end{bmatrix}_c \tag{2.9}$$

is diagonal, and the u^i $(i = 1, 2, \ldots, n)$ are the principal coordinates. Here, and in the following, subscripts on H denote derivatives with respect to the corresponding variables.

If all the *stability coefficients* h_i $(i = 1, 2, \ldots, n)$ are non-zero and positive, the equilibrium state c is stable. If any of these coefficients is negative, the state is unstable. On the other hand, if the point c represents the primary critical point, then, r $(r \leq n)$ elements (eigenvalues) of eqn (2.9) vanish simultaneously, and c is said to be an *r-fold* (*coincident*) critical point; *simple* (*distinct*) critical points are the special cases in which $r = 1$. Also, note that the equilibrium condition (eqn 2.2) is now replaced by

$$\operatorname{grad}_{\boldsymbol{u}} H(\boldsymbol{u}, \boldsymbol{\mu}) = 0. \tag{2.10}$$

Suppose c is an r-fold critical point; then

$$h_a = 0 \quad \text{where} \quad a = 1, 2, \ldots, r;\ r < n, \tag{2.11}$$

and

$$h_s > 0 \quad \text{where} \quad s = r + 1, \ldots, n. \tag{2.12}$$

The notation u^a and u^s suggest themselves for the corresponding coordinates.

In order to examine whether or not the potential function (eqn 2.8) has a relative minimum at c, consider the variations of the potential function along an arbitrary path (in u^i-space) of the form

$$u^i(\xi) = u^{i,\xi}\xi + \tfrac{1}{2}u^{i,\xi\xi}\xi^2 + \ldots \tag{2.13}$$

where ξ is a parameter, $\xi = 0$ at c, and $u^{i,\xi} \triangleq \partial u^i/\partial \xi|_{\xi=0}$, etc.

The variation of the potential function with respect to the path of eqn (2.13) is expressed as

$$h(\xi) = H\{u^i(\xi), 0\} - H_c(0, 0). \tag{2.14}$$

Then

$$\frac{\mathrm{d}h}{\mathrm{d}\xi} = H_i u^{i,\xi}, \qquad u^{i,\xi} \triangleq \frac{\partial u^i}{\partial \xi}, \tag{2.15}$$

where the summation convention is adopted, and will be employed throughout the book unless stated otherwise. By virtue of eqn (2.10), one has

$$\left.\frac{\mathrm{d}h}{\mathrm{d}\xi}\right|_c = 0.$$

A second differentiation of eqn (2.14) yields

$$\frac{\mathrm{d}^2h}{\mathrm{d}\xi^2} = H_{ij}u^{i,\xi}u^{j,\xi} + H_iu^{i,\xi\xi}, \tag{2.16}$$

and evaluation at c results in

$$\left.\frac{\mathrm{d}^2h}{\mathrm{d}\xi^2}\right|_c = h_a(u^{a,\xi})^2 + h_s(u^{s,\xi})^2, \tag{2.17}$$

or, using eqns (2.11) and (2.12),

$$\left.\frac{\mathrm{d}^2h}{\mathrm{d}\xi^2}\right|_c = h_s(u^{s,\xi})^2 > 0, \tag{2.18}$$

which indicates that the second variation of the potential function,

$$\delta^2 h = \frac{1}{2}\left.\frac{\mathrm{d}^2h}{\mathrm{d}\xi^2}\right|_c \xi^2,$$

along the path of the form of eqn (2.13), which involves at least one u^s, is positive. On the other hand, $\delta^2 h$ with respect to a path which is entirely in the sub-space spanned by the critical coordinates u^a vanishes. It is, therefore, evident that higher order variations with respect to an infinite number of paths, which are *initially* in u^a-space, must now be examined. In view of eqn (2.11), this can best be done with respect to a surface directly, rather than a path, by expressing u^s in terms of u^a and considering the surface

$$u^s(u^a) = u^{s,a}u^a + \tfrac{1}{2}u^{s,ab}u^au^b + \ldots, \tag{2.19}$$

where $u^{a,b} = u^{a,bc} = \ldots = 0$, all $u^{s,a} = 0$, and $(a, b, c, \ldots)$ are used to denote the critical coordinates.

Equipped with this information, return now to the initial formulation and express the change in the potential function along the surface (2.19) as

$$h(u^a) = H\{u^s(u^a), u^a; 0\} - H_c, \tag{2.20}$$

which yields

$$\frac{\partial h}{\partial u^a} = H_su^{s,a} + H_a, \qquad (a = 1, 2, \ldots, r), \tag{2.21}$$

and

$$\left.\frac{\partial h}{\partial u^a}\right|_c = 0, \tag{2.22}$$

since $H_s = H_a = 0$ at c.

A second differentiation with respect to u^b $(b = 1, 2, \ldots, r)$ leads to

$$\left.\frac{\partial^2 h}{\partial u^a\,\partial u^b}\right|_c = h_s u^{s,a} u^{s,b} = 0, \tag{2.23}$$

since it has been assumed that $u^{s,a} = 0$ along the surface (2.19). It is noted that eqn (2.23), in fact, confirms the earlier observation that for any $u^{s,a} \neq 0$ the second variation

$$\delta^2 h = \left.\frac{\partial^2 h}{\partial u^a\,\partial u^b}\right|_c u^a u^b \tag{2.24}$$

will be positive. In other words, eqn (2.24) vanishes in the directions identified by $u^{s,a} = 0$ $(a = 1, 2, \ldots, r)$ only.

Similarly, a third differentiation of eqn (2.20) leads to

$$\left.\frac{\partial^3 h}{\partial u^a\,\partial u^b\,\partial u^c}\right|_c = H_{abc},$$

which should vanish if the potential function is to have a minimum. In other words,

$$H_{abc} = 0, \qquad (a, b, c = 1, 2, \ldots, r), \tag{2.25}$$

is a necessary condition for stability of the state c.

If this condition is satisfied, one proceeds to a fourth differentiation which yields

$$\left.\frac{\partial^4 h}{\partial u^a\,\partial u^b\,\partial u^c\,\partial u^d}\right|_c = H_{abcd} + 6H_{scd}u^{s,ab} + 3h_s u^{s,ab} u^{s,cd}. \tag{2.26}$$

It is observed that the second derivatives, $u^{s,ab}$, of the surface (eqn 2.19) have now appeared in the expression (2.26). However, $u^{s,ab}$ can readily be determined (Huseyin 1972*a*) by minimizing the fourth variation $\delta^4 h$ with respect to these unknown derivatives. Indeed,

$$\frac{\partial(\delta^4 h)}{\partial(u^{t,ab})} = \frac{1}{4!}(H_{tcd} + h_{\underline{t}} u^{t,cd}) u^c u^d = 0, \tag{2.27}$$

where t takes values ranging from $r + 1$ to n and the bar under t suspends the summation convention. Hence, one has the derivatives

$$u^{t,cd} = \frac{H_{tcd}}{h_t}. \tag{2.28}$$

Introducing eqn (2.28) into the fourth variation leads to

$$\delta^4 h = \frac{1}{4!}(H_{abcd} - 3H_{sab}H_{scd}/h_s) u^a u^b u^c u^d, \tag{2.29}$$

which is positive for all $u^a \neq 0$ if and only if the expression in parentheses is positive definite; that is,

$$\tilde{H}_{abcd} \triangleq (H_{abcd} - 3H_{sab}H_{scd}/h_s) > 0. \tag{2.30}$$

If eqn (2.29) admits negative values for some combination of u^a, the potential function has no minimum at c and the state is unstable. It is noted that eqn (2.30) is a sufficient condition for stability; if it turns out that $\tilde{H}_{abcd}$ is positive semi-definite rather than positive definite, one has to proceed to higher order variations according to the established procedure.

One way of ascertaining the positive definiteness of a quartic form like eqn (2.29) is to express it as a quadratic form. Assume, for example, that a general, two-variable quartic form is given as

$$y(x_1, x_2) = \frac{1}{4!}(ax_1^4 + 4bx_1^3x_2 + 6cx_1^2x_2^2 + 4dx_1x_2^3 + ex_2^4). \tag{2.31}$$

Here, y is a quartic form in x_1 and x_2, and it is possible to express it as

$$y(x_1, x_2) = \frac{1}{4!}\langle z, Az\rangle \tag{2.32}$$

where the vector $z = (x_1^2 \quad x_1x_2 \quad x_2^2)'$ and A is a symmetric matrix given by

$$A = \begin{bmatrix} a & 2b & c \\ 2b & 4c & 2d \\ c & 2d & e \end{bmatrix}. \tag{2.33}$$

The symmetric matrix A and the quadratic form (eqn 2.32) are positive definite if

$$a > 0,$$

$$\begin{vmatrix} a & 2b \\ 2b & 4c \end{vmatrix} > 0,$$

and (2.34)

$$\begin{vmatrix} a & 2b & c \\ 2b & 4c & 2d \\ c & 2d & e \end{vmatrix} > 0.$$

The results of this section are summarized in Table 2.1. Note that in the special case of $r = 1$, that is, in the case of a distinct critical point, one has

$$H_{abc} = H_{111} \quad \text{and} \quad \tilde{H}_{1111} = H_{1111} - 3\sum_s (H_{s11})^2/h_s$$

Table 2.1. Summary of results of Section 2.1

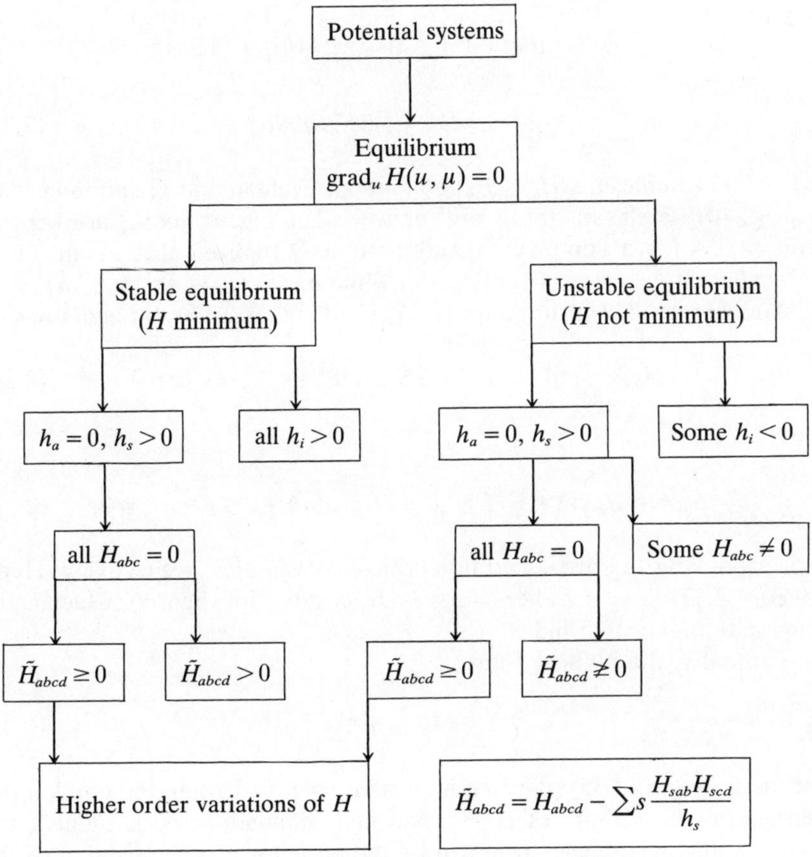

Problem 2.1. Derive the sufficient condition (eqn 2.30) for the stability of a critical equilibrium state by continuing the analysis in Section 2.1 on the basis of the path (eqn 2.13) rather than shifting to the reference surface (eqn 2.19). Discuss the procedure (see Huseyin 1975).

Problem 2.2. The paths with respect to which the variations of the potential function should be examined for a decision on stability can be identified in advance; establish an appropriate basis for such an identification and construct a systematic procedure.

2.2 Classification of critical conditions

Consider now the expansion of the potential function (eqn 2.8) into Taylor's series around an r-fold critical point c characterized by eqns

(2.11) and (2.12),

$$H(u^i, \mu^\alpha) = \frac{1}{2!}[h_a(u^a)^2 + h_s(u^s)^2 + 2H_{i\alpha}u^i\mu^\alpha]$$

$$+\frac{1}{3!}[H_{ijk}u^iu^ju^k + 3H_{ij\alpha}u^iu^j\mu^\alpha + \ldots] + \ldots, \qquad (2.35)$$

where the coefficients $H_{i\alpha}$, H_{ijk}, etc. are all evaluated at c, and in view of eqn (2.10), irrelevant terms are omitted. The eigenvalue h_a are kept in eqn (2.35) for a complete picture, but it is realized that at an r-fold critical point $h_a = 0$ $(a = 1, 2, \ldots, r)$ while $h_s > 0$ $(s = r+1, \ldots, n)$.

Applying $\text{grad}_u H = 0$ to eqn (2.35) yields the equilibrium equations

$$H_{a\alpha}\mu^\alpha + \tfrac{1}{2}(H_{aij}u^iu^j + 2H_{ai\alpha}u^i\mu^\alpha + \ldots) + \ldots = 0 \qquad (2.36)$$

and

$$h_{\underline{s}}u^s + H_{s\alpha}\mu^\alpha + \tfrac{1}{2}(H_{sij}u^iu^j + 2H_{si\alpha}u^i\mu^\alpha + \ldots) + \ldots = 0, \qquad (2.37)$$

corresponding to critical and non-critical coordinates, respectively. Here, in eqn (2.37), and elsewhere a bar under a subscript (here s) suspends the summation over that index.

Evidently, the bilinear form

$$H_{a\alpha}u^a\dot{\mu}^\alpha \qquad (2.38)$$

of the function (2.35) plays an important role in shaping the equilibrium surface in the vicinity of c (Huseyin and Mandadi 1977*a*). Indeed, the rank 'k' of this bilinear form can be the basis for a generalization of the concepts of *general* and *special* critical points introduced in a thesis (Huseyin 1967) in connection with *distinct* critical points. Thus, *if the rank of the bilinear form* (*eqn* 2.38) *k is equal to r, the critical point will be called 'general'*, and *if the rank k is less than r, it will be called 'special'*. The significance of the rank associated with eqn (2.38) lies in the fact that the equilibrium surface around c is normally a *proper* (*improper*) surface if the rank is r ($<r$), and c is *general* (*special*). Here, the term 'proper' refers to a surface which is smooth with continuously turning tangent planes and well-defined normals; this is not so in the *special* case. This definition reduces to the one given for distinct critical points (Huseyin 1967) for $r = 1$.

It can be shown that by introducing appropriate transformations of the variables u^a and μ^α, the bilinear form (eqn 2.38) can be replaced by a canonical form such that the matrix $H_{a\alpha}$ is given by one of the following canonical matrices:

$$[\boldsymbol{I} \quad \boldsymbol{0}] \quad \text{if the rank} \quad k=r<m, \tag{2.39}$$

$$\begin{bmatrix} \boldsymbol{I} \\ \boldsymbol{0} \end{bmatrix} \quad \text{if the rank} \quad k=m<r, \tag{2.40}$$

$$\begin{bmatrix} \boldsymbol{I} & \boldsymbol{0} \\ \boldsymbol{0} & \boldsymbol{0} \end{bmatrix} \quad \text{if the rank} \quad k<\min(r,m) \tag{2.41}$$

where $\boldsymbol{I}$ is a $k \times k$ identity matrix, and $\boldsymbol{0}$ represents a null matrix.

According to the definitions given above, eqn (2.39) is associated with a *general* critical point, and it will be shown (Section 2.3.2) that the transformation leading to matrix (2.39) can be accomplished by a suitable change of coordinates from $(\boldsymbol{u}, \boldsymbol{\mu})$ to $(\boldsymbol{v}, \boldsymbol{\varphi})$ such that the Taylor's expansion of the resulting function $\Pi(\boldsymbol{v}, \boldsymbol{\varphi})$ is in the form of

$$\Pi(v^i, \varphi^\alpha) = \frac{1}{2!}[\Pi_s(v^s)^2 + 2\delta_{a\varepsilon}v^a\varphi^\varepsilon + 2\Pi_{s\alpha}v^s\varphi^\alpha] + \frac{1}{3!}(\Pi_{ijk}v^iv^jv^k + 2\Pi_{ab\nu}v^av^b\varphi^\nu) + \dots, \tag{2.42}$$

where $\delta_{a\varepsilon}$ is the Kronecker delta and $\varepsilon = 1, 2, \dots, r$. It is noted that the transformation to the function Π results in a significant split in the parameters μ^α such that the set φ^ε $(\varepsilon = 1, 2, \dots, r)$ is associated with a linear form in v^a, while the remaining set φ^ν $(\nu = r+1, \dots, m)$ first appears in conjunction with a quadratic form in v^a. The former set of parameters will, therefore, be called *primary* and the latter set *secondary*. It will be seen that the primary parameters span the *normal space* of the equilibrium manifold, while the secondary parameters, together with the v^a, constitute a basis for the *tangent space*.

2.3 Equilibrium surface via the multiple-parameter perturbation technique

2.3.1 *Regular (non-critical) equilibrium states*

Consider first a non-critical point c on the equilibrium surface at which the Hessian is non-singular and all $h_i \neq 0$; such states are called *regular*. It follows from the fundamental existence theorem on implicit functions that a unique, single-valued equilibrium surface, given as $\boldsymbol{u}(\boldsymbol{\mu})$, passes through c. The asymptotic equations of this unique surface in the vicinity of c can be obtained through the multiple-parameter perturbation technique (Huseyin 1973, 1975). Thus, substituting the assumed solution $\boldsymbol{u}(\boldsymbol{\mu})$ into the equilibrium equation (2.10) results in the identically vanishing functions

$$H_i[u^j(\mu^\alpha), \mu^\alpha] \equiv 0. \tag{2.43}$$

Repeated differentiations of eqn (2.43) with respect to the independent parameters μ^α ($\alpha = 1, 2, \ldots, m$), and evaluation at c, yield

$$H_{ij}u^{j,\alpha} + H_{i\alpha} = 0, \tag{2.44}$$

$$(H_{ijk}u^{k,\beta} + H_{ij\beta})u^{j,\alpha} + H_{ij}u^{j,\alpha\beta} + H_{ij\alpha}u^{j,\beta} + H_{i\alpha\beta} = 0, \tag{2.45}$$

etc.

These equations can be solved successively to obtain the first order, second order, etc., surface derivatives $u^{j,\alpha}$, $u^{j,\alpha\beta}$, etc. to construct the surface through c as

$$u^i = u^{i,\alpha}\mu^\alpha + \frac{1}{2!}u^{i,\alpha\beta}\mu^\alpha\mu^\beta + \ldots \tag{2.46}$$

To this end, eqn (2.44) yields

$$u^{i,\alpha} = -\frac{H_{i\alpha}}{h_i}, \tag{2.47}$$

and the first-order equations of the equilibrium surface take the form

$$h_i u^i + H_{i\alpha}\mu^\alpha = 0,$$

which is, in fact, an m-dimensional plane in R^{n+m}, indicating a one-to-one correspondence between the parameters μ^α and u^i. This plane can be envisaged as a small local portion of the equilibrium surface around c, and for further information about the surface one has to substitute for the first derivatives (eqn 2.47) into eqn (2.45) to obtain the curvatures $u^{j,\alpha\beta}$, and so on.

A *regular* point c is structurally stable since the Hessian $|\partial^2 H/\partial u^i\, \partial u^j|_c$ does not vanish, and the behaviour of the system in the vicinity of a regular point is quite simple as demonstrated above. If, however, some of the stability coefficients vanish, the equilibrium surface may take a complicated shape in the vicinity of such a critical point. This will be explored next.

2.3.2 *General critical points*

Suppose now that c is a *general*, r-fold critical point described by

$$\text{grad}_{\boldsymbol{u}}\, H(\mathbf{0}, \mathbf{0}) = \mathbf{0},$$

$$h_a = 0; \quad a = 1, 2, \ldots, r \quad \text{and} \quad m > r.$$

Introduce next the non-singular transformations

$$\boldsymbol{u} = \boldsymbol{R}\boldsymbol{v},$$

where

$$\boldsymbol{R}=\begin{bmatrix}\boldsymbol{T} & \boldsymbol{0}\\ \boldsymbol{0} & \boldsymbol{I}\end{bmatrix},\qquad \boldsymbol{u}=\begin{bmatrix}u^a\\ u^s\end{bmatrix},\qquad \boldsymbol{v}=\begin{bmatrix}v^a\\ v^s\end{bmatrix},$$

and

$$\boldsymbol{\mu}=\boldsymbol{Q}\boldsymbol{\varphi},$$

such that the resulting function

$$\Pi(\boldsymbol{v},\boldsymbol{\varphi})\equiv H(\boldsymbol{R}\boldsymbol{v},\boldsymbol{Q}\boldsymbol{\varphi}) \tag{2.48}$$

has the properties

$$\begin{gathered}\Pi_{ij}=0 \quad \text{for} \quad i\neq j,\\ \Pi_{ii}\triangleq\Pi_i:\begin{cases}\Pi_a=0 & \text{for} \quad a=1,\ldots,r,\\ \Pi_s>0 & \text{for} \quad s=r+1,\ldots,n,\end{cases}\end{gathered} \tag{2.49}$$

and

$$[\Pi_{a\alpha}]=[\Pi_{a\varepsilon}\Pi_{a\nu}]=[\delta_{a\varepsilon}0] \quad \text{where} \quad \varepsilon=1,2,\ldots,r;\ \nu=r+1,\ldots,m.$$

Also note that $u^s=v^s$ and the eigenvalues of the Hessian remain invariant, i.e. $\Pi_s=h_s$.

It is assumed that the equilibrium equations can be solved simultaneously to obtain the equilibrium surface around c in a parametric form

$$\boldsymbol{v}=\boldsymbol{v}(\boldsymbol{\sigma}),\qquad \boldsymbol{\varphi}=\boldsymbol{\varphi}(\boldsymbol{\sigma}), \tag{2.50}$$

where $\boldsymbol{\sigma}$ represents a set of m unspecified parameters, $\boldsymbol{\sigma}=\boldsymbol{0}$ giving the critical point c. According to the multiple-parameter perturbation technique (Huseyin 1973, 1975), these perturbation parameters must be chosen such that the functions (eqn 2.50) are single-valued. It can be demonstrated that, in view of the transformations introduced here, the basic variables v^a and φ^ν are suitable for the role of $\boldsymbol{\sigma}$, and if this is recognized and employed right at the outset, the analysis becomes simpler.

Thus, substituting the assumed solutions $v^s(v^a,\varphi^\nu)$ and $\varphi^\varepsilon(v^a,\varphi^\nu)$ back into the equilibrium equations yields the identities

$$\Pi_i[v^j(v^a,\varphi^\nu),\varphi^\alpha(v^a,\varphi^\nu)]\equiv 0 \tag{2.51}$$

where $v^j(v^a,\varphi^\nu)$ and $\varphi^\alpha(v^a,\varphi^\nu)$ reduce simply to v^a and φ^ν for $j=a$ and $\alpha=\nu$, respectively.

Successive differentiations of eqn (2.51) with respect to the independent variables, v^a and φ^ν, generate a sequence of perturbation equations. Each step of the sequence, then, yields a set of linear equations of the surface derivatives which can be solved and substituted in the next set of equations. This procedure produces enough derivatives to construct the asymptotic equations of the surface.

To this end, the first and second-order equations are generated by differentiating eqn (2.51) with respect to v^a and φ^ν as follows:

$$\Pi_{ij} v^{j,a} + \Pi_{i\alpha} \varphi^{\alpha,a} = 0, \tag{2.52a}$$

$$\Pi_{ij} v^{j,\nu} + \Pi_{i\alpha} \varphi^{\alpha,\nu} = 0, \tag{2.52b}$$

$$(\Pi_{ijk} v^{k,b} + \Pi_{ij\alpha} \varphi^{\alpha,a}) v^{j,a} + \Pi_{ij} v^{j,ab} + (\Pi_{ij\alpha} v^{j,b} + \Pi_{i\alpha\beta} \varphi^{\beta,b}) \varphi^{\alpha,a} + \Pi_{i\alpha} \varphi^{\alpha,ab} = 0, \tag{2.53a}$$

$$(\Pi_{ijk} v^{k,\nu} + \Pi_{ij\alpha} \varphi^{\alpha,\nu}) v^{j,a} + \Pi_{ij} v^{j,a\nu} + (\Pi_{ij\alpha} v^{j,\nu} + \Pi_{i\alpha\beta} \varphi^{\beta,\nu}) \varphi^{\alpha,a} + \Pi_{i\alpha} \varphi^{\alpha,a\nu} = 0, \tag{2.53b}$$

and

$$(\Pi_{ijk} v^{k,\xi} + \Pi_{ij\alpha} \varphi^{\alpha,\xi}) v^{i,\nu} + \Pi_{ij} v^{i,\nu\xi} + (\Pi_{ij\alpha} v^{j,\xi} + \Pi_{i\alpha\beta} \varphi^{\beta,\xi}) \varphi^{\alpha,\nu} + \Pi_{i\alpha} \varphi^{\alpha,\nu\xi} = 0, \tag{2.53c}$$

where the sets of indices $(i, j, k, \ldots)$, $(a, b, c, \ldots)$, $(\alpha, \beta, \gamma, \ldots)$, and $(\nu, \xi, \ldots)$ range from 1 to n, 1 to r, 1 to m, and $(r+1)$ to m, respectively. It is understood that superscripts on v and φ after commas indicate differentiation with respect to the corresponding variables, and subscripts on Π denote differentiation with respect to the corresponding v and/or φ. This convention will apply throughout.

Evaluating eqn (2.52a) at c yields

$$0 + \delta_{a\varepsilon} \varphi^{\varepsilon,a} = 0 \quad \text{for} \quad i = a$$

and

$$\Pi_{\underline{s}} v^{s,a} + \Pi_{s\alpha} \varphi^{\alpha,a} = 0 \quad \text{for} \quad i = s$$

which result in

$$\varphi^{\varepsilon,a} = 0 \quad \text{and} \quad v^{s,a} = 0, \tag{2.54}$$

upon noting that $\varphi^{\nu,a} = 0$ since v^a and φ^ν are independent variables.

Similarly, evaluating eqn (2.52b) at c leads to

$$0 + \delta_{a\varepsilon} \varphi^{\varepsilon,\nu} = 0 \quad \text{for} \quad i = a$$

and

$$\Pi_{\underline{s}} v^{s,\nu} + \Pi_{s\varepsilon} \varphi^{\varepsilon,\nu} = 0 \quad \text{for} \quad i = s$$

which yield

$$\varphi^{\varepsilon,\nu} = 0 \quad \text{and} \quad v^{s,\nu} = -\frac{\Pi_{s\nu}}{\Pi_s}. \tag{2.55}$$

Evaluating the second-order perturbation equation (2.53a) at c and using the derivatives (2.54) and (2.55) leads to

$$\Pi_{iab} + \Pi_{\underline{i}} v^{i,ab} + \Pi_{i\varepsilon} \varphi^{\varepsilon,ab} = 0, \tag{2.56a}$$

$$\Pi_{ias} v^{s,\nu} + \Pi_{ia\nu} + \Pi_{\underline{i}} v^{i,a\nu} + \Pi_{i\varepsilon} \varphi^{\varepsilon,a\nu} = 0, \tag{2.56b}$$

and

$$\Pi_{ist}v^{s,v}v^{t,\xi}+\Pi_{is\xi}v^{s,v}+\Pi_{\underset{\sim}{i}}v^{i,v\xi}+\Pi_{isv}v^{s,\xi}+\Pi_{iv\xi}+\Pi_{i\varepsilon}\varphi^{\varepsilon,v\xi}=0, \quad (2.56c)$$

Eqn (2.56*a*) reduces to

$$\Pi_{cab}+\delta_{c\varepsilon}\varphi^{\varepsilon,ab}=0 \quad \text{for} \quad i=c,$$

and

$$\Pi_{sab}+\Pi_{\underset{\sim}{s}}v^{s,ab}+\Pi_{s\varepsilon}\varphi^{\varepsilon,ab}=0 \quad \text{for} \quad i=s,$$

which result in

$$\varphi^{\varepsilon,ab}=-\delta^{\varepsilon c}\Pi_{cab}\triangleq F_{\varepsilon ab} \tag{2.57}$$

and

$$v^{s,ab}=-\frac{F_{sab}}{\Pi_{\underset{\sim}{s}}}$$

where $F_{sab}\triangleq\Pi_{sab}-\Pi_{s\varepsilon}\,\delta^{\varepsilon c}\Pi_{cab}$.

Similarly, eqn (2.56*b*) yields

$$\varphi^{\varepsilon,av}=F_{\varepsilon av}\triangleq-\delta^{\varepsilon b}\{\Pi_{bav}-(\Pi_{bas}\Pi_{sv})/\Pi_{\underset{\sim}{s}}\} \quad \text{for} \quad i=b. \tag{2.58}$$

Eqn (2.56*c*) results in

$$\begin{aligned}\varphi^{\varepsilon,v\xi}=F_{\varepsilon v\xi}\triangleq-\delta^{\varepsilon a}\{\Pi_{av\xi}&-(\Pi_{as\xi}\Pi_{sv}+\Pi_{asv}\Pi_{s\xi})/\Pi_{\underset{\sim}{s}}\\&+(\Pi_{ast}\Pi_{sv}\Pi_{t\xi})/\Pi_{\underset{\sim}{s}}\Pi_{\underset{\sim}{t}}\}.\end{aligned} \tag{2.59}$$

The derivatives (2.54), (2.55), (2.57), (2.58), and (2.59) may now be used to construct the asymptotic equations of the equilibrium surface in the vicinity of the compound general critical point *c* as

$$\varphi^{\varepsilon}=\tfrac{1}{2}F_{\varepsilon ab}v^{a}v^{b}+F_{\varepsilon av}v^{a}\varphi^{v}+\tfrac{1}{2}F_{\varepsilon v\xi}\varphi^{v}\varphi^{\xi} \tag{2.60}$$

and

$$\Pi_{\underset{\sim}{s}}v^{s}+\Pi_{sv}\varphi^{v}+\tfrac{1}{2}F_{sab}v^{a}v^{b}=0. \tag{2.61}$$

It is observed that eqn (2.60) is the projection of the equilibrium surface onto the (u^{a},φ^{α}) sub-space, and provides significant quantitative information concerning the local behaviour of the system. The second fundamental tensors associated with *r* orthogonal normals, for example, are now represented by

$$\boldsymbol{F}^{(\varepsilon)}=\begin{bmatrix}F_{(\varepsilon)ab} & F_{(\varepsilon)av}\\ F_{(\varepsilon)va} & F_{(\varepsilon)v\xi}\end{bmatrix}, \qquad (\varepsilon=1,2,\ldots,r), \tag{2.62}$$

which provides information about the geometric invariants of the surface—such as principal curvatures, principal directions. Indeed, the eigenvalues and eigenvectors of eqn (2.62) are the principal curvatures and principal directions, respectively.

It may turn out that, due to the symmetry properties of the system, all the cubic coefficients Π_{cab} vanish simultaneously, resulting in $\varphi^{\varepsilon,ab} = 0$ and requiring a further step in the sequence of perturbations to determine higher order derivatives of the surface. Indeed, the third-order perturbation yield the derivative

$$\varphi^{\varepsilon,abc} = -\delta^{\varepsilon d}(\Pi_{dcab} - 3\Pi_{sda}\Pi_{sbc}/\Pi_s) \triangleq F_{\varepsilon cab}, \tag{2.63}$$

and the corresponding first-order equations of the equilibrium surface follow immediately as

$$\varphi^{\varepsilon} = \frac{1}{3!} F_{\varepsilon cab} v^a v^b v^c + F_{\varepsilon a\nu} v^a \varphi^{\nu} + \tfrac{1}{2} F_{\varepsilon\nu\xi} \varphi^{\nu} \varphi^{\xi} \tag{2.64}$$

and

$$\Pi_s v^s + \Pi_{s\nu}\varphi^{\nu} + \Pi_{sab} v^a v^b = 0. \tag{2.65}$$

A general critical point at which all $\Pi_{abc} = 0$ so that the associated equilibrium surface takes the above form will be called *singular general*, following the terminology introduced for simple critical points (Huseyin 1975).

Clearly, if more coefficients vanish at c, rendering the first-order equations of the equilibrium surface invalid, the perturbation procedure must continue so that higher order surface derivatives are generated as may be required for construction of the asymptotic equations of the surface.

Eqns (2.60) and (2.64) will be analysed in more detail, and the behaviour of the system as reflected by these equations under specified conditions will be explored in the following sections.

Problem 2.3. By differentiating the functions (2.51) with respect to u^a for a third time and evaluating at the critical point c, show that $\varphi^{\varepsilon,abc}$ is given by eqn (2.63) if all $\Pi_{cab} = 0$.

2.4 Simple general points and elementary catastrophes

2.4.1 General points of order 2

Suppose now that the critical point c is simple such that $r = 1$; it then follows that only one stability coefficient vanishes at c (i.e. $\Pi_1 = 0$), and eqn (2.60) reduces to

$$\varphi^1 = \tfrac{1}{2} F_{111}(v^1)^2 + F_{11\nu} v^1 \varphi^{\nu} + \tfrac{1}{2} F_{1\nu\xi} \varphi^{\nu} \varphi^{\xi} \tag{2.66}$$

where $F_{11\nu}$ and $F_{1\nu\xi}$ are obtained from eqns (2.58) and (2.59), respectively, by substituting for $\varepsilon = 1$ and $a = b = 1$ in the formulas. Similarly, eqn (2.61) takes the form

$$\Pi_s v^s + \Pi_{sv}\varphi^v + \tfrac{1}{2}F_{s11}(v^1)^2 = 0. \qquad (2.67)$$

Consider next the arbitrary control path in parameter space, described by

$$\varphi^\alpha(\rho) = \varphi^{\alpha,\rho}\rho + \tfrac{1}{2}\varphi^{\alpha,\rho\rho}\rho^2 + \ldots; \qquad \varphi^\alpha(0) = 0, \qquad (2.68)$$

where $\varphi^{\alpha,\rho}$, $\varphi^{\alpha,\rho\rho}$, etc. are derivatives (constants) with respect to ρ, evaluated at c $(\rho = 0)$, and ρ is now a variable parameter. Then, introducing eqn (2.68) into eqn (2.66) yields the first-order approximation

$$\varphi^{1,\rho}\rho = \tfrac{1}{2}F_{111}(v^1)^2 \quad \text{if} \quad \varphi^{1,\rho} \neq 0, \qquad (2.69)$$

which is immediately recognized as the *fold catastrophe*. On a plot of v^1 against the independent parameter ρ, one obtains a *limit point*. Clearly, the behaviour of the system does not change qualitatively for different control paths provided $\varphi^{1,\rho} \neq 0$, and it may be suggested, therefore, that the general critical point of order 2 (fold catastrophe) is essentially a one-parameter phenomenon (i.e. of *co-dimension* one). This, however, is a misleading conclusion in the sense that one may be led to believe that the equilibrium surface (eqn 2.66) is *parabolic* and there is no need for a second parameter at all. Actually, the properties of the *critical zone* (fold zone) as well as the entire equilibrium surface can only be explored fully by retaining sufficient independent parameters. Furthermore, if the control path (eqn 2.68) is chosen in such a way that it lies entirely in the subspace spanned by φ^v, eqn (2.66) yields

$$v^1 = \frac{1}{F_{111}}[-F_{11v}\varphi^{v,\xi} \pm \surd\{(F_{11v}\varphi^{v,\rho})^2 - F_{111}F_{1v\xi}\varphi^{v,\rho}\varphi^{\xi,\rho}\}]\rho, \quad (2.70)$$

which indicates that an asymmetric bifurcation phenomenon takes place at the critical point c provided the expression under the square-root is positive. In other words, the same critical point may appear as a *limit* or *bifurcation* point on different plots (Huseyin 1970*a*, 1970*b*). This phenomenon is of course linked to topological properties of the surface described by eqn (2.66). Indeed, the equilibrium surface (eqn 2.66) around a simple general point of order 2 is described as *synclastic*, *anticlastic*, or *parabolic* according to whether the matrix

$$[F_{111}F_{1v\xi} - F_{11v}F_{11\xi}] \qquad (2.71)$$

is positive definite, negative definite, or null, respectively (Fig. 2.1).

It can be seen from the fundamental tensor (eqn 2.62), which is now represented by

$$\boldsymbol{F}^{(1)} = \begin{bmatrix} F_{111} & F_{11v} \\ F_{1v1} & F_{1v\xi} \end{bmatrix}, \qquad (2.72)$$

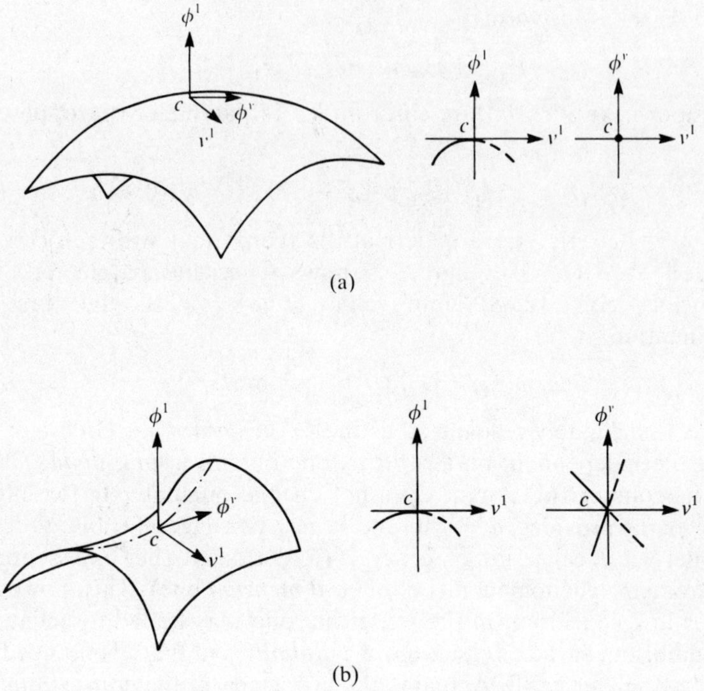

Fig. 2.1 General critical point of order 2 showing (a) synclastic surface and (b) anticlastic surface.

that the principal curvatures of the surface (2.66) have the same signs in the synclastic case, while the sign of the curvature with respect to v^1 is always the opposite of those with respect to φ^ν. In fact, the rank, index, and signature of the symmetric matrix (2.72) are the same as those of matrix

$$\begin{bmatrix} F_{111} & 0 \\ 0 & G \end{bmatrix} \quad \text{where} \quad \boldsymbol{G} = F_{1\nu\xi} - \frac{1}{F_{111}} F_{11\nu} F_{11\xi}, \tag{2.73}$$

and it follows that the eigenvalues of

$$\frac{1}{F_{111}} [F_{111} F_{1\nu\xi} - F_{11\nu} F_{11\xi}] \tag{2.74}$$

and F_{111} have the same signs if matrix (2.71) is positive definite. On the other hand, if matrix (2.71) is negative definite, one observes that the eigenvalues of matrix (2.71) and F_{111} will always have opposite signs (Fig. 2.1). It is also noted that the primary parameter φ^1 is normal to the surface at c while v^1 and φ^ν are in tangential directions.

In order to verify matrix (2.73), consider the non-singular transformation matrix

$$\boldsymbol{P} = \begin{bmatrix} 1 & -\dfrac{F_{11\nu}}{F_{111}} \\ 0 & I_{m-1} \end{bmatrix}$$

where I_{m-1} is an $(m-1)$-dimensional identity matrix. The result (eqn 2.73) follows from the congruence transformation

$$\boldsymbol{P}'\boldsymbol{F}^{(1)}\boldsymbol{P} = \begin{bmatrix} F_{111} & 0 \\ 0 & G \end{bmatrix}$$

where $\boldsymbol{P}'\boldsymbol{F}^{(1)}\boldsymbol{P}$ and $\boldsymbol{F}^{(1)}$ are subject to Sylvester's law of inertia (Huseyin 1978*b*).

The critical zone, on the other hand, can be obtained readily (Huseyin 1970*b*) by differentiating eqn (2.66) with respect to v^1 as

$$\overset{*}{v}^1 = -\frac{F_{11\nu}}{F_{111}}\overset{*}{\varphi}^\nu,$$

which together with eqn (2.66) yields

$$\overset{*}{\varphi}^1 = \frac{1}{2F_{111}}(F_{111}F_{1\nu\xi} - F_{11\nu}F_{11\xi})\overset{*}{\varphi}^\nu\overset{*}{\varphi}^\xi, \tag{2.75}$$

where a star denotes a critical variable.

The stability boundary of the system in the vicinity of the critical point c is given by eqn (2.75). Several interesting properties concerning the stability boundary can now be established (Huseyin 1975); however, only one such property will be discussed here. Suppose the system is described by one behaviour variable, v^1, and the potential function is linear in the parameters (which is the case in many applications); then eqn (2.66) takes the form

$$\varphi^1 = \tfrac{1}{2}F_{111}(v^1)^2 + F_{11\nu}v^1\varphi^\nu \tag{2.76}$$

and eqn (2.75) reduces to

$$\overset{*}{\varphi}^1 = -\frac{1}{2F_{111}}(F_{11\nu}\overset{*}{\varphi}^\nu)^2. \tag{2.77}$$

It is now observed that the curvatures of the stability boundary,

$$\overset{*}{\varphi}^{1,\nu\nu} = -\frac{(F_{11\nu})^2}{F_{111}}, \tag{2.78}$$

and the curvature of the equilibrium surface along v^1,

$$\varphi^{1,11} = F_{111}, \tag{2.79}$$

always have opposite signs. This implies that the stability boundary cannot have convexity with regard to the region of instability, a result that may have practical consequences.

For $\nu = 2$, that is for a two-parameter system, it is easy to see that if the equilibrium surface is stable on one side of the critical zone—which takes the form of a critical line for $\nu = 2$—it has to be unstable on the other side. This result can be generalized to systems with multiple parameters (Huseyin 1970*b*). Stable and unstable paths in Fig. 2.1 are shown by full and dashed lines, respectively. It is also noted that the critical point c itself—as well as all the other critical states on the critical zone—is unstable since $\Pi_{111} \neq 0$ and the function Π cannot have a minimum at c (Section 2.1).

Problem 2.4. The critical zone described by eqns (2.74) and (2.75) can also be obtained by observing that the stability determinant $\Delta(v^i, \varphi^\alpha) \equiv |\Pi_{ij}(v^k, \varphi^\alpha)|$ vanishes at critical points. Verify eqns (2.74) and (2.75) on the basis of this fact.

2.4.2 *General points of order 3 (singular general points)*

A topologically distinct phenomenon arises when $\Pi_{111} = 0$ at the critical point c. The asymptotic equations of the equilibrium surface around such a *singular critical point* are obtained directly from eqns (2.64) and (2.65) as

$$\varphi^1 = \frac{1}{3!} F_{1111}(v^1)^3 + F_{11\nu} v^1 \varphi^\nu + \tfrac{1}{2} F_{1\nu\xi} \varphi^\nu \varphi^\xi \tag{2.80}$$

and

$$\Pi_{\underset{s}{s}} v^s + \Pi_{s\nu} \varphi^\nu + F_{s11}(v^1)^2 = 0 \tag{2.81}$$

where

$$F_{1111} = -\left\{\Pi_{1111} - 3 \sum_s (\Pi_{s11})^2 / \Pi_{\underset{s}{s}}\right\}. \tag{2.82}$$

It is noted that the behaviour of the system remains qualitatively the same for all control paths $\varphi^\nu(\rho)$ in the φ^ν-space. If one considers φ^2 as a representative parameter for all φ^ν instead of introducing a new variable parameter ρ, the response characteristic will again be qualitatively similar. Consider, therefore, the two-parameter system

$$\varphi^1 = \frac{1}{3!} F_{1111}(v^1)^3 + F_{112} v^1 \varphi^2 + \tfrac{1}{2} F_{122} (\varphi^2)^2 \tag{2.83}$$

to identify the salient features of the equilibrium surface. Additional transformations may be introduced to further simplify eqn (2.83) and facilitate comparisons with elementary catastrophes. Indeed, introducing for example the transformation

$$v^1 = \frac{1}{2}\frac{F_{122}}{F_{112}}\varphi^2\left\{-1 + \frac{1}{4!}F_{1111}\frac{(F_{122})^2}{(F_{112})^3}\varphi^2\right\} + w^1 \tag{2.84}$$

and keeping to a first-order approximation results in

$$\varphi^1 = \frac{1}{3!}F_{1111}(w^1)^3 + F_{112}w^1\varphi^2. \tag{2.85}$$

Note that this transformation preserves original parameters without mixing them. A simpler way of transforming eqn (2.83) into a form similar to eqn (2.85) consists of defining a new parameter by $\bar{\varphi}^1 = \varphi^1 - \frac{1}{2}F_{122}(\varphi^2)^2$, which results in

$$\bar{\varphi}^1 = \tfrac{1}{3}F_{1111}(v^1)^3 + F_{112}v^1\varphi^2.$$

This equation or eqn (2.85) describes an equilibrium surface associated with the cusp catastrophe (Riemann–Hugoniot catastrophe), and was obtained through other methods earlier (Huseyin 1972*b*, 1977). It is interesting to note that its initial derivation precedes the advent of catastrophe theory, as do the developments discussed in Section 2.4.1 (fold catastrophe). Furthermore, the analysis here is quite general in the sense that it does not rely on the relatively restrictive framework of imperfection sensitivity formulation.

The critical line is given by

$$\tfrac{1}{2}F_{1111}(\overset{*}{w}{}^1)^2 + F_{112}\overset{*}{\varphi}{}^2 = 0 \tag{2.86}$$

and

$$\overset{*}{\varphi}{}^2 = -\frac{1}{2}\frac{(F_{1111})^{1/3}}{F_{112}}(3\overset{*}{\varphi}{}^1)^{2/3}. \tag{2.87}$$

The stability boundary (eqn 2.87) is the projection of the critical line onto the parameter space, and takes the form of a cusp (Fig. 2.2). The critical point *c* appears as a singular point on the stability boundary where the convexity of the curve changes abruptly, and this property of the boundary provided the motivation for calling the critical point *c* ‘singular’. It is clear, however, that the critical line on the equilibrium surface is smooth. Furthermore, the orientation of the cusp in the original parameter space is generally tilted (Fig. 2.3). The transformed parameters φ^1 and φ^2 are identified as the shock and normal parameters of the catastrophe theory, respectively. The significance of the transformation $\boldsymbol{\mu} = \boldsymbol{Q}\boldsymbol{\varphi}$ lies in the fact that the interesting features of the equilibrium surface and the stability boundary can only be brought to focus through this transformation. Another interesting phenomenon is

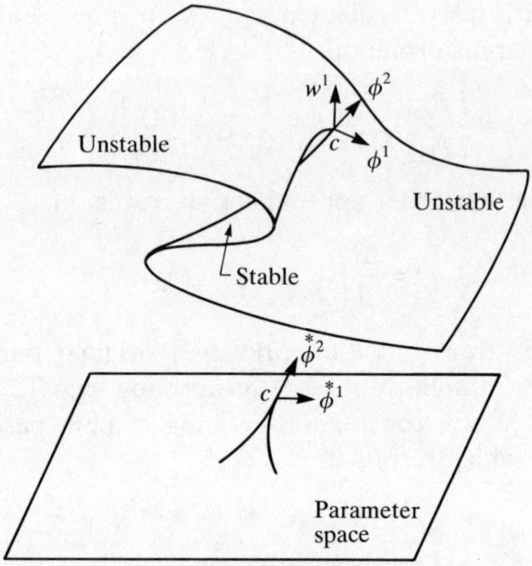

Fig. 2.2 Equilibrium surface in the vicinity of a singular general point (cusp catastrophe).

revealed when one sets $\varphi^1 = 0$ in eqn (2.85), yielding

$$w^1 = 0 \quad \text{or} \quad \frac{1}{3!} F_{1111}(w^1)^2 + F_{112}\varphi^2 = 0, \tag{2.88}$$

which indicates a symmetric bifurcation phenomenon. The equilibrium path $w^1 = 0$ loses its stability at the critical point c where a totally stable (Fig. 2.4) or unstable (Fig. 2.5) post-critical path branches off from this point. Note that the equilibrium surface depicted in Fig. 2.2 corresponds to the latter phenomenon (Fig. 2.5), and an analogous surface corresponding to Fig. 2.4 can readily be produced by observing the stability properties and orientation of axes. Indeed, the situation depicted in Fig. 2.4 is a cross-section of a surface similar to the one in Fig. 2.2, but with reversed stability properties on each of the three surface sheets. The stability distribution on the equilibrium surface follows from the criteria discussed in Section 2.1 and has been explored in detail (Huseyin 1975).

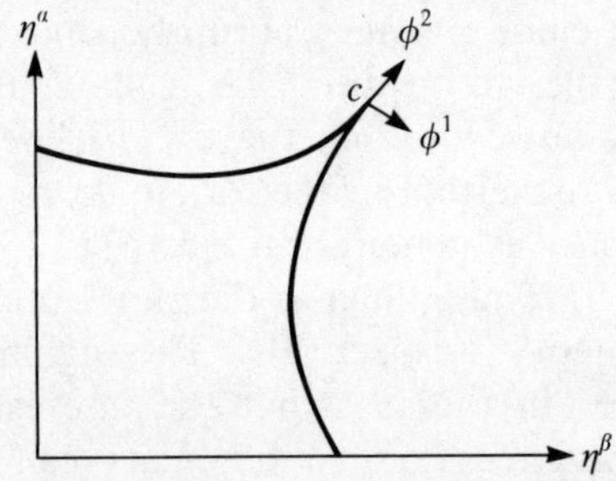

Fig. 2.3 Stability boundary in η-space.

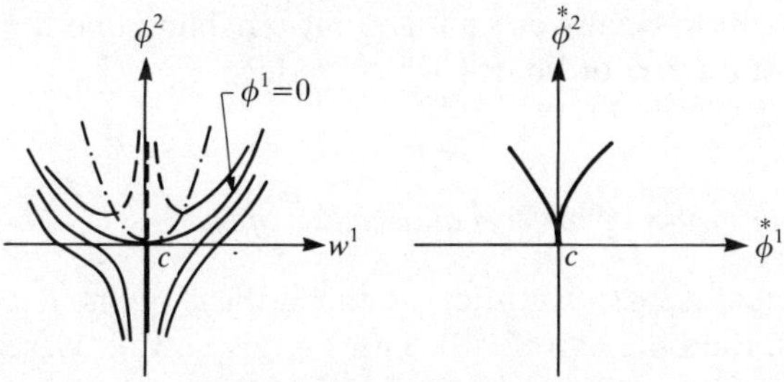

Fig. 2.4 Stable symmetric point.

Note that the critical point c is stable or unstable according to whether the coefficient

$$\Pi_{1111} - 3\sum_{s} (\Pi_{s11})^2/\Pi_s \tag{2.89}$$

is positive or negative, respectively. This conclusion is reached after observing that the criteria in Table 2.1, given in terms of the derivatives of H, may simply be replaced by the derivatives of Π on the basis of the transformation (2.48). It is not difficult to see that if the critical point c is stable (unstable), the post-critical path in eqn (2.88) is also stable (unstable). Similarly, the sheets of the equilibrium surface at two flanks which join to form a single sheet are stable (unstable) while the interior sheet is unstable (stable) if the critical point c is stable (unstable). Instabilities occur at the critical fold lines of the surface, resulting in jumps from one sheet to another. On the other hand, if an equilibrium path lies entirely on the single sheet and two flank sheets it is either totally stable or unstable; i.e. if a control path is followed in parameter space such that the states develop gradually on a smooth equilibrium path which starts at one flank and reaches the other one through the single sheet, passing around c and avoiding the critical folds, the system does not experience any change in its stability properties.

Problem 2.5. Derive the first-order equations of the equilibrium surface in the vicinity of a singular critical point by observing that a symmetric

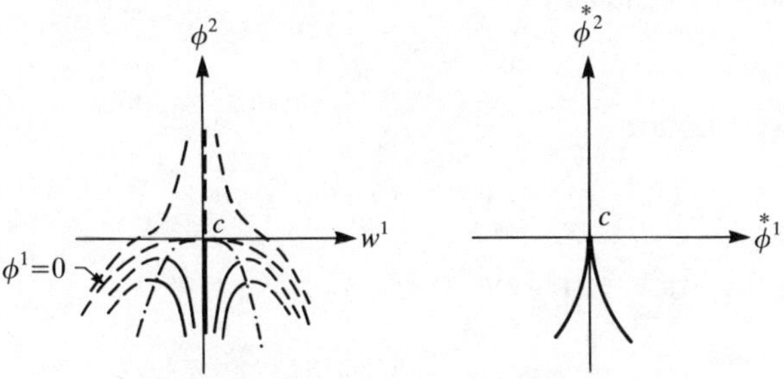

Fig. 2.5 Unstable symmetric point.

point of bifurcation occurs at such a point, enabling one to introduce the sliding and rotating coordinates which reduce eqn (2.83) to eqn (2.85) (see Huseyin 1977, p. 241).

2.4.3 General points of order 4 and higher order

If the coefficient (2.82) vanishes at the critical point c, together with $\Pi_{111} = \Pi_1 = 0$, then the eqn (2.80) ceases to give a valid description of the equilibrium surface in the vicinity of c. Furthermore, the general results of Section 2.3.2 are no longer capable of yielding higher order terms necessary for constructing a first-order approximation for the surface in this case. It is clear from the preceding two sections, however, that the non-critical coordinates and higher order terms in the parameters can be effectively eliminated from the final equation by suitable transformations—while the effect of these coordinates is reflected in the coefficients of the equation. Consider, therefore, the simple case of one behaviour variable and assume that the potential function is linear in the parameters. Furthermore, it will be assumed here that the transformation of parameters, $\boldsymbol{\mu} = \boldsymbol{Q}\boldsymbol{\varphi}$, is chosen in a different way, such that

$$\begin{gathered} \Pi_{1\alpha} = \delta_{1\alpha}, \qquad \Pi_{11\nu} = \delta_{2\nu} \qquad (\nu \geq 2), \\ \Pi_{111\nu} = \delta_{3\mu} \ (\mu \geq 3) \quad \text{and} \quad \Pi_{111\xi} = \delta_{4\xi} \ (\xi \geq 4), \end{gathered} \tag{2.90}$$

in the transformed function which is again denoted by $\Pi(v^i, \varphi^\alpha)$.

By following the familiar perturbation procedure, the equilibrium surface in the vicinity of the critical point c, under the new conditions, is constructed (Huseyin 1977) as

$$\varphi^1 + \frac{1}{4!} d(v^1)^4 + \frac{1}{3!} (v^1)^3 \varphi^4 + \frac{1}{2!} (v^1)^2 \varphi^3 + v^1 \varphi^2 = 0, \tag{2.91}$$

where d is a constant.

On the other hand, any polynomial

$$v^m + \alpha_m v^{m-1} + \ldots + \alpha_1 \tag{2.92}$$

may be brought to the form

$$w^m + \sum_{k=0}^{m-2} \beta_{k+1} w^k \tag{2.93}$$

by the transformation

$$w = v + \frac{1}{m} \alpha_m. \tag{2.94}$$

Applying this transformation to eqn (2.91) yields

$$\varphi^1 + \frac{1}{4!} a(w)^4 + \frac{1}{2!} b(w)^2 \bar{\varphi}^3 + cw\bar{\varphi}^2 = 0, \tag{2.95}$$

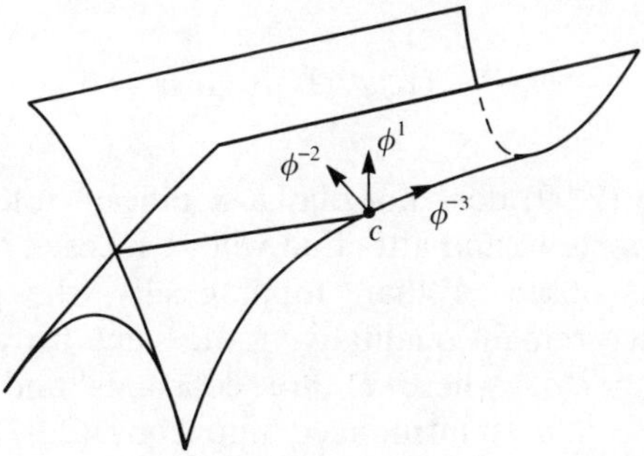

Fig. 2.6 Swallow tail.

where a, b, c are constants, w is the new behaviour variable, and $\bar{\varphi}^2$ and $\bar{\varphi}^3$ are the new parameters. It is now recognized that eqn (2.95) describes the swallow's tail of the elementary catastrophes. It is interesting to note that the transformation (eqn 2.94) has reduced the number of *significant parameters* to three. The stability boundary in the three-dimensional parameter space is obtained by eliminating w between eqn (2.95) and

$$\frac{1}{3!}a(w)^3 + bw\bar{\varphi}^3 + c\bar{\varphi}^2 = 0, \tag{2.96}$$

and obviously has a complicated shape. It is depicted in Fig. 2.6 (see Thom 1975).

Higher order surfaces may be generated by continuing the pattern established here. For instance, if the coefficient of $(w)^4$ in eqn (2.95) vanishes as well, together with all the other coefficients that vanish at the general point of order 4, an equilibrium surface of order 5 emerges as the next structurally stable form, which is identified as the *butterfly* in catastrophe theory. This, however, will not be pursued here since the increasing complexity inevitably results in a loss of geometric intuition.

2.5 Two-fold general critical points

2.5.1 Equilibrium surface and connection with umbilics

Consider now a coincident critical point c under the condition $r = 2$, that is, a two-fold critical point. The asymptotic equations of the equilibrium surface (2.60) and (2.61), derived for an r-fold critical point, reduce to

$$\begin{aligned}\varphi^1 = {} & \tfrac{1}{2}F_{111}(v^1)^2 + F_{112}v^1v^2 + \tfrac{1}{2}F_{122}(v^2)^2 \\ & + F_{11\nu}v^1\varphi^\nu + F_{12\nu}v^2\varphi^\nu + \tfrac{1}{2}F_{1\nu\xi}\varphi^\nu\varphi^\xi,\end{aligned} \tag{2.97}$$

$$\begin{aligned}\varphi^2 = {} & \tfrac{1}{2}F_{211}(v^1)^2 + F_{212}v^1v^2 + \tfrac{1}{2}F_{222}(v^2)^2 \\ & + F_{21\nu}v^1\varphi^\nu + F_{22\nu}v^2\varphi^\nu + \tfrac{1}{2}F_{2\nu\xi}\varphi^\nu\varphi^\xi,\end{aligned} \tag{2.98}$$

and

$$\Pi_s v^s + \Pi_{sv}\varphi^v + \tfrac{1}{2}F_{sab}v^a v^b = 0, \tag{2.99}$$

where $a, b = 1, 2$.

As before, eqn (2.99) does not play a significant role in describing the behaviour of the system, and attention will be focused on eqns (2.97) and (2.98). It is first observed that, topologically, the properties of the equilibrium surface remain qualitatively the same for all parameter rays of the form $\varphi^v = l^v \rho$, where l^v are constants and ρ is a variable parameter. If $\varphi^v = l^v\rho$ is introduced into eqns (2.97) and (2.98), one obtains

$$\varphi^1 = \tfrac{1}{2}F_{111}(v^1)^2 + F_{112}v^1v^2 + \tfrac{1}{2}F_{122}(v^2)^2 + (F_{11}v^1 + F_{12}v^2)\rho + \tfrac{1}{2}F_1\rho^2 \tag{2.100}$$

$$\varphi^2 = \tfrac{1}{2}F_{211}(v^1)^2 + F_{212}v^1v^2 + \tfrac{1}{2}F_{222}(v^2)^2 + (F_{21}v^1 + F_{22}v^2)\rho + \tfrac{1}{2}F_2\rho^2, \tag{2.101}$$

where

$$F_{11} = F_{11v}l^v, \qquad F_{12} = F_{21} = F_{12v}l^v, \qquad F_1 = F_{v\xi}l^v l^\xi,$$
$$F_{22} = F_{22v}l^v, \qquad F_2 = F_{2v\xi}l^v l^\xi.$$

These equations may be assumed to be derivable from the potential function

$$F = \frac{1}{3!}F_{11}(v^1)^3 + \tfrac{1}{2}F_{112}(v^1)^2v^2 + \tfrac{1}{2}F_{122}v^1(v^2)^2 + \frac{1}{3!}F_{222}(v^2)^3$$
$$+ \tfrac{1}{2}[F_{11}(v^1)^2 + 2F_{12}v^1v^2 + F_{22}(v^2)^2]\rho - v^1\varepsilon^1 - v^2\varepsilon^2, \tag{2.102}$$

by differentiating F with respect to v^1 and v^2, respectively. Here ε^1, ε^2, and ρ are the new parameters defined by $\varepsilon^1 = \varphi^1 - \tfrac{1}{2}F_1\rho^2$ and $\varepsilon^2 = \varphi^2 - \tfrac{1}{2}F_2\rho^2$. The potential function (2.102) describes the local behaviour of the system while the original potential $\Pi(u, \varphi)$ is an all-embracing function characterizing the global behaviour.

The homogeneous cubic polynomial of F in v^1 and v^2 may be factored into three linear forms as

$$P(v^1, v^2) = (a_1v^1 + b_1v^2)(a_2v^1 + b_2v^2)(a_3v^1 + b_3v^2), \tag{2.103}$$

where the coefficient vectors (a_1, b_1), (a_2, b_2), and (a_3, b_3) may serve as a basis to classify topologically distinct equilibrium surfaces (Zeeman 1976, Poston and Stewart 1976). Thus, consider the situation in which all the vectors are pairwise linearly independent; two distinct cases may be identified:

Case 1. All the (a_i, b_i) *are real*. In this case it can be demonstrated that appropriate transformations reduce eqn (2.103) to the simple form

$$P(x, y) = x^3 - xy^2 \tag{2.104}$$

which is associated with the organizing centre of elliptic umbilic. To this end, consider first the transformation

$$\begin{aligned} a_1 v^1 + b_1 v^2 &= p \\ a_2 v^2 + b_2 v^2 &= q. \end{aligned} \tag{2.105}$$

Then eqn (2.103) takes the form

$$P(p, q) = pq(ap + bq) \tag{2.106}$$

where a and b are given in terms of a_i and b_i. Next, introduce

$$\begin{bmatrix} p \\ \\ q \end{bmatrix} = \begin{bmatrix} \dfrac{(ab)^{1/3}}{a} & 0 \\ \\ 0 & \dfrac{(ab)^{1/3}}{b} \end{bmatrix} \begin{bmatrix} r \\ \\ s \end{bmatrix} \tag{2.107}$$

to obtain

$$P(r, s) = rs(r + s). \tag{2.108}$$

Finally, the transformation

$$\begin{bmatrix} r \\ s \end{bmatrix} = 2^{-1/3} \begin{bmatrix} 1 & 1 \\ 1 & -1 \end{bmatrix} \begin{bmatrix} x \\ y \end{bmatrix}$$

yields eqn (2.104).

Case 2. Two of the vectors (a_i, b_i) *are complex conjugate*. Suppose eqn (2.103) is in the form

$$P(v^1, v^2) = (a_1 v^1 + b_1 v^2)(a_2 v^1 + b_2 v^2)(\bar{a}_2 v^1 + \bar{b}_2 v^2) \tag{2.109}$$

where $\bar{a}_2$ and $\bar{b}_2$ are complex conjugates of a_2 and b_2, respectively. Then the product of the last two forms is positive definite and a transformation may be introduced to bring this quadratic form to a sum of squares while canonizing the linear form $(a_1 v^1 + b_1 v^2)$ simultaneously. Thus, in new variables (t, u), eqn (2.109) may be expressed as $P(t, u) = t(t^2 + r^2)$ which is then transformed into

$$P(x, y) = x^3 + y^3, \tag{2.110}$$

by observing that $(t + r)^3 + (t - r)^3 = 2t(t^2 + 3r^2)$ where the coefficients 2 and 3 can readily be eliminated by scalar transformations as demonstrated in the case of the elliptic umbilic.

The cubic form (eqn 2.110) is the organizing centre of the hyperbolic umbilic of catastrophe theory.

If all the vectors (a_i, b_i) are not linearly independent, a number of other situations may arise. Consider, for instance, the case in which two of the vectors, say, the second and third vectors are not linearly independent; it can then be demonstrated that fourth-order terms in v^1 should be brought into picture and this case is related to the parabolic umbilic with an organizing centre given by $f(x, y) = x^2y + y^4$. This implies that eqn (2.64) should replace eqn (2.60) as a first-order approximation for the equilibrium surface in the vicinity of such a critical point. The cases in which all the vectors (a_i, b_i) are linearly dependent or the cubic form vanishes identically, require more than four independent parameters for a structurally stable universal unfolding.

In numerical applications, another way of ascertaining the character of the equilibrium surface (2.60) is to determine the principal curvatures of the surface at the critical point and to compare these curvatures with those of umbilics. The principal curvatures associated with the equilibrium surface (2.60) are given by the eigenvalues of eqn (2.62). On the other hand, the second fundamental tensors associated with the elliptic umbilic (Thom 1975)

$$V = x^3 - 3xy^2 + \rho(x^2 + y^2) - \varepsilon_1 x - \varepsilon_2 y, \tag{2.111}$$

(where ρ, ε_1, and ε_2 are the parameters) are represented by

$$\begin{bmatrix} 6 & 0 & 2 \\ 0 & -6 & 0 \\ 2 & 0 & 0 \end{bmatrix} \quad \text{and} \quad \begin{bmatrix} 0 & -6 & 0 \\ -6 & 0 & 2 \\ 0 & 2 & 0 \end{bmatrix} \tag{2.112}$$

which have the sets of eigenvalues $(-6, 3 \pm \sqrt{13})$ and $(0, \pm 2\sqrt{10})$.

Similarly, the second fundamental tensors associated with the hyperbolic umbilic

$$V = x^3 + y^3 + \rho xy - \varepsilon_1 x - \varepsilon_2 y \tag{2.113}$$

are represented by

$$\begin{bmatrix} 6 & 0 & 0 \\ 0 & 0 & 1 \\ 0 & 1 & 0 \end{bmatrix} \quad \text{and} \quad \begin{bmatrix} 0 & 0 & 1 \\ 0 & 6 & 0 \\ 1 & 0 & 0 \end{bmatrix} \tag{2.114}$$

with the eigenvalues $(-1, 1, 6)$ and $(-1, 1, 6)$.

2.5.2 *Critical zone*

It may not be possible to express the stability boundary in parameter space explicitly in a simple general form. In order to locate the critical zone on the equilibrium surface associated with the relatively more manageable system (2.102), it is first noted that the equilibrium equations in this case may be expressed as

$$\varepsilon^1 = \tfrac{1}{2}F_{1ab}v^a v^b + F_{1a}v^a \rho \tag{2.115}$$

and

$$\varepsilon^2 = \tfrac{1}{2}F_{2ab}v^a v^b + F_{2a}v^a \rho \tag{2.116}$$

where $a, b = 1, 2$.

Next, consider the fundamental criticality condition, that is, the vanishing Hessian

$$\Delta(v, \rho, \varepsilon^i) \triangleq \left| \frac{\partial^2 F}{\partial v^a \, \partial v^b} \right| = 0. \tag{2.117}$$

The critical points on the equilibrium surface described by eqns (2.115) and (2.116) may be obtained by solving these equations together with the criticality condition (2.117), simultaneously. In general, the solution of three equations with five unknowns may be expressed with the aid of two independent parameters in a parametric form. First, note that eqn (2.117) yields

$$f_{ab}v^a v^b + f_a v^a \rho + f\rho^2 = 0, \tag{2.118}$$

where

$$\begin{aligned} f_{ab} &= F_{11a}F_{22b} - F_{12a}F_{12b}, \\ f_a &= F_{11a}F_{22} + F_{11}F_{22a} - 2F_{12a}F_{12}, \end{aligned} \tag{2.119}$$

and

$$f = F_{11}F_{22} - F_{12}^2.$$

Substituting for ρ in eqns (2.115) and (2.116) results in

$$\begin{aligned} \varepsilon^1 &= \tfrac{1}{2}F_{1ab}v^a v^b + \frac{F_{1a}v^a}{2f}[-f_b v^b \pm \surd\{(f_b f_c - 4ff_{bc})v^b v^c\}], \\ \varepsilon^2 &= \tfrac{1}{2}F_{2ab}v^a v^b + \frac{F_{2a}v^a}{2f}[-f_b v^b \pm \surd\{(f_b f_c - 4ff_{bc})v^b v^c\}], \\ \rho &= \frac{1}{2f}[-f_a v^a \pm \surd\{(f_a f_b - 4ff_{ab})v^a v^b\}], \end{aligned} \tag{2.120}$$

where the variables v^a $(a = 1, 2)$ play the role of independent parameters. Obviously, it does not seem feasible to eliminate the v^a from these equations so that an explicit relationship for the critical surface can be obtained. The parametric form (2.120), however, has computational advantages since the critical surface may be constructed numerically by evaluating eqn (2.120) for specific sets of v^a. In applications, a good estimate of the boundary emerges if a sufficient number of points is used.

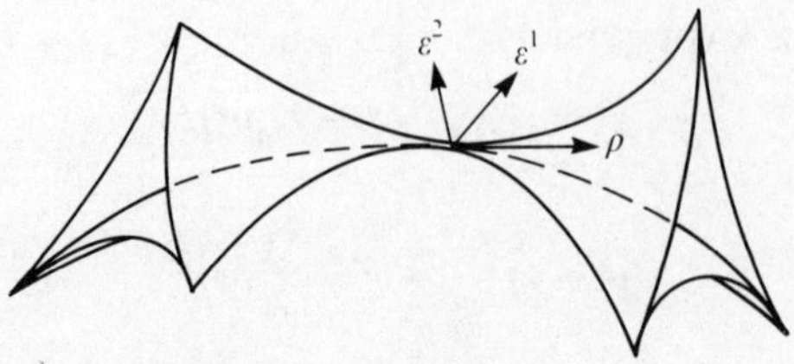

Fig. 2.7 Elliptic umbilic.

The critical surfaces in parameter space associated with elliptic and hyperbolic umbilics have been constructed by several authors with the aid of a digital computer; these surfaces are sketched in Figs. 2.7 and 2.8, respectively.

2.6 Special critical points and imperfection sensitivity

Recall now the definitions given in Section 2.2 concerning the types of critical points which were classified into two broad classes as *general* and *special*. The potential function referred to a *general* critical point has as many *primary* parameters as the nullity of the Hessian, and the rest of the parameters are *secondary* or may even be of higher order. In the case of *special* points, the number of *primary* parameters is less than the nullity of the Hessian matrix, and a variety of situations may arise. An interesting case, which occurs in many investigations in the field of elastic stability, is concerned with a potential function without any primary parameters; in this case the rank of the matrix $H_{a\alpha}$ is zero. Consider, for example, a perfectly flat thin plate subjected to two independent uniformly distributed in-plane compression forces along its perpendicular edges. It is expected, of course, that this system will have small imperfections, either in the structure itself or in the external loading. However, an idealized model which initially ignores such imperfections is invaluable in a quantitative analysis. In fact, the most convenient method of analysis aimed at exploring the effect of imperfections is based on the perfect model whose loss of stability is often associated with *special* critical points. In this example cited above, for instance, the potential energy of the idealized model—referred to a critical point—does not have

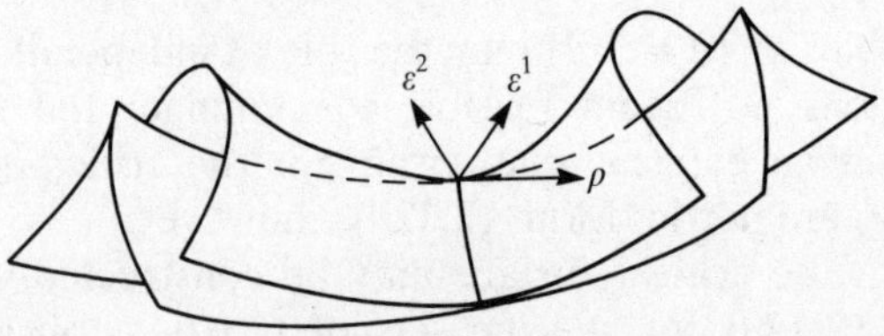

Fig. 2.8 Hyperbolic umbilic.

primary parameters. Addition of imperfections to the perfect model, however, alters the mathematical description of the system, rendering a *special* critical point *general* provided a sufficient number of imperfection parameters—in *primary* form—is incorporated.

As remarked earlier, a variety of situations may develop in connection with special points. The rest of this section will be devoted to analyses of some interesting cases which arise as a result of some sort of symmetry in the system. The perfect models to be considered exhibit bifurcation points where a single-valued fundamental equilibrium path is intersected by a post-critical path(s). These systems may, therefore, be formulated in a different way for convenience (Huseyin 1975).

Consider the original potential system described by the potential function (2.1), but with a single parameter only; i.e. assume that $\eta \in R$. By solving the equilibrium equations

$$\frac{\partial}{\partial q^i} V(q^j, \eta) = 0 \tag{2.121}$$

simultaneously, one may express the fundamental equilibrium path as $q^i_F(\eta)$. The potential function may now be referred to this single-valued path by introducing the transformation

$$q^i = q^i_F(\eta) + T_{ij}(\eta)u^j, \qquad \boldsymbol{T}(\eta)\boldsymbol{T}'(\eta) = \boldsymbol{I}, \tag{2.122}$$

which keeps the quadratic form of the energy in generalized coordinates diagonalized along the fundamental path $q^i_F(\eta)$. It is noted that the transformation (eqn 2.122) is essentially different from eqn (2.6) which is valid at the critical point c only. Here, the principal coordinates u^i slide along the fundamental path and rotate so that the quadratic form of the resulting new function in u^i is kept diagonalized. Indeed, introducing eqn (2.122) into the potential function $V(q^i, \eta)$ leads to

$$S(u^i, \eta) \equiv V[q^i_F + T_{ij}(\eta)u^j, \eta], \tag{2.123}$$

The new potential function $S(u^i, \eta)$ possesses the properties

$$S_i(0, \eta) = S_i'(0, \eta) = S_i''(0, \eta) = \ldots = 0 \tag{2.124}$$

and

$$S_{ij}(0, \eta) = S_{ij}'(0, \eta) = S_{ij}''(0, \eta) = \ldots = 0 \quad \text{for} \quad i \neq j, \tag{2.125}$$

where the subscripts and primes on S denote partial differentiation with respect to the u^i and η, respectively.

Let c be a two-fold critical point on $q^i_F(\eta)$, at $\eta = \eta_c$, where the Hessian S_{ij} has nullity two so that $S_{aa} \triangleq S_a = 0$ for $a = 1, 2$; and $S_{ss} \triangleq S_s \neq 0$ for $s = 3, 4, \ldots, n$. It is expected that the behaviour of a slightly imperfect system in the vicinity of c can be determined as a perturbation of the perfect system. Assume now that imperfections associated with the principal coordinates u^1 and u^2 are represented by ε^1 and ε^2, respectively, such that the potential function of the imperfect system $S(u^i, \eta, \varepsilon^\alpha)$ has a canonical bilinear form $S_{a\alpha}u^a\varepsilon^\alpha$ where

$$[S_{a\alpha}] = \begin{bmatrix} 1 & 0 \\ 0 & 1 \end{bmatrix}, \qquad (\alpha = 1, 2), \tag{2.126}$$

The solutions of the equilibrium equations

$$S_i(u^j, \eta, \varepsilon^\alpha) = 0 \tag{2.127}$$

are assumed to be in the parametric form

$$\begin{aligned} \varepsilon^\alpha &= \varepsilon^\alpha(u^a, \eta) \qquad (\alpha = 1, 2;\ a = 1, 2), \\ u^s &= u^s(u^a, \eta), \qquad (s = 3, 4, \ldots, n), \end{aligned} \tag{2.128}$$

and the basic identity of the perturbation procedure is obtained by introducing eqn (2.128) back into the potential function $S(u^i, \eta, \varepsilon^\alpha)$ as

$$S_i[u^j(u^a, \eta), \varepsilon^\alpha(u^a, \eta), \eta] \equiv 0 \tag{2.129}$$

where $u^j(u^a, \eta)$ reduce to u^a for $j = a$.

A sequence of perturbation equations can now be generated by successive differentiations of eqn (2.129) with respect to the independent variables u^a and η. Thus, the first-order equations are

$$\begin{aligned} S_{ij}u^{j,a} + S_{i\alpha}\varepsilon^{\alpha,a} + S_{ia} &= 0, \\ S_{ij}u^{j,\prime} + S_{i\alpha}\varepsilon^{\alpha,\prime} + S_i' &= 0, \end{aligned}$$

which, upon evaluation at c, yield

$$\varepsilon^{\alpha,a} = 0, \qquad \varepsilon^{\alpha,\prime} = 0, \qquad u^{s,a} = 0, \quad \text{and} \quad u^{s,\prime} = 0$$

where a prime after a comma indicates differentiation with respect to η. Continuing with the perturbation procedure yields the surface derivatives necessary to construct the equilibrium surface (Huseyin 1975, 1977). Here, however, attention will be focused on the effect of the inherent symmetry properties of the system. Two cases will be considered (Mandadi and Huseyin 1978).

Case 1. Systems with symmetry in one critical coordinate. Let the potential function S be symmetric in the coordinate u^2 so that $S_{211} =$

$S_{222}=0$. The second and third perturbations then yield the derivatives

$$\varepsilon^{1,11}=-S_{111},\qquad \varepsilon^{1,22}=-S_{122},\qquad \varepsilon^{1,1'}=-S'_{11},$$

$$\varepsilon^{1,2'}=0,\qquad \varepsilon^{1,''}=0,\qquad \varepsilon^{2,2'}=-S'_{22},\qquad \varepsilon^{2,1'}=0,$$

$$\varepsilon^{2,''}=0,\qquad u^{s,1'}=\frac{S_{21}S'_{11}}{S_s},\qquad u^{s,2'}=\frac{S_{s2}S'_{22}}{S_s},$$

$$u^{s,''}=0,\qquad u^{s,11}=\frac{1}{S_s}(S_{s11}-S_{s1}S_{111}),$$

$$\varepsilon^{2,11}=0,\qquad \varepsilon^{2,22}=0.$$

With the aid of the above surface derivatives, one constructs the asymptotic equations of the equilibrium surface as

$$\varepsilon^1+\tfrac{1}{2}S_{111}(u^1)^2+\tfrac{1}{2}S_{122}(u^2)^2+S'_{11}u^1\mu=0, \tag{2.130}$$

$$\varepsilon^2+S_{122}u^1u^2+S'_{22}u^2\mu=0, \tag{2.131}$$

where $\eta=\eta_c+\mu$. The non-critical coordinates u^s do not play an active role in shaping the behaviour characteristics of the system and the corresponding equations are not, therefore, constructed.

The structure of the eqns (2.130) and (2.131) has an unmistakable resemblance to that of the *elliptic umbilic,* and indeed this will be the case provided the coefficients satisfy certain conditions.

In order to explore certain interesting situations in more detail, let $\varepsilon^2=0$. Eqn (2.131) then yields

$$u^2=0\quad\text{or}\quad S_{122}u^1+S'_{22}\mu=0. \tag{2.132}$$

which, in conjunction with eqn (2.130), results in two surfaces

$$\left.\begin{aligned} u^2=0,\\ \varepsilon^1+\tfrac{1}{2}S_{111}(u^1)^2+S'_{11}u^1\mu=0,\end{aligned}\right\} \tag{2.133}$$

and

$$\left.\begin{aligned} S_{122}u^1+S'_{22}\mu=0,\\ \varepsilon^1+\tfrac{1}{2}S_{111}(u^1)^2+\tfrac{1}{2}S_{122}(u^2)^2+S'_{11}u^1\mu=0.\end{aligned}\right\} \tag{2.134}$$

Interestingly, the surface (2.133) is related to general points of order 2 (i.e., fold catastrophe) which are simple critical points as discussed in Section 2.4.1. The system may be envisaged as developing initially in the sub-space $(u^1-\mu-\varepsilon^1)$ until it loses its stability upon reaching the critical line (locus of limit points or fold)

$$S_{111}u^1+S'_{11}\mu=0. \tag{2.135}$$

Unlike the simple case, however, bifurcations may occur before the critical fold line (eqn 2.135) is reached due to the presence of a second surface (2.134). Indeed, when the intersection of the two surfaces (2.133)

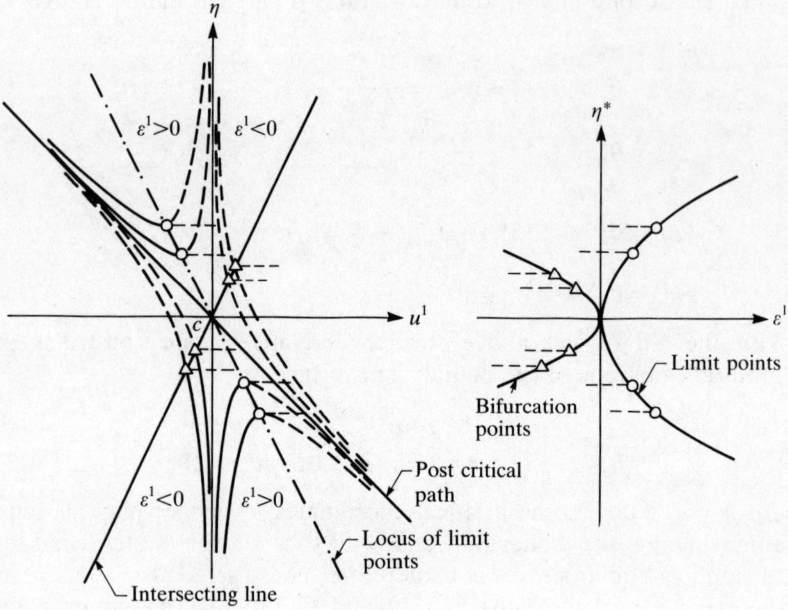

Fig. 2.9 Stability behaviour for $\alpha < \frac{1}{2}$.

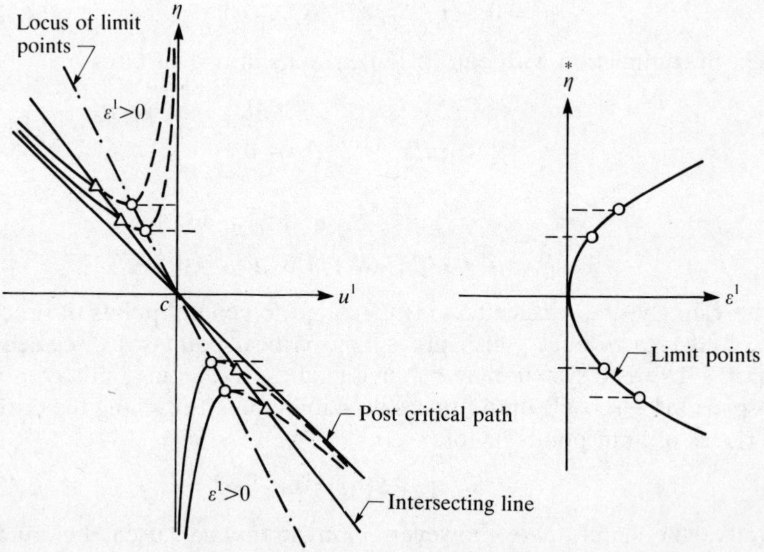

Fig. 2.10 Stability behaviour for $\frac{1}{2} < \alpha < 1$.

and (2.134) is considered, it is seen that the straight line in surface (2.134) may intersect the curved surface (2.133) in different ways, depending on the coefficients involved. The asymptotes of the hyperbolas associated with the surface (2.133) for constant values of ε^1 are given by

$$u^1 = 0 \quad \text{and} \quad \mu = -\frac{S_{111}}{2S'_{11}} u^1 \tag{2.136}$$

which are, of course, recognized as the fundamental and post-critical equilibrium paths of the perfect system, and the intersection can take place in several regions defined by these asymptotes. Let

$$\alpha \triangleq \frac{S'_{11} S_{122}}{S'_{22} S_{111}}; \tag{2.137}$$

the following four distinct cases can be identified:

i) $\alpha < \frac{1}{2}$.

It then follows from eqn (2.136) and the intersecting line

$$S_{122} u^1 + S'_{22} \mu = 0 \tag{2.138}$$

that the intersection takes place outside the asymptotes (eqn 2.136), involving the stable equilibrium paths. In other words, a particular equilibrium path corresponding to a given positive value of ε^1 loses stability at a limit point as in the case of simple points; however, the stable branch of the hyperbola (for $\varepsilon^1 < 0$) is now intersected, resulting in bifurcations that could not have occurred if c were a simple point (Fig. 2.9).

ii) $\frac{1}{2} < \alpha < 1$.

In this case, the line of intersection lies between the post-critical path and the critical line (2.135), as depicted in Fig. 2.10. Bifurcations now occur after the limit points, and this case is not, therefore, of much practical significance.

iii) $\alpha = 1$.

The line of intersection coincides with the critical line (2.135), indicating that limit points and bifurcation phenomena take place simultaneously (Fig. 2.11).

iv) $\alpha > 1$.

This condition implies

$$-\frac{S_{122}}{S'_{22}} < -\frac{S_{111}}{S'_{11}}$$

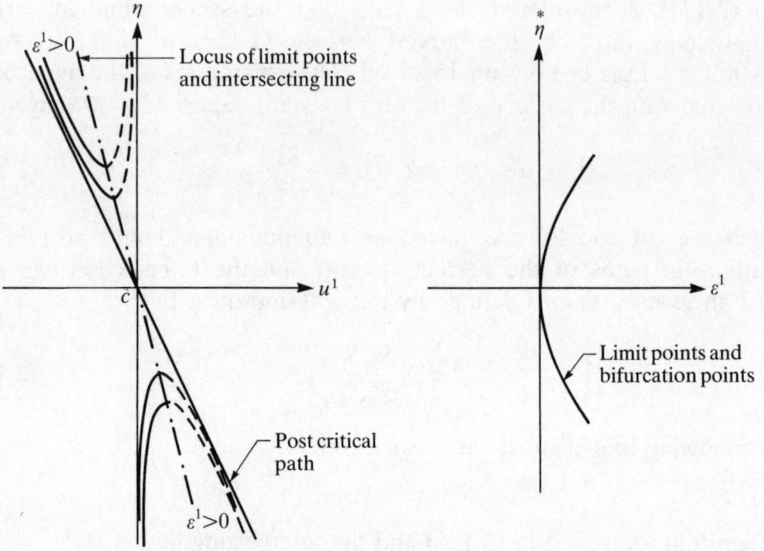

Fig. 2.11 Stability behaviour for $\alpha = 1$.

which ensures that the line of intersection lies between the fundamental path ($u^1 = 0$) and the critical line (eqn 2.135). It then follows that initially stable equilibrium paths corresponding to $\varepsilon^1 > 0$ lose their stability via bifurcation before the limit points are reached as η is increased gradually (Fig. 2.12).

On the basis of these four distinct cases, one observes that the behaviour of a slightly imperfect system in the vicinity of a coincident critical point can be quite different from that associated with simple critical points. Firstly, in contrast with simple points, there may not be a totally stable path in the vicinity of a coincident critical point; i.e. the equilibrium paths corresponding to $\varepsilon^1 > 0$ and $\varepsilon^1 < 0$ may both lose stability at a certain value of the parameter. The loss of stability may take place at a limit or bifurcation point in the former case (Figs. 2.10 and 2.12), while it can only occur at a point of bifurcation in the latter case (Fig. 2.9). Secondly, a loss of stability may take place at much lower bifurcation loads, *thus indicating higher imperfection sensitivity in the case of coincident critical points.*

The stability boundaries (here, imperfection sensitivity curves) corresponding to limit and bifurcation points can be obtained readily. Indeed, introducing eqn (2.135) into eqn (2.133) yields

$$\overset{*}{\mu} = \pm \frac{(2S_{111})^{1/2}}{S'_{11}} (\varepsilon^1)^{1/2}. \tag{2.139}$$

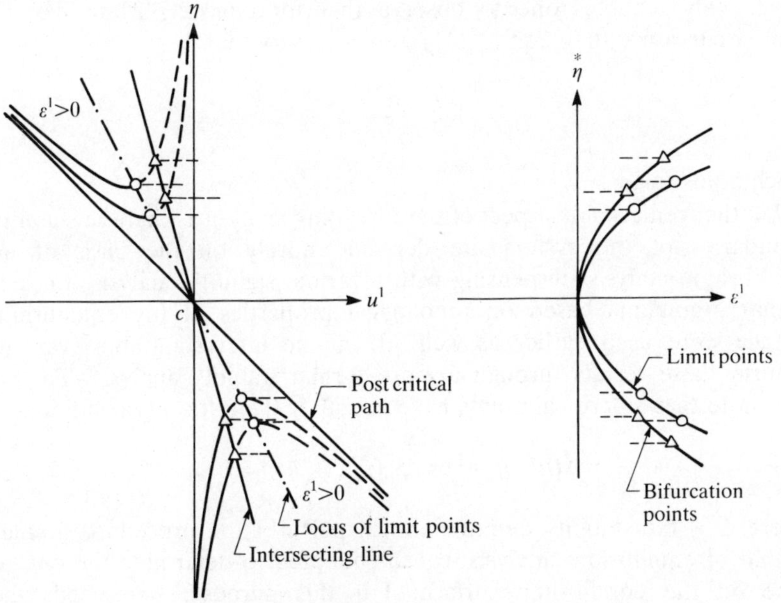

Fig. 2.12 Stability behaviour for $\alpha > 1$.

The stability boundary corresponding to bifurcation points is likewise obtained by eliminating u^1 between eqn (2.138) and eqn (2.133), as

$$\overset{*}{\mu} = \pm \frac{S_{122}}{S'_{22}} \left[\frac{2}{S_{111}(2\alpha - 1)} \right]^{1/2} (\varepsilon^1)^{1/2}. \tag{2.140}$$

These parabolic relationships are shown in Figs. 2.9 to 2.12 for the four cases discussed above. In Fig. 2.9, the parabola (2.140) has negative curvature since $\alpha < \frac{1}{2}$, while its curvature is positive in all other cases. Obviously, if the line of intersection and the locus of limit points (critical line) coincide (Case iii), the two parabolas (2.139 and 2.140) become identical. It is also expected from the equilibrium analysis that the curvature of the parabola (2.140) is smaller than that of the parabola (2.139) in Case iv. This can be shown independently on the basis of the condition $\alpha > 1$. In fact, expressing eqns (2.139) and (2.140) as

$$\overset{*}{\mu}{}^2 = \frac{2(S_{122})^2}{(S'_{22})^2 S_{111} \alpha^2} \varepsilon^1 \tag{2.141}$$

and

$$\overset{*}{\mu}{}^2 = \frac{2(S_{122})^2}{(S'_{22})^2 S_{111}(2\alpha - 1)} \varepsilon^1, \tag{2.142}$$

respectively, enables one to observe that for eqn (2.142) to have a smaller curvature than eqn (2.141) it is necessary that

$$1 < \frac{\alpha^2}{2\alpha - 1},$$

which holds for $\alpha > 1$.

Another remarkable aspect of the foregoing analysis is that the stability boundaries of the system are derived entirely on the basis of an equilibrium analysis, dispensing with a formal stability analysis. In fact, similar arguments based on topological properties of the equilibrium surface were used earlier as well. It will be interesting, however, to confirm these results through a more formal stability analysis. To this end, note that all critical points have to satisfy the criticality condition

$$\Delta(u^k, \eta, \varepsilon^1) = |S_{ij}(u^k, \eta, \varepsilon^1)| = 0 \tag{2.143}$$

where Δ is the stability determinant. A perturbation procedure, similar to that of equilibrium analysis, can be adopted to determine the critical zone on the equilibrium surface. For this purpose, one needs the derivatives of the stability determinant Δ which are obtained by differentiating eqn (2.143), by rows or columns, successively with respect to its arguments and evaluating at c. Thus,

$$\begin{gathered} \Delta_i = 0, \qquad \Delta^1 = 0, \qquad \Delta' = 0, \\ \Delta_{ij} = 2(S_{11i}S_{22j} - S_{12i}S_{12j}) \prod_{s=3}^{n} S_s, \\ \Delta'_i = (S_{11i}S'_{22} + S'_{11}S_{22i}) \prod_{s=3}^{n} S_s, \\ \Delta'' = 2S'_{11}S'_{22} \prod_{s=3}^{n} S_s \end{gathered} \tag{2.144}$$

where the notation follows the earlier pattern.

In view of the $(n+1)$ equations, $S_i = 0$, and $\Delta = 0$, the critical zone should be expressed in terms of a single parameter (Huseyin 1975*a*). By choosing u^1 as this parameter, the solutions of the $(n+1)$ equations are expressed as

$$u^j = u^j(u^1), \qquad \eta = \eta(u^1), \quad \text{and} \quad \varepsilon^1 = \varepsilon^1(u^1), \tag{2.145}$$

which are then substituted back into $S_i = 0$ and $\Delta = 0$ to yield the identities

$$\begin{gathered} S_i[u^j(u^1), \eta(u^1), \varepsilon^1(u^1)] \equiv 0, \\ \Delta[u^j(u^1), \eta(u^1), \varepsilon^1(u^1)] \equiv 0. \end{gathered} \tag{2.146}$$

Following the familiar pattern leads to

$$S_{ij}u^{j,1} + S'_i\eta^{,1} + S^1_i\varepsilon^{1,1} = 0,$$
$$\Delta_j u^{j,1} + \Delta'\eta^{,1} + \Delta^1\varepsilon^{1,1} = 0 \tag{2.147}$$

where $\eta^{,1} \triangleq \partial\eta/\partial u^1$. Evaluations at c result in

$$u^{s,1} = 0, \qquad \varepsilon^{1,1} = 0. \tag{2.148}$$

Similarly, a second perturbation yields two sets of derivatives

$$\eta^{,1} = -\frac{S_{111}}{S'_{11}}, \qquad \varepsilon^{1,11} = S_{111}, \tag{2.149}$$

and

$$\eta^{,1} = -\frac{S_{122}}{S'_{22}}, \qquad \varepsilon^{1,11} = -\left(S_{111} - \frac{2S'_{11}S_{122}}{S'_{22}}\right). \tag{2.150}$$

Using Taylor's expansions results in the critical zones

$$\mu = -\frac{S_{111}}{S'_{11}}u^1, \qquad \varepsilon^1 = \tfrac{1}{2}S_{111}(u^1)^2 \tag{2.151}$$

and

$$\mu = -\frac{S_{122}}{S'_{22}}u^1, \qquad \varepsilon^1 = -\frac{1}{2}\left(S_{111} - 2\frac{S'_{11}S_{122}}{S'_{22}}\right)(u^1)^2. \tag{2.152}$$

By eliminating u^1 in eqns (2.151) and (2.152) one obtains the stability boundaries (2.139) and (2.140), respectively.

Case 2. Systems with symmetry in both critical coordinates. If the potential function S is symmetric in both u^1 and u^2, all the cubic coefficients vanish, and only even derivatives of S with respect to u^1 and u^2 may be non-zero. A third perturbation of eqn (2.129) yields

$$\varepsilon^{1,111} = -\bar{S}_{1111}, \qquad \varepsilon^{1,112} = \varepsilon^{1,222} = 0,$$
$$\varepsilon^{1,122} = -\bar{S}_{1122}, \qquad \varepsilon^{2,222} = -\bar{S}_{2222}, \tag{2.153}$$
$$\varepsilon^{2,111} = \varepsilon^{2,122} = 0, \qquad \varepsilon^{2,112} = -\bar{S}_{1122},$$

where

$$\bar{S}_{aabb} = S_{aabb} - \sum_s \frac{S_{saa}S_{sbb}}{S_s}, \qquad (a \neq b)$$

$$\bar{S}_{aaaa} = S_{aaaa} - 3\sum_s \frac{(S_{saa})^2}{S_s}, \qquad (a, b = 1, 2).$$

The equilibrium surface in critical coordinates is then described by

$$\varepsilon^1 + \frac{1}{3!}\bar{S}_{1111}(u^1)^3 + \tfrac{1}{2}\bar{S}_{1122}u^1(u^2)^2 + S'_{11}u^1\mu = 0 \tag{2.154}$$

$$\varepsilon^2 + \frac{1}{3!}\bar{S}_{2222}(u^2)^3 + \tfrac{1}{2}\bar{S}_{1122}(u^1)^2u^2 + S'_{22}u^2\mu = 0. \tag{2.155}$$

Consider again the situation arising when $\varepsilon^2 = 0$; it is easily observed that the equilibrium surface disintegrates into two surfaces given by

$$\begin{gathered} u^2 = 0, \\ \varepsilon^1 + \frac{1}{3!}\bar{S}_{1111}(u^1)^3 + S'_{11}u^1\mu = 0, \end{gathered} \tag{2.156}$$

and

$$\begin{gathered} \bar{S}_{2222}(u^2)^2 + 3\{\bar{S}_{1122}(u^1)^2 + 2S'_{22}\mu\} = 0, \\ \varepsilon^1 + \frac{1}{3!}\left\{\bar{S}_{1111} - 9\frac{(\bar{S}_{1122})^2}{\bar{S}_{2222}}\right\}(u^1)^3 + 2\left\{3S'_{11} - 9\frac{\bar{S}_{1122}S'_{22}}{\bar{S}_{2222}}\right\}u^1\mu = 0. \end{gathered} \tag{2.157}$$

The surface (2.156) is associated with a singular general point (i.e. cusp catastrophe) as discussed in Section 2.4.2 (see eqn 2.85). One may again envisage the behaviour of the system initially on the surface (2.156) until it loses its stability upon reaching the critical line

$$\tfrac{1}{2}\bar{S}_{1111}(u^1)^2 + S'_{11}\mu = 0. \tag{2.158}$$

However, while the equilibrium paths in the vicinity of a singular general point can only become unstable at limit points described by eqn (2.158), in this case the effect of a 2-fold critical point is manifested in the appearance of a second surface (eqn 2.157) which may result in bifurcations on the neighbouring equilibrium paths.

In analogy with the previous case, consider the intersection of the surfaces (2.156) and (2.157) in u^1-μ space; this is readily obtained as

$$\tfrac{1}{2}\bar{S}_{1122}(u^1)^2 + S'_{22}\mu = 0 \tag{2.159}$$

by setting $u^2 = 0$ in eqn (2.157). Define now the quantities

$$a = -2\frac{S'_{11}}{\bar{S}_{1111}} \quad \text{and} \quad b = -2\frac{S'_{22}}{\bar{S}_{1122}}. \tag{2.160}$$

The asymptotes associated with the equilibrium surface (2.156)—that is the equilibrium paths of the perfect system on the surface (2.156)—are then given by

$$u^1 = 0, \qquad (u^1)^2 = 3a\mu, \tag{2.161}$$

and the intersecting parabola (2.159) by

$$(u^1)^2 = b\mu. \tag{2.162}$$

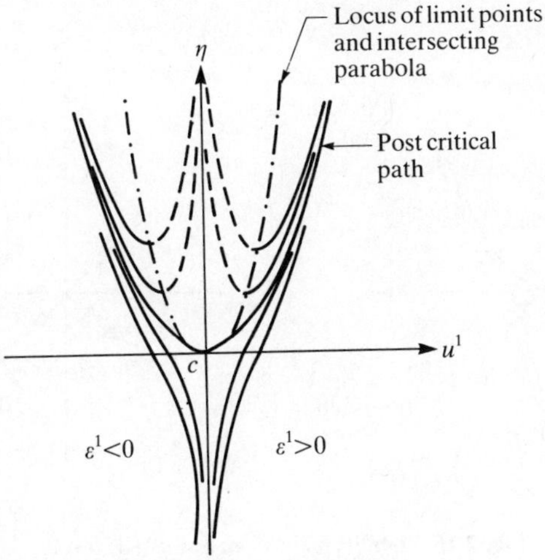

Fig. 2.13 Stability behaviour for $a>0$, $b>0$, $b<3a$.

Four interesting situations arise:

i) $a>0$, $b>0$.

In this case both the post-critical path (2.161) and the intersecting parabola (2.162) have positive curvatures. If, in addition, $b<3a$, the parabola (2.162) intersects the complementary paths, and for $b=a$ bifurcation points coincide with the critical line (2.158) as depicted in Fig. 2.13. If, on the other hand, $b>3a$, intersections take place between the post-critical path and the u^1 axis, and initially stable paths lose stability at bifurcation points above the critical point c (Fig. 2.14). Induced

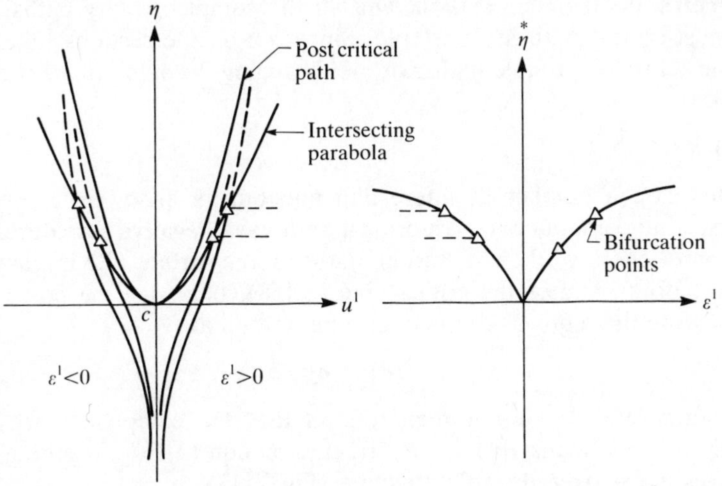

Fig. 2.14 Stability behaviour for $a>0$, $b>0$, $b>3a$.

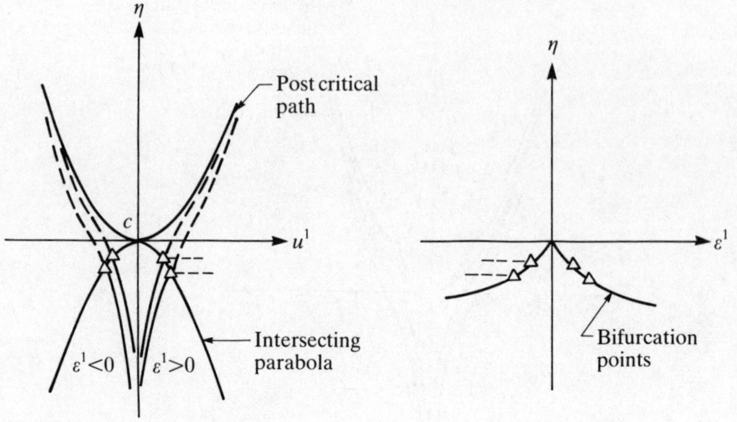

Fig. 2.15 Stability behaviour for $a>0$, $b<0$.

branching states are described by the surface (2.157). This is a phenomenon that cannot be exhibited in the vicinity of a simple critical point.

ii) $a>0$, $b<0$.

In this case the stable equilibrium paths—which are entirely stable in the simple case—are intersected by the parabola (2.162) below the u^1 axis, thus losing their stability at bifurcation points at values of $\eta<\eta_c$ (Fig. 2.15).

iii) $a<0$, $b>0$.

Under the condition $a<0$, the perfect system exhibits an unstable symmetric point of bifurcation, and all the complementary paths above the post-critical path are unstable. Since $b>0$, intersections take place on these totally unstable paths without changing the stability picture (Fig. 2.16).

iv) $a<0$, $b<0$.

In this case a number of interesting phenomena arise, since both the intersecting parabola and post-critical path have negative curvatures, and the intersections of the equilibrium paths on the surface (2.156) may take place before or after the critical line (2.158), or above the post-critical path. Note that eqn (2.158) may also be written as

$$(u^1)^2 = a\mu, \tag{2.163}$$

and suppose $|b|>3\,|a|$; it then follows that the complementary paths which are totally unstable are intersected by eqn (2.162), resulting in no changes in the overall stability picture (Fig. 2.17).

Now let $|a|<|b|<3\,|a|$; the two equilibrium surfaces then intersect

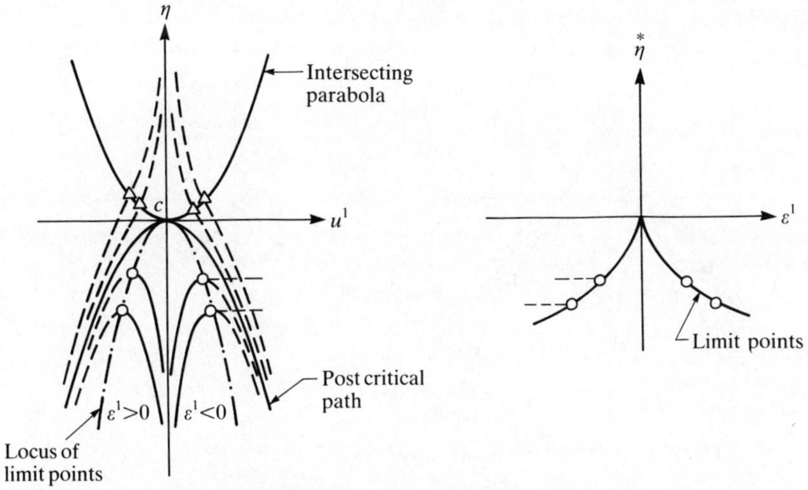

Fig. 2.16 Stability behaviour paths for $a<0$, $b>0$.

along a parabola which lies between the critical line (2.163) and post-critical path (2.161). Thus, the limit points (fold line) on the surface (2.156) play a predominant role as if c were a simple critical point (Fig. 2.18).

Finally, consider the situation under the condition $|b|<|a|$. The equilibrium paths on the surface (2.156) now lose their stability at bifurcation points before reaching the critical fold line as depicted in Fig. 2.19. The bifurcation boundary assumes a pre-dominant role in this case, and implies higher imperfection sensitivities.

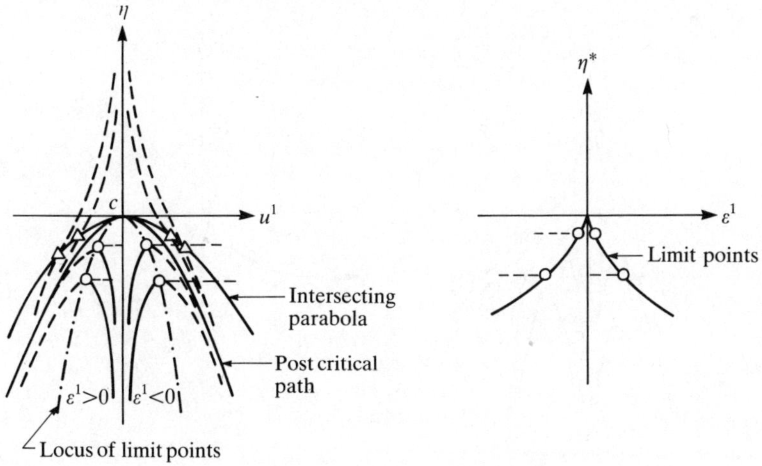

Fig. 2.17 Stability behaviour paths for $a<0$, $b<0$, $|b|>3\,|a|$.

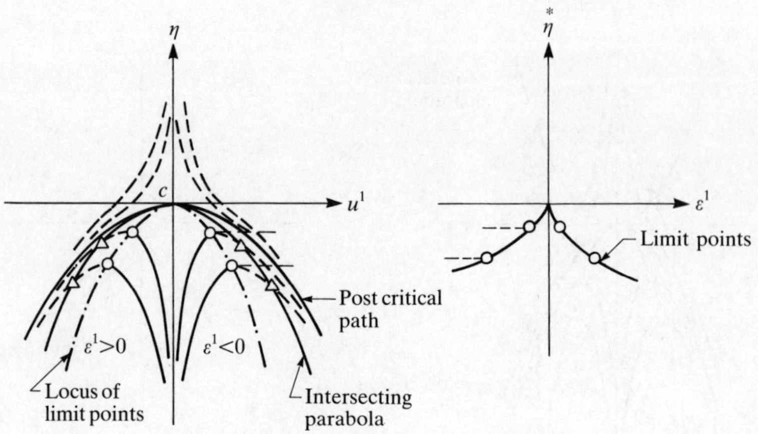

Fig. 2.18 Stability behaviour paths for $a<0$, $b<0$, $|a|<|b|<3|a|$.

In all the cases discussed above, the stability boundary consists of points of bifurcation or limit points. The bifurcation boundary is obtained by eliminating u^1 between eqns (2.156) and (2.159) and is

$$\overset{*}{\eta} = \eta_c - \frac{1}{2}\left\{\frac{36\bar{S}_{1122}}{S'_{22}(\bar{S}_{1111}S'_{22} - 3\bar{S}_{1122}S'_{11})^2}\right\}^{1/3} (\varepsilon^1)^{2/3}. \tag{2.164}$$

The stability boundary consisting of limit points is, of course, the projection of the critical fold line (2.158) onto the parameter space η-ε^1, and follows from the elimination of u^1 between eqns (2.156) and (2.158):

$$\overset{*}{\eta} = \eta_c - \frac{1}{2}\frac{(9\bar{S}_{1111})^{1/3}}{S'_{11}}(\varepsilon^1)^{2/3}. \tag{2.165}$$

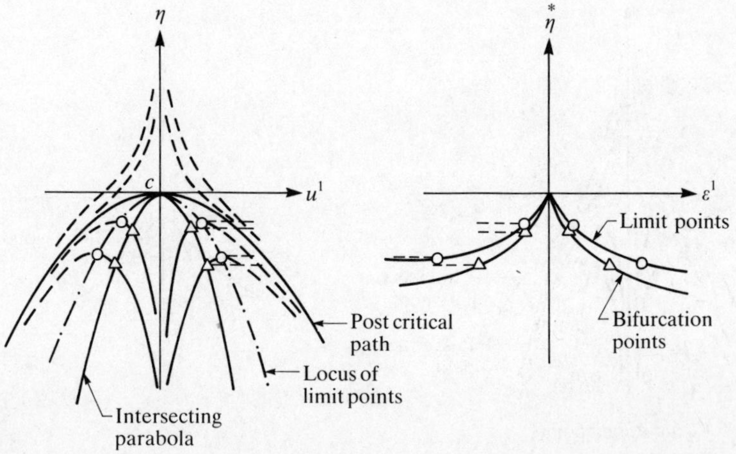

Fig. 2.19 Stability behaviour for $a<0$, $b<0$, $|b|<|a|$.

Problem 2.6. Verify eqns (2.164) and (2.165) through a formal perturbation analysis based on the stability determinant (2.143) by adopting the procedure described in the analysis of systems with symmetry in one critical coordinate only.

2.7 Concluding remarks

In view of the fact that potential systems have been explored relatively thoroughly in a number of excellent monographs, this chapter has been kept to a concise form. In recent years attention has focused mainly in non-potential systems, which will be explored in the following chapters. Before proceeding to do so, however, it should be emphasized once more that Koiter's original work (1945, 1963, 1965) constitutes a cornerstone in the historical development of the non-linear concepts underlying static instabilities encountered not only in the field of elastic stability but in the entire domain of potential systems. His ideas inspired many researchers—notably at University College, London and Harvard University (e.g., Budiansky and Hutchinson 1966, Hutchinson and Budiansky 1966). Some of the earlier work (Britvec and Chilver 1963, Thompson 1963, Roorda 1965, Sewell 1965, Chilver 1967) at University College has led the way to an extensive study of imperfection sensitivities of distinct and coincident critical points. In this context, when a general theory concerning the non-linear instability behaviour of multiple-parameter systems was submitted to London University in 1967 in a Ph.D. thesis (Huseyin 1967) and initial publications appeared (Huseyin 1969, 1970*a*, *b*, *c*, *d*; 1971*a*, *b*), there was no word of catastrophe theory or any other non-linear, multiple-parameter stability theory. Initial motivation for this work had come from a theorem due to Papkovich (1934) concerning the convexity of the stability boundary of linear conservative systems. Incidentally, Papkovich's theorem became widely available through Renton's article (1967), and although it was generally believed that the proof of the theorem had to go back to the nineteen thirties, this author has learned the exact reference (Papkovich 1934) rather recently through a private communication with Professor Koiter who had discovered the convexity property independently. It was after the publication of the monograph on *Non-Linear Theory of Elastic Stability* (Huseyin 1975) that catastrophe theory became widely known, and an explosion of activities in a variety of fields and directions was set in motion. Indeed, interest and controversy surrounding catastrophe theory has been so intense and baffling that it will be an overwhelming task to try to describe what has actually transpired, even now that the dust seems to be settling and interests shifting in other directions. Certainly, it is not the objective here to give an elaborate account of these developments—other than what has already been said—nor is there a desire to get involved in the controversy. Extensive lists of references can be found in a recent book by Thompson (1982*a*), and a review article by Stewart (1981). It has to be noted that Arnold (1974) has a list of 'Lagrangian singularities', similar to elementary catastrophes, defined in the context of Lagrangian manifolds.

Several relatively recent books contain related material (Abraham and Marsden 1978, Marsden and Hughes 1983, Ioos and Joseph 1980, Chow and Hale 1982, Guckenheimer and Holmes 1983, Sattinger 1973 and Keller and Antman 1969). In addition, Haken's work on synergetics (1978, 1983), and Santilli's book *Foundations of Theoretical Mechanics* (1983) provide interesting viewpoints of two theoretical physicists. A comprehensive article by Knops and Wilkes in *Handbuch der Physik* (1973) should also be consulted.

Returning to the historical developments motivated by problems of elastic stability, it is noted that a substantial amount of work has been done, and it is not easy to provide a full bibliography. In addition to those already cited, only a few general studies will be mentioned here. Following Koiter's work, some of the initial investigations of compound branching and the associated sensitivities were carried out by Chilver (1967), Supple (1967), Ho (1972, 1974), Johns (1974), Huseyin (1975), Keener (1976), Huseyin and Mandadi (1977*b*), Mandadi and Huseyin (1978), Hunt (1977, 1981), Thompson and Gaspar (1977), Hansen (1977), Reis and Roorda (1979), Samuels (1979, 1982), Golubitsky and Shaeffer (1979), and Potier-Ferry (1979, 1981*a*, *b*). Discussions concerning the relationship to elementary catastrophes include the papers by Thompson (1982*a*, *b*), Gaspar, Huseyin, and Mandadi (1978), Huseyin (1977, 1978*a*), Sewell (1977), Troger (1976), Chillingworth (1975), and books by Leipholz (1980) and Gilmore (1981).

3

STATIC INSTABILITY OF AUTONOMOUS SYSTEMS

3.1 One-parameter systems

The approach in this and the following chapters differs from that adopted in Chapter 2 where the governing equations of the equilibrium surface, derived at the outset, were in a sufficiently general form capable of yielding information for a variety of more specific cases. This was partly due to a self-imposed constraint to keep the analysis of potential systems to as concise a form as possible. The treatment of non-potential systems in this and the following chapters, however, will commence with relatively simple situations and progress systematically through increasing complexities. As discussed in the introduction, the analysis of general behaviour patterns will be carried out within the framework of an autonomous model which enables one to examine static and dynamic instabilities as well as the stability of particular states on the basis of a unified mathematical formulation. Also note that this formulation does not necessarily exclude potential systems.

Consider an autonomous system characterized by

$$\frac{\mathrm{d}\boldsymbol{z}}{\mathrm{d}t} = \boldsymbol{Z}(\boldsymbol{z}, \eta), \qquad \boldsymbol{z} \in R^n, \qquad \eta \in R, \tag{3.1}$$

where the vector function $\boldsymbol{Z}(\boldsymbol{z}, \eta)$ is assumed to be real and smooth—at least in a region G of interest. The equilibrium states of eqn (3.1) satisfy

$$\boldsymbol{Z}(\boldsymbol{z}, \eta) = \boldsymbol{0}, \tag{3.2}$$

and the point sets in the *state-parameter space* R^{n+1}, corresponding to the solutions of eqn (3.2), constitute one-dimensional manifolds which will be called equilibrium paths as before.

Consider an equilibrium path which is initially stable so that the eigenvalues of the Jacobian evaluated at a *regular* equilibrium state c on this portion of the equilibrium path all have negative real parts. By virtue of the implicit function theorem it can then be concluded that there exists a unique single-valued equilibrium path, $\boldsymbol{z}(\eta)$, through the *regular* point c—so that $\boldsymbol{z}_c = \boldsymbol{z}(\eta_c)$—and the path is stable in the immediate vicinity of c.

The asymptotic equations of this unique path may be obtained readily by perturbing the identity

$$\boldsymbol{Z}[\boldsymbol{z}(\eta),\eta]\equiv 0, \tag{3.3}$$

following the procedure described in Section 2.3.1.

As the parameter η is varied slowly it is conceivable that a real eigenvalue of the Jacobian crosses the origin of the complex plane at a *critical* value of the parameter, $\eta=\eta_c$, and becomes real positive for $\eta>\eta_c$ while the remaining eigenvalues continue to have negative real parts. It may also happen that several real or complex conjugate eigenvalues cross the origin simultaneously and become real positive, resulting in *static* (*divergence*) *instability*. The critical divergence point c, at which the Jacobian becomes singular, is said to be *simple* if the *nullity* of the Jacobian matrix at c is one. Although critical points with repeated (coincident) zeros are customarily described as *compound critical*—as in the case of potential systems—the post-critical behaviour is essentially influenced by the dimension of the eigenvector space (Huseyin 1978*b*), and here the term *r-fold compound* will be reserved for critical points at which the Jacobian matrix has *nullity* r.

In order to facilitate the following analyses, an appropriate canonical form of the Jacobian matrix

$$[Z_{ij}]=\left[\frac{\partial Z_i}{\partial z^j}\right]_c \tag{3.4}$$

will be required. In this regard it is first recognized that the Jacobian matrix (3.4) is, in general, asymmetric, and unlike the quadratic form of a potential function, it cannot always be diagonalized. If the eigenvectors are linearly independent, diagonalization is possible. On the other hand, it can be shown that the eigenvectors are linearly independent if the eigenvalues are distinct. A number of other canonical forms, however, are always possible and the Jordan form is one of them (e.g., Huseyin 1978*b*). Nevertheless, one of the objectives here is to keep the entire analysis real, and Jordan blocks will be used in conjunction with real blocks corresponding to complex conjugate eigenvalues. Formal proofs and derivations can be found in standard books on matrices (e.g. Gantmacher 1959, Hirsch and Smale 1974), but certain transformations and underlying concepts and procedures will be made plausible here for convenience.

Let $\boldsymbol{A}$ be a real square matrix; if λ is a complex eigenvalue of $\boldsymbol{A}$, its complex conjugate $\bar{\lambda}$ is also an eigenvalue of $\boldsymbol{A}$, and the same rule applies to the corresponding eigenvectors $\boldsymbol{q}$ and $\bar{\boldsymbol{q}}$ since

$$(\boldsymbol{A}-\lambda\boldsymbol{I})\boldsymbol{q}=\boldsymbol{0} \quad \text{and} \quad (\boldsymbol{A}-\bar{\lambda}\boldsymbol{I})\bar{\boldsymbol{q}}=\boldsymbol{0}. \tag{3.5}$$

Suppose that one of the eigenvalues of $\boldsymbol{A}$, say λ_1, is repeated r times, and the dimension of the corresponding eigenvector space remains one.

It is, of course, possible that the number of linearly independent eigenvectors may vary between one and r; nevertheless, the treatment of such situations follows the pattern to be described below. Thus, if $\boldsymbol{q}_1$ is the eigenvector corresponding to the r-fold eigenvalue λ_1, then

$$(\boldsymbol{A} - \lambda_1 \boldsymbol{I})\boldsymbol{q}_1 = 0, \tag{3.6}$$

and one defines a set of linearly independent eigenvectors by

$$(\boldsymbol{A} - \lambda_1 \boldsymbol{I})\boldsymbol{q}_j = \boldsymbol{q}_{j-1}, \qquad (j = 2, 3, \ldots, r). \tag{3.7}$$

It is noticed that if λ_1 is complex, then $\bar{\lambda}_1$ is also r-fold and corresponding relations similar to eqns (3.6) and (3.7) may be constructed. The vectors $\boldsymbol{q}_j$ are often referred to as the *generalized eigenvectors*.

To help fix ideas, consider the simple case of a (2×2) matrix $\boldsymbol{A}$ with repeated eigenvalue λ_1; by virtue of eqns (3.6) and (3.7) one has

$$\begin{aligned}[\boldsymbol{A}\boldsymbol{q}_1 \quad \boldsymbol{A}\boldsymbol{q}_2] &= [\lambda_1 \boldsymbol{q}_1 \quad (\boldsymbol{q}_1 + \lambda_1 \boldsymbol{q}_2)] \\ &= [\boldsymbol{q}_1 \quad \boldsymbol{q}_2]\begin{bmatrix}\lambda_1 & 1 \\ 0 & \lambda_1\end{bmatrix}\end{aligned} \tag{3.8}$$

which may be written as

$$\boldsymbol{A}\boldsymbol{Q} = \boldsymbol{Q}\boldsymbol{J} \quad \text{where} \quad \boldsymbol{J} = \begin{bmatrix}\lambda_1 & 1 \\ 0 & \lambda_1\end{bmatrix} \quad \text{and} \quad \boldsymbol{Q} = [\boldsymbol{q}_1 \quad \boldsymbol{q}_2], \tag{3.9}$$

or

$$\boldsymbol{J} = \boldsymbol{Q}^{-1}\boldsymbol{A}\boldsymbol{Q}.$$

This procedure can readily be generalized to an arbitrary n-square matrix $\boldsymbol{A}$. Indeed, it can be demonstrated that there exists a non-singular matrix $\boldsymbol{Q}$ such that

$$\boldsymbol{Q}^{-1}\boldsymbol{A}\boldsymbol{Q} = \begin{bmatrix}\boldsymbol{J}_{n_1}(\lambda_1) & & & 0 \\ & \boldsymbol{J}_{n_2}(\lambda_2) & & \\ & & \ddots & \\ 0 & & & \boldsymbol{J}_{n_k}(\lambda_k)\end{bmatrix} \tag{3.10}$$

where λ_i are the eigenvalues of A—distinct or coincident—and

$$n_1 + n_2 + \ldots + n_k = n.$$

The J_{n_i} are called *Jordan blocks* whose main diagonals consist of the eigenvalues and the diagonal above each main diagonal consists of ones; e.g. if $n_2 = 3$

$$\boldsymbol{J}_{n_2} = \begin{bmatrix}\lambda_2 & 1 & 0 \\ 0 & \lambda_2 & 1 \\ 0 & 0 & \lambda_2\end{bmatrix}.$$

If $n_1 = n_2 = \ldots = n_k = 1$ then, the Jordan canonical form becomes a diagonal matrix. The maximum number of eigenvectors corresponding to a repeated eigenvalue is the number of Jordan blocks in which it appears. Thus if, for example,

$$\boldsymbol{J} = \begin{bmatrix} \lambda_1 & 0 & 0 \\ 0 & \lambda_1 & 1 \\ 0 & 0 & \lambda_1 \end{bmatrix},$$

there exists two independent eigenvectors corresponding to λ_1. The number of *Jordan blocks* in which an eigenvalue appears is the *index of the eigenvalue*. It is clear that the *multiplicity* of an eigenvalue cannot be less than its index. If the multiplicity is equal to the *index* for all eigenvalues of $\boldsymbol{A}$, then $\boldsymbol{A}$ is diagonalizable.

Next, consider again the simple case of a (2×2) matrix, and suppose that its eigenvalues consist of a complex conjugate pair, $\lambda_1 = \alpha + \mathrm{i}\omega$ and $\lambda_2 = \alpha - \mathrm{i}\omega$. It then follows that

$$[\boldsymbol{A} - (\alpha + \mathrm{i}\omega)\boldsymbol{I}](\boldsymbol{q}^R + \mathrm{i}\boldsymbol{q}^I) = \boldsymbol{0}$$

and (3.11)

$$[\boldsymbol{A} - (\alpha - \mathrm{i}\omega)\boldsymbol{I}](\boldsymbol{q}^R - \mathrm{i}\boldsymbol{q}^I) = \boldsymbol{0},$$

where $\boldsymbol{q}^R$ and $\boldsymbol{q}^I$ are the real and imaginary parts of the eigenvector $\boldsymbol{q}$, respectively. It is easy to see that eqns (3.11) yields

$$\boldsymbol{A}\boldsymbol{q}^R = \alpha\boldsymbol{q}^R - \omega\boldsymbol{q}^I$$

and (3.12)

$$\boldsymbol{A}\boldsymbol{q}^I = \omega\boldsymbol{q}^R + \alpha\boldsymbol{q}^I.$$

One may then construct the relationship

$$\begin{aligned} [\boldsymbol{A}\boldsymbol{q}^R \quad \boldsymbol{A}\boldsymbol{q}^I] &= [(\alpha\boldsymbol{q}^R - \omega\boldsymbol{q}^I) \quad (\omega\boldsymbol{q}^R + \alpha\boldsymbol{q}^I)] \\ &= [\boldsymbol{q}^R \quad \boldsymbol{q}^I]\begin{bmatrix} \alpha & \omega \\ -\omega & \alpha \end{bmatrix} \end{aligned} \tag{3.13}$$

which leads to $\boldsymbol{A}\boldsymbol{Q} = \boldsymbol{Q}\boldsymbol{B}$, or

$$\boldsymbol{B} = \boldsymbol{Q}^{-1}\boldsymbol{A}\boldsymbol{Q} \tag{3.14}$$

where

$$\boldsymbol{B} = \begin{bmatrix} \alpha & \omega \\ -\omega & \alpha \end{bmatrix} \quad \text{and} \quad \boldsymbol{Q} = [\boldsymbol{q}^R \quad \boldsymbol{q}^I]. \tag{3.15}$$

Note that if the vectors in $\boldsymbol{Q}$ are rearranged, one may also have

$$\boldsymbol{B} = \begin{bmatrix} \alpha & -\omega \\ \omega & \alpha \end{bmatrix}.$$

As an illustrative example, suppose now that $\boldsymbol{A}$ has an r-fold zero eigenvalue, a pair of distinct complex conjugate eigenvalues and two equal pairs with a repeated eigenvector. Then, combining and generalizing the transformations discussed above, one may form a non-singular transformation matrix $\boldsymbol{Q}$ as

$$\boldsymbol{Q} = [\boldsymbol{q}_1, \ldots, \boldsymbol{q}_r, \boldsymbol{q}_{r+1}^R, \boldsymbol{q}_{r+1}^I, q_{r+3}^R, \boldsymbol{q}_{r+3}^I, \boldsymbol{q}_{r+5}^R, \boldsymbol{q}_{r+5}^I] \tag{3.16}$$

where $\boldsymbol{q}_1, \ldots, \boldsymbol{q}_r$ are the eigenvectors or generalized eigenvectors corresponding to vanishing eigenvalues, $\boldsymbol{q}_{r+1}$ and $\boldsymbol{q}_{r+3}$ are two linearly independent eigenvectors and $\boldsymbol{q}_{r+5}$ is the generalized eigenvector corresponding to the repeated complex pairs. On the basis of these assumptions, it is not difficult to verify that

$$\boldsymbol{AQ} = \boldsymbol{QC} \quad \text{or} \quad \boldsymbol{C} = \boldsymbol{Q}^{-1}\boldsymbol{AQ} \tag{3.17}$$

where

$$\boldsymbol{C} = \left[\begin{array}{c|cc|cc|cc} \boldsymbol{J} & \multicolumn{2}{c|}{\boldsymbol{0}} & \multicolumn{2}{c|}{\boldsymbol{0}} & \multicolumn{2}{c}{\boldsymbol{0}} \\ \hline \boldsymbol{0} & \alpha_{r+1} & \omega_{r+1} & \multicolumn{2}{c|}{\boldsymbol{0}} & \multicolumn{2}{c}{\boldsymbol{0}} \\ & -\omega_{r+1} & \alpha_{r+1} & & & & \\ \hline \boldsymbol{0} & \multicolumn{2}{c|}{\boldsymbol{0}} & \alpha_{r+3} & \omega_{r+3} & 1 & 0 \\ & & & -\omega_{r+3} & \alpha_{r+3} & 0 & 1 \\ \hline \boldsymbol{0} & \multicolumn{2}{c|}{\boldsymbol{0}} & & & \alpha_{r+5} & \omega_{r+5} \\ & & & & & -\omega_{r+5} & \alpha_{r+5} \end{array}\right], \tag{3.18}$$

where

$$\begin{cases} \alpha_{r+3} = \alpha_{r+5}, \\ \omega_{r+3} = \omega_{r+5}, \end{cases}$$

and $\boldsymbol{J}$ is the Jordan canonical form corresponding to r repeated zero eigenvalues which consists of the Jordan blocks $\boldsymbol{J}_1$, $\boldsymbol{J}_2$, etc., associated with linearly independent eigenvectors. If the number of linearly independent eigenvectors is r, then there are r Jordan blocks, implying that $\boldsymbol{J}$ is actually diagonal.

Clearly, it is not difficult to extend the concepts underlying the transformations above to more general situations.

3.1.1 *Simple critical points (divergence)*

Case 1. A real eigenvalue vanishes. Suppose now that a real eigenvalue of the Jacobian goes through zero at a critical value of the parameter, $\eta = \eta_c$, while the remaining eigenvalues continue to have negative real parts (Fig. 3.1). Introduce the transformations

$$\boldsymbol{z} = \boldsymbol{z}_c + \boldsymbol{Qy} \quad \text{and} \quad \eta = \eta_c + \mu \tag{3.19}$$

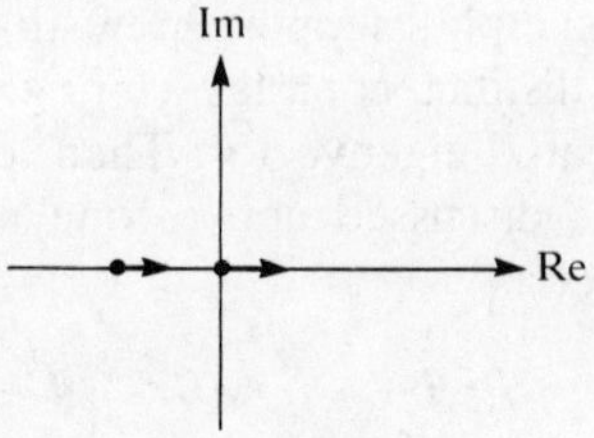

Fig. 3.1 Case 1. Real eigenvalue goes through zero at a critical value of $\eta = \eta_c$, while remaining eigenvalues continue to have negative real parts.

to replace eqn (3.1) by

$$\frac{\mathrm{d}\boldsymbol{y}}{\mathrm{d}t} = \boldsymbol{Y}(\boldsymbol{y}, \mu) \tag{3.20}$$

such that the Jacobian is now in the canonical form

$$[Y_{ij}] = \left[\frac{\partial \boldsymbol{Y}}{\partial \boldsymbol{y}}\right]_c = \mathrm{diag}\,[\boldsymbol{J}, \boldsymbol{K}_3, \boldsymbol{K}_5, \ldots] \tag{3.21}$$

where

$$\boldsymbol{J} = \begin{bmatrix} 0 & 0 \\ 0 & \alpha_2 \end{bmatrix}, \qquad \boldsymbol{K}_m = \begin{bmatrix} \alpha_m & -\omega_m \\ \omega_m & \alpha_m \end{bmatrix}, \qquad (m = 3, 5, \ldots), \tag{3.22}$$

and $\alpha_2 < 0$, $\alpha_m < 0$. The matrices $\boldsymbol{K}_m$ correspond to complex conjugate eigenvalues $(\alpha_m \pm \mathrm{i}\omega_m)$ which are assumed to be distinct without loss of generality. Furthermore, the block $\boldsymbol{J}$ is chosen as a (2×2) matrix in order to avoid another subscript for real eigenvalues. In fact, incorporating r non-zero, real eigenvalues $(\alpha_r < 0)$ in J does not pose any difficulty, but the presence of such eigenvalues does not have an essential influence on the behaviour characteristics of the system (Huseyin 1981*a*).

It will now be assumed that the equilibrium path(s) in the vicinity of the critical point c, satisfying

$$\boldsymbol{Y}(\boldsymbol{y}, \mu) = \boldsymbol{0}, \tag{3.23}$$

can be expressed in the parametric form

$$\boldsymbol{y} = \boldsymbol{y}(\sigma), \qquad \mu = \mu(\sigma), \tag{3.24}$$

where the functions $\boldsymbol{y}(\sigma)$ and $\mu(\sigma)$ are single-valued and may be expanded into Taylor's series in terms of the scalar parameter σ.

Introducing eqn (3.24) back into eqn (3.23) results in the identities

$$\boldsymbol{Y}[\boldsymbol{y}(\sigma), \mu(\sigma)] \equiv \boldsymbol{0} \tag{3.25}$$

which will form a basis for generating explicit information about the path(s) through c.

Thus, expressing eqn (3.25) as

$$Y_i[y^j(\sigma), \mu(\sigma)] \equiv 0, \tag{3.26}$$

and differentiating successively with respect to σ yields

$$Y_{ij}\dot{y}^j + Y'_i\dot{\mu} = 0, \tag{3.27}$$

$$(Y_{ijk}\dot{y}^k + Y'_{ij}\dot{\mu})\dot{y}^j + Y_{ij}\ddot{y}^j + (Y'_{ij}\dot{y}^j + Y''_i\dot{\mu})\dot{\mu} + Y'_i\ddot{\mu} = 0, \tag{3.28}$$

and

$$Y_{ijkl}\dot{y}^j\dot{y}^k\dot{y}^l + 3Y'_{ijk}\dot{y}^j\dot{y}^k\dot{\mu} + 3Y_{ijk}\ddot{y}^j\dot{y}^k + 3Y''_{ij}\dot{y}^j\dot{\mu}^2 + 3Y'_{ij}\ddot{y}^j\dot{\mu} + 3Y'_{ij}\dot{y}^j\ddot{\mu}$$
$$+ Y_{ij}\dddot{y}^j + Y'''_i\dot{\mu}^3 + 3Y''_i\dot{\mu}\ddot{\mu} + Y'_i\dddot{\mu} = 0, \tag{3.29}$$

where a dot denotes differentiation with respect to σ; this notation will be adopted henceforth for convenience. Primes and subscripts on Y's denote differentiation with respect to μ and the corresponding y^i, respectively, and the summation convention applies as before.

If the parameter σ is specified in some suitable way, the analysis may be simplified considerably. Indeed, it will be assumed here that σ measures the arc length of the equilibrium path through the point c. The unit tangent vector to the path is then given by $(n+1)$ components, $\dot{y}$ and $\dot{\mu}$, which satisfy the relation

$$(\dot{y}^1)^2 + \ldots + (\dot{y}^n)^2 + (\dot{\mu})^2 = 1. \tag{3.30}$$

The perturbation equations (3.27) to (3.29) will be solved in conjunction with the auxiliary equation (3.30).

Thus, evaluating eqn (3.27) at the critical point c and taking into account eqn (3.21) yield

$$\dot{y}^2 = 0, \quad \dot{\mu} = 0, \quad \text{and} \quad \dot{y}^s = 0, \quad (s = 3, 4, \ldots, n), \tag{3.31}$$

provided

$$Y'_1 \neq 0. \tag{3.32}$$

Indeed, the coefficient Y'_1 plays a significant role in shaping the topological properties of the equilibrium path through c, and the situation in which $Y'_1 = 0$ will be discussed later.

Proceeding now under the condition (3.32), and using eqns (3.31) and the auxiliary equation (3.30), yields

$$\dot{y}^1 = 1. \tag{3.33}$$

Evaluating the second perturbation equation (2.28) at c and using these results leads to

$$Y_{111} + Y'_1\ddot{\mu} = 0, \tag{3.34}$$

$$Y_{211} + \alpha_2\ddot{y}^2 + Y'_2\ddot{\mu} = 0, \tag{3.35}$$

and

$$\begin{bmatrix} \alpha_m & -\omega_m \\ \omega_m & \alpha_m \end{bmatrix} \begin{bmatrix} \ddot{y}^m \\ \ddot{y}^{m+1} \end{bmatrix} = -\begin{bmatrix} Y_{m11} + Y'_m\ddot{\mu} \\ Y_{(m+1)11} + Y'_{m+1}\ddot{\mu} \end{bmatrix},$$

$$(m = 3, 5, \ldots, n-1), \quad (3.36)$$

which result in the path derivatives

$$\ddot{\mu} = -\frac{Y_{111}}{Y'_1}, \qquad \ddot{y}^2 = \frac{1}{\alpha_2}\left(Y_{111}\frac{Y'_2}{Y'_1} - Y_{211}\right) \triangleq b_2,$$

$$\ddot{y}^m = \frac{\alpha_m k_m + \omega_m k_{m+1}}{\alpha_m^2 + \omega_m^2} \triangleq b_m, \quad (3.37)$$

and

$$\ddot{y}^{m+1} = \frac{\alpha_m k_{m+1} - \omega_m k_m}{\alpha_m^2 + \omega_m^2} \triangleq b_{m+1}$$

where

$$k_m = Y_{111}\frac{Y'_m}{Y'_1} - Y_{m11}.$$

The derivative (3.33) has already indicated at the end of the first perturbation that the equilibrium path through c lies in the direction of y^1. This implies that y^1 could have been chosen as σ in the beginning; nevertheless, this information was not available in advance, and a wrong choice would have resulted in inconsistencies, breaking down the analysis. One could also miss completely a possible path bifurcating into a perpendicular direction. It is, therefore, essential to keep σ unidentified until sufficient information about directions of prospective path(s) emerges in the analysis, unless such information is available *a priori* on the basis of a previous analysis or other experience.

The first-order equations of the equilibrium path in the vicinity of c can now be constructed by using Taylor's expansion and the derivatives (3.31), (3.33), and (3.37) as

$$\mu = -\frac{1}{2}\frac{Y_{111}}{Y'_1}(y^1)^2, \quad (3.38)$$

$$y^t = \tfrac{1}{2}b_t(y^1)^2, \qquad (t = 2, 3, \ldots, n). \quad (3.39)$$

The local equilibrium path defined by eqns (3.38) and (3.39) is actually a portion of a space curve which describes an initially stable equilibrium path and reaches a maximum (or a minimum) at the critical point c as the parameter is varied. A plot of μ against y^1 yields a smooth extremum (Fig. 3.2), while on a plot of μ against any y^t the same curve appears as a sharp cusp as may be deduced from eqns (3.38) and (3.39). Clearly, these

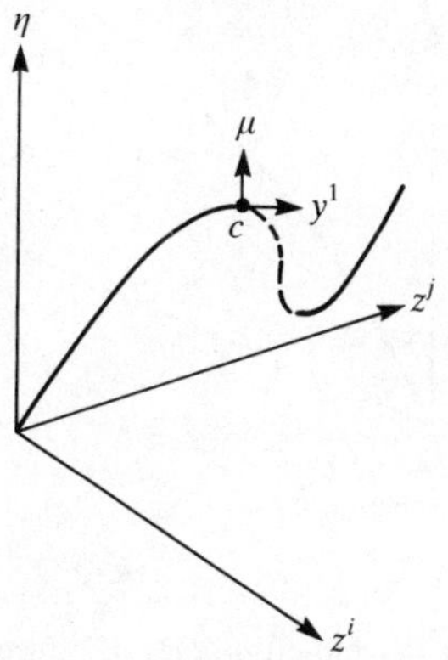

Fig. 3.2 Local equilibrium path defined by eqns (3.38) and (3.39).

projections represent different views of a space curve from various angles in the vicinity of a *limit point*, as in the case of potential systems. The vanishing eigenvalue α_1 of the Jacobian actually changes sign from negative to positive as it goes through c, and *the originally stable path becomes unstable at a limit point*. In Fig. 3.2, a continuous line represents a stable path and a broken line an unstable path. The system under consideration will then exhibit a sudden jump as the parameter η attains its critical value η_c beyond which the equilibrium path becomes unstable and the system seeks another stable position. In many applications the unstable path gains its stability upon passing through another critical limit point c' (an extremum) and the jump takes place to a point b on this branch as depicted in Fig. 3.3. If the parameter now decreases gradually, the system follows the stable equilibrium path from b to c' and exhibits a jump at c' back to the original stable branch.

Consider now the case in which condition (3.32) does not hold, i.e. the situation when

$$Y_1' = 0. \tag{3.40}$$

If the perturbation equations (3.27) and (3.28) are evaluated at the critical point c under this condition, the first-order relationship linking

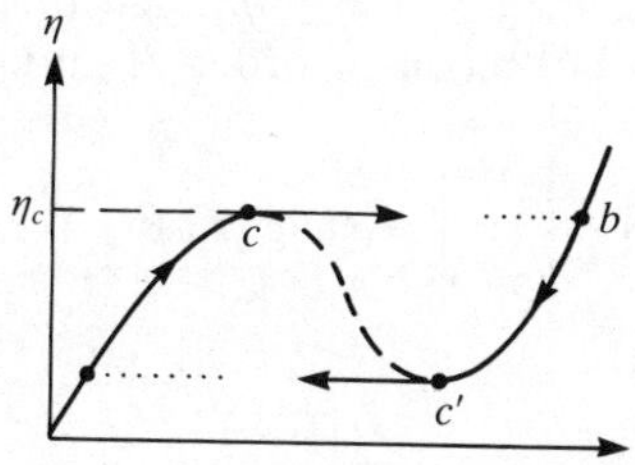

Fig. 3.3 Equilibrium path showing jump to a point b.

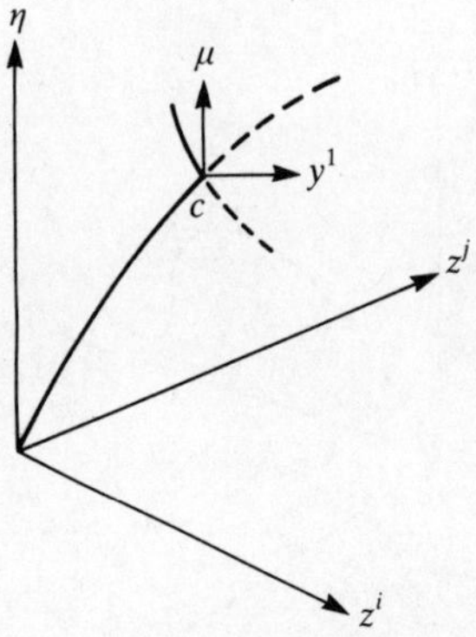

Fig. 3.4 Asymmetric point of bifurcation.

the critical coordinate to the parameter μ takes the form

$$\tfrac{1}{2}(y^1)^2 + dy^1\mu + \tfrac{1}{2}e\mu^2 = 0 \qquad (3.41)$$

where the coefficients d and e are certain expressions in terms of the derivatives of Y evaluated at c. This equation may be solved for y^1 to yield

$$y^1 = \{-d \pm \sqrt{(d^2 - e)}\}\mu \qquad (3.42)$$

which indicates that c is now a point of *bifurcation* where the initial equilibrium path is intersected by a post-critical path, provided $(d^2 - e) > 0$ (Fig. 3.4). The analyses leading to this type of *asymmetric point of bifurcation* as well as *symmetric* and other more degenerate bifurcation phenomena can be simplified considerably by modifying the formulation and introducing what may be called 'sliding coordinates' along the initial path. This will be done in the following sections where the stability distribution on the equilibrium paths in the neighbourhood of a point of bifurcation will also be examined. Before proceeding with the analysis of various points of bifurcation, however, other conditions giving rise to limit points will be explored.

Case 2. A Jordan block of order 2. Suppose now that a pair of eigenvalues crosses the origin at $\eta = \eta_c$ such that the canonical form of the Jacobian has a Jordan block of order 2, and is assumed to be in the form

$$[Y_{ij}] = \left[\frac{\partial \boldsymbol{Y}}{\partial \boldsymbol{y}}\right]_c = \operatorname{diag}[\boldsymbol{J}, \boldsymbol{K}_3, \boldsymbol{K}_5, \ldots] \qquad (3.43)$$

where

$$\boldsymbol{J} = \begin{bmatrix} 0 & 1 \\ 0 & 0 \end{bmatrix}, \qquad (3.44)$$

and $\boldsymbol{K}_m$ ($m = 3, 5, \ldots$) are as given by eqn (3.22). There may, of course, exist negative real eigenvalues which are not included in eqn (3.43) for simplicity.

Evaluating eqn (3.27) at c with the aid of eqn (3.43) yields

$$\dot{y}^2 + Y'_1\dot{\mu} = 0 \quad \text{for} \quad i = 1,$$
$$Y'_2\dot{\mu} = 0 \quad \text{for} \quad i = 2,$$

and (3.45)

$$\begin{bmatrix} \alpha_m & -\omega_m \\ \omega_m & \alpha_m \end{bmatrix}\begin{bmatrix} \dot{y}^m \\ \dot{y}^{m+1} \end{bmatrix} + \begin{bmatrix} Y'_m \\ Y'_{m+1} \end{bmatrix}\dot{\mu} = 0, \qquad (m = 3, 5, \ldots, n-1)$$

which result in

$$\dot{\mu} = 0, \qquad \dot{y}^2 = 0, \quad \text{and} \quad \dot{y}^s = 0, \qquad (s = 3, 4, \ldots, n) \qquad (3.46)$$

provided

$$Y'_2 \neq 0. \qquad (3.47)$$

The condition (3.47) replaces condition (3.32) in this case, and eqn (3.30) again yields

$$\dot{y}^1 = 1. \qquad (3.48)$$

By evaluating the second perturbation equation (2.28) at c and using eqns (3.46) and (3.48), one obtains

$$Y_{111} + \ddot{y}^2 + Y'_1\ddot{\mu} = 0,$$
$$Y_{211} + Y'_2\ddot{\mu} = 0,$$

and (3.49)

$$\begin{bmatrix} \alpha_m & -\omega_m \\ \omega_m & \alpha_m \end{bmatrix}\begin{bmatrix} \ddot{y}^m \\ \ddot{y}^{m+1} \end{bmatrix} = -\begin{bmatrix} Y_{m11} + Y'_m\ddot{\mu} \\ Y_{(m+1)11} + Y'_{m+1}\ddot{\mu} \end{bmatrix}, \qquad (m = 3, 5, \ldots, n-1)$$

which yield

$$\ddot{\mu} = -\frac{Y_{211}}{Y'_2}, \qquad \ddot{y}^2 = Y_{211}\frac{Y'_1}{Y'_2} - Y_{111} \triangleq c_2,$$
$$\ddot{y}^m = c_m, \quad \text{and} \quad \ddot{y}^{m+1} = c_{m+1}, \qquad (3.50)$$

where c_m and c_{m+1} are certain expressions similar to b_m and b_{m+1} in eqn (3.37), except for k_m which has to be replaced by

$$l_m = Y_{211}\frac{Y'_m}{Y'_2} - Y_{111},$$

The asymptotic equations of the equilibrium path through the critical point c can now be constructed as (Mandadi and Huseyin 1979)

$$\mu = -\frac{1}{2}\frac{Y_{211}}{Y'_2}(y^1)^2, \qquad (3.51)$$

$$y^t = \tfrac{1}{2}c_t(y^1)^2, \qquad (t = 2, 3, \ldots, n). \qquad (3.52)$$

It is immediately observed that the equilibrium path described by these equations is topologically identical to that described by eqns (3.38) and (3.39). In other words, the critical point c is again a *limit point* as depicted in Fig. 3.2. Similarities with Case 1 continue to hold when the condition (3.47) is violated. Indeed, it can be demonstrated that if $Y_2' = 0$, the critical point c emerges as a point of bifurcation illustrated in Fig. 3.4.

In this case, however, the trajectories of the vanishing eigenvalues of the Jacobian in the complex λ-plane are not immediately discernable as in the case of a single real eigenvalue crossing the origin of the λ-plane and becoming positive. As a matter of fact, in the presence of more than one independent parameter the crossing may be into the right half-plane, suggesting *dynamic instability*. This situation will be discussed further when multiple-parameter systems are considered in Section 3.2.2 (see Case 2). It can be demonstrated, however, that under the conditions prescribed in this section, complex conjugate eigenvalues become real upon crossing the origin of the λ-plane, and a stable path can exhibit *static instability* only. To this end, consider the expansion of the Jacobian in the vicinity of c,

$$Y_{ij}(y^k, \mu) = Y_{ij} + Y_{ijk} y^k + Y'_{ij}\mu + \dots, \tag{3.53}$$

where Y_{ij} is given by eqn (3.43).

Evaluating eqn (3.53) on the equilibrium path described by eqns (3.51) and (3.52), and keeping to a first-order approximation shows that the variation of eigenvalues in the vicinity of c may be traced with the aid of a single parameter—the critical coordinate y^1:

$$Y_{ij}(y^1) = Y_{ij} + Y_{ij1} y^1 + 0\{(y^1)^2\} + \dots. \tag{3.54}$$

Since the non-vanishing eigenvalues are assumed to remain in the left half-plane, and by virtue of the structure of eqn (3.54), one may simply examine the eigenvalues of the matrix

$$Y_{ij}(y^1) = \begin{bmatrix} 0 & 1 \\ 0 & 0 \end{bmatrix} + \begin{bmatrix} Y_{111} & Y_{121} \\ Y_{211} & Y_{221} \end{bmatrix} y^1 \tag{3.55}$$

which are given by

$$\begin{vmatrix} Y_{111} y^1 - \lambda & 1 + Y_{121} y^1 \\ Y_{211} y^1 & Y_{221} y^1 - \lambda \end{vmatrix} = 0. \tag{3.56}$$

Ignoring the higher order terms leads to the characteristic equation

$$\lambda^2 - (Y_{111} + Y_{221}) y^1 \lambda - Y_{211} y^1 = 0. \tag{3.57}$$

The eigenvalues are then given by

$$\lambda_{1,2}(y^1) = \tfrac{1}{2}(Y_{111} + Y_{221}) y^1 \pm \tfrac{1}{2}\surd\{(Y_{111} + Y_{221})^2 (y^1)^2 + 4 Y_{211} y^1\} \tag{3.58}$$

which reduces to

$$\lambda_{1,2}(y^1) = \tfrac{1}{2}(Y_{111} + Y_{221})y^1 \pm (Y_{211}y^1)^{1/2} \tag{3.59}$$

in the vicinity of c.

If $(Y_{111} + Y_{221}) > 0$ and $Y_{211} > 0$, then eqn (3.59) yields a complex conjugate pair with a negative real part for $y^1 < 0$. At $y^1 = 0$, the pair crosses the origin, $\lambda_{1,2}(0) = 0$, and becomes real for $y^1 > 0$. Clearly, for small magnitudes of y^1, one of these real roots is positive—indicating divergence instability—while the other one is negative. The trajectories of the eigenvalues are shown in Fig. 3.5*a*. Suppose now that $(Y_{111} + Y_{221}) = 0$ and $Y_{211} > 0$. It is then inferred from eqn (3.59) that an imaginary pair approaches the origin, coalesces there, and then the two roots depart in opposite directions on the real axis, as y^1 takes on values ranging from negative to positive (Fig. 3.5*b*). It is not difficult to verify that *under no circumstances may a stable path become flutter unstable, and a stable path loses its stability by divergence upon passing through a limit point*.

Problem 3.1. In analogy with the aforegoing analysis, consider the case of a Jordan block of order r such that the matrix $\boldsymbol{J}$ in eqn (3.43) is r-square,

$$\boldsymbol{J} = \begin{bmatrix} 0 & 1 & & & \\ & 0 & 1 & & \\ & & 0 & \cdot & \\ & & & \cdot & \cdot \\ & & & \cdot & 1 \\ & & & & 0 \end{bmatrix}. \tag{3.60}$$

Demonstrate analytically that if the condition $Y'_r \neq 0$ is satisfied, the equilibrium path through c is in the form shown in Fig. 3.2, and c is again a limit point. Derive the asymptotic equations of the equilibrium path in the vicinity of the critical point c. What happens if $Y'_r = 0$?

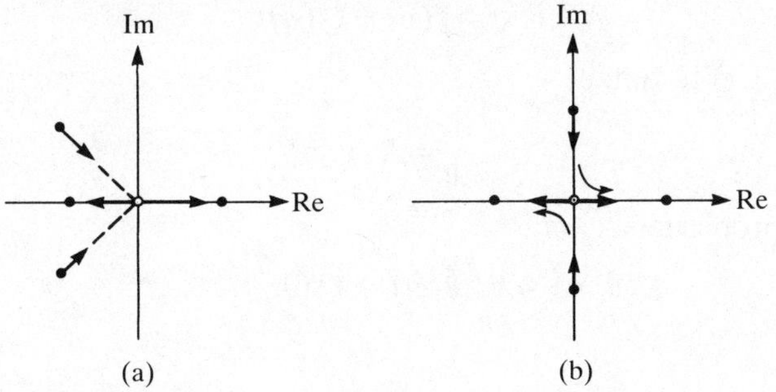

Fig. 3.5 Trajectories of eigenvalues for (a) $(Y_{111} + Y_{221}) > 0$, $Y_{211} > 0$ and (b) $(Y_{111} + Y_{221}) = 0$ and $Y_{211} > 0$.

3.1.2 Simple bifurcation points

Equilibrium path configurations and the stability distribution in the vicinity of a critical bifurcation point can be studied more conveniently by introducing further transformations.

Let $\boldsymbol{z}=\boldsymbol{f}(\eta)$ be an initial equilibrium path associated with system (3.1), and assume that the path is single-valued in the vicinity of a bifurcation point c on it. A coordinate system $\boldsymbol{w}$ can then be attached to this path by the transformation

$$\boldsymbol{z}=\boldsymbol{f}(\eta)+\boldsymbol{Q}\boldsymbol{w}, \qquad |\boldsymbol{Q}|\neq 0, \tag{3.61}$$

such that system (3.1) is transformed into

$$\frac{\mathrm{d}\boldsymbol{w}}{\mathrm{d}t}=\boldsymbol{W}(\boldsymbol{w},\eta) \tag{3.62}$$

with the properties

$$\boldsymbol{W}(\boldsymbol{0},\eta)=\boldsymbol{W}'(\boldsymbol{0},\eta)=\boldsymbol{W}''(\boldsymbol{0},\eta)=\ldots=\boldsymbol{0} \tag{3.63}$$

and

$$\left[\frac{\partial \boldsymbol{W}}{\partial \boldsymbol{w}}\right]_c=\operatorname{diag}\,[\boldsymbol{J},\boldsymbol{K}_3,\boldsymbol{K}_5,\ldots] \tag{3.64}$$

where $\boldsymbol{J}$ is again a Jordan canonical form associated with vanishing eigenvalues, and $\boldsymbol{K}_m$ $(m=3,5,\ldots)$ are (2×2) blocks as in eqn (3.22).

If only one real eigenvalue passes through zero at c, a further transformation can be introduced. Indeed, in this case the transformation matrix $\boldsymbol{Q}$ can be chosen as a function of η such that when η varies in the neighbourhood of $\eta=\eta_c$, the canonical form (3.64) of the Jacobian matrix along the initial path $\boldsymbol{f}(\eta)$ is preserved. In other words, for each value of η in the vicinity of c, an appropriate transformation matrix $\boldsymbol{Q}(\eta)$ is formed in such a way that the Jacobian matrix of the resulting system evaluated on the initial path has the block-diagonal form of eqn (3.64) with $\boldsymbol{J}$ always diagonal. Thus, introduce

$$\boldsymbol{z}=\boldsymbol{f}(\eta)+\boldsymbol{Q}(\eta)\boldsymbol{x} \tag{3.65}$$

into eqn (3.1) to obtain

$$\frac{\mathrm{d}\boldsymbol{x}}{\mathrm{d}t}=\boldsymbol{X}(\boldsymbol{x},\eta) \tag{3.66}$$

with the properties

$$\boldsymbol{X}(\boldsymbol{0},\eta)=\boldsymbol{X}'(\boldsymbol{0},\eta)=\boldsymbol{X}''(\boldsymbol{0},\eta)=\ldots=\boldsymbol{0} \tag{3.67}$$

and

$$\left[\frac{\partial \boldsymbol{X}}{\partial \boldsymbol{x}}\right]_{\boldsymbol{x}=\boldsymbol{0}}=\operatorname{diag}\,[\boldsymbol{J},\boldsymbol{K}_3,\boldsymbol{K}_5,\ldots] \tag{3.68}$$

where

$$\boldsymbol{J}=\begin{bmatrix}\alpha_1(\eta) & 0\\ 0 & \alpha_2(\eta)\end{bmatrix},\qquad \boldsymbol{K}_m=\begin{bmatrix}\alpha_m(\eta) & -\omega_m(\eta)\\ \omega_m(\eta) & \alpha_m(\eta)\end{bmatrix},\tag{3.69}$$

$(m=3,5,\ldots)$, $\alpha_1(\eta_c)=0$, and $\alpha_t(\eta_c)<0$ for $t\neq 1$.

It follows that all the off-block-diagonal elements and their derivatives with respect to η vanish along the initial path $f(\eta)$. This property may be expressed by

$$X_{ij}(0,\eta)=X'_{ij}(0,\eta)=X''_{ij}(0,\eta)=\ldots=0\tag{3.70}$$

for $i\neq j$ provided $j\neq i+1$ when $i=3,5,\ldots$ and $j\neq i-1$ when $i=4,5,\ldots$. This transformation was originally introduced (Huseyin 1981*a*) to facilitate the stability analyses of the equilibrium paths.

According to this formulation, the initial path is defined by $\boldsymbol{x}=\boldsymbol{0}$, and the possible paths branching off the critical point c are expressed in the parametric form

$$\boldsymbol{x}=\boldsymbol{x}(\sigma),\qquad \eta=\eta(\sigma),\tag{3.71}$$

as before.

Introducing eqn (3.71) into the equilibrium equations $X_i(x^j,\eta)=0$ yields the identities

$$X_i[x^j(\sigma),\eta(\sigma)]\equiv 0,\tag{3.72}$$

which will be used to generate asymptotic solutions intrinsically by successive differentiations. Thus, first, second, and third perturbations yield

$$X_{ij}\dot{x}^j+X'_i\dot{\eta}=0,\tag{3.73}$$

$$(X_{ijk}\dot{x}^k+X'_{ij}\dot{\eta})\dot{x}^j+X_{ij}\ddot{x}^j+(X'_{ij}\dot{x}^j+X''_i\dot{\eta})\dot{\eta}+X'_i\ddot{\eta}=0,\tag{3.74}$$

$$X_{ijkl}\dot{x}^j\dot{x}^k\dot{x}^l+3(X'_{ijk}\dot{x}^j\dot{x}^k\dot{\eta}+X_{ijk}\ddot{x}^j\dot{x}^k+X''_{ij}\dot{x}^j\dot{\eta}^2+X'_{ij}\ddot{x}^j\dot{\eta}+X'_{ij}\dot{x}^j\ddot{\eta}+X_i\dot{\eta}\ddot{\eta})$$
$$+X_{ij}\dddot{x}^j+X'''_i\dot{\eta}^3+X'_i\dddot{\eta}=0,\tag{3.75}$$

where a dot again denotes differentiation with respect to σ. If the arc length of the equilibrium path (3.71), measured from the critical point c, is described by σ, one has

$$(\dot{x}^1)^2+\ldots+(\dot{x}^n)^2+(\dot{\eta})^2=1.\tag{3.76}$$

Several cases will now be explored:

Case 1. Asymmetric points of bifurcation. Evaluating eqn (3.73) at c with the aid of eqns (3.67), (3.68), and (3.70) results in

$$\dot{x}^2=0\quad\text{and}\quad \dot{x}^s=0,\qquad (s=3,4,\ldots,n).\tag{3.77}$$

Similarly, evaluation of the second perturbation equation (3.74) with the aid of eqns (3.67), (3.68), (3.70), and (3.77) leads to two solutions:

$$\dot{x}^1 = \dot{x}^2 = \ldots = \dot{x}^m = \dot{x}^{m+1} = \ldots = 0$$

or

$$\dot{\eta} = -\frac{X_{111}}{2X'_{11}}\dot{x}^1, \qquad \ddot{x}^2 = -\frac{X_{211}}{\alpha_2}(\dot{x}^1)^2, \tag{3.78}$$

$$\begin{bmatrix} \alpha_m & -\omega_m \\ \omega_m & \alpha_m \end{bmatrix}\begin{bmatrix} \ddot{x}^m \\ \ddot{x}^{m+1} \end{bmatrix} = -\begin{bmatrix} X_{m11} \\ X_{(m+1)11} \end{bmatrix}(\dot{x}^1)^2.$$

The first set of derivatives yields the initial path

$$x^1 = x^2 = x^3 = \ldots = 0 \tag{3.79}$$

as expected, whereas eqns (3.78) yield the post-critical path

$$\begin{aligned} \mu &= -\frac{X_{111}}{2X'_{11}}x^1, \\ x^2 &= -\frac{X_{211}}{2\alpha_2}(x^1)^2 \triangleq -\tfrac{1}{2}a_2(x^1)^2, \\ x^m &= -\frac{1}{2}\frac{\alpha_m X_{m11} + \omega_m X_{(m+1)11}}{\alpha_m^2 + \omega_m^2}(x^1)^2 \triangleq -\tfrac{1}{2}a_m(x^1)^2, \\ x^{m+1} &= -\frac{1}{2}\frac{\alpha_m X_{(m+1)11} - \omega_m X_{m11}}{\alpha_m^2 + \omega_m^2}(x^1)^2 \triangleq -\tfrac{1}{2}a_{m+1}(x^1)^2, \end{aligned} \tag{3.80}$$

upon setting $\sigma = x^1$ and $\eta = \eta_c + \mu$. The post-bifurcation path (3.80) has a finite slope and intersects the initial path at c, indicating that c is an *asymmetric point of bifurcation* (Fig. 3.6). An exchange of stabilities occurs at c, in complete analogy with potential systems, as will be shown in the following section.

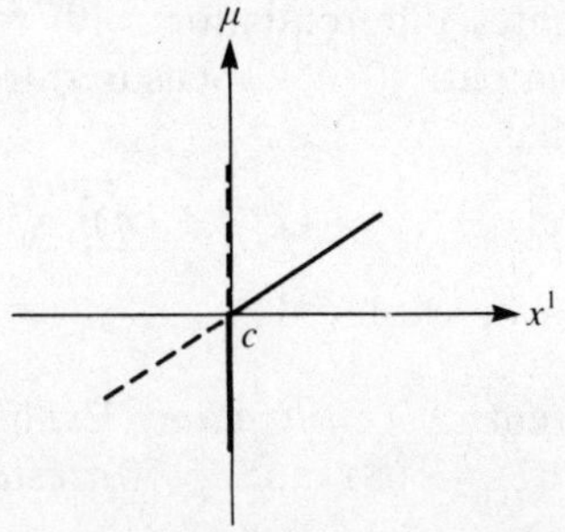

Fig. 3.6 Case 1: asymmetric point of bifurcation at c.

Case 2. Symmetric points of bifurcation. It may turn out that certain key coefficients vanish at c, resulting in a different form for the post-critical path. Indeed, eqn (3.80) indicates that the case in which

$$X_{111} = 0 \tag{3.81}$$

deserves further attention. Under this condition, one has to proceed to a third perturbation in order to generate more derivatives for a first-order approximation. To this end, one evaluates eqn (3.75) at c to obtain the second derivative

$$\begin{aligned} \eta^{,11} \equiv \ddot{\eta} &= -\frac{1}{3X'_{11}}\left(X_{111} - 3\frac{X_{121}X_{211}}{\alpha_2} - 3X_{1s1}a_s\right) \\ &= -\frac{1}{3X'_{11}}(X_{1111} - 3X_{1t1}a_t) \\ &\triangleq -\frac{1}{3X'_{11}}\bar{X}_{1111} \triangleq -a_0 \end{aligned} \tag{3.82}$$

where $\eta^{,11} = \partial^2\eta/\partial(x^1)^2|_c$, and x^1 serves as σ.

By using this derivative, the first-order equations of the post-critical path are expressed as (Huseyin 1981*a*)

$$\begin{aligned} \mu &= -\tfrac{1}{2}a_0(x^1)^2, \\ x^t &= -\tfrac{1}{2}a_t(x^1)^2, \qquad (t = 2, 3, \ldots, n), \end{aligned} \tag{3.83}$$

which represent a space curve in the μ-x^i space, intersecting the initial path at $\eta = \eta_c$. According to eqn (3.80), the slope of this path is zero, and the critical point is now a *symmetric point of bifurcation*. Depending on the curvature a_0 it may be a *stable* or *unstable symmetric point of bifurcation*. Stability properties will be analysed in the next section. It is also noted that in the vicinity of c, post-critical equilibrium states exist only either for $\eta > \eta_c$ or $\eta < \eta_c$ (Fig. 3.7).

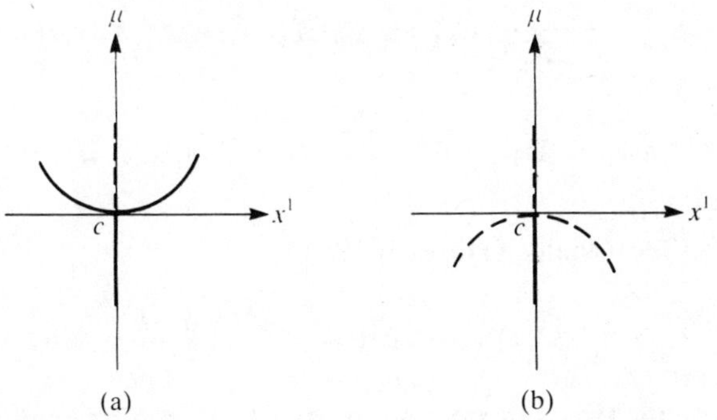

Fig. 3.7 Symmetric point of bifurcation, (a) stable and (b) unstable.

Case 3. A critical point of tri-furcation. In some mathematical models, and often due to symmetry properties incorporated in the model, additional key coefficients may vanish at c, rendering the critical point even more degenerate than asymmetric and symmetric bifurcations.

Suppose now that

$$X_{111} = X'_{11} = 0 \tag{3.84}$$

at c; in this case, the slope $\partial\mu/\partial x^1|_c$ in eqn (3.80) is indeterminate, and the first perturbation equation (3.73) yields, upon evaluation at c,

$$\dot{\eta}^2 = 0 \quad \text{and} \quad \dot{\eta}^s = 0, \qquad (s = 3, 4, \ldots, n). \tag{3.85}$$

The second perturbation equation (3.74), when evaluated at c under the condition (3.84), yields

$$\ddot{x}^t = -a_t(\dot{x}^1)^2 \qquad (t = 2, 3, \ldots, n), \tag{3.86}$$

where the a_t are defined in eqn (3.80).

Finally, the third perturbation equation (3.75) results in

$$X_{1111}(\dot{x}^1)^3 + 3\{X'_{111}(\dot{x}^1)^2\dot{\eta} + X_{1t1}\ddot{x}^t\dot{x}^1 + X''_{11}\dot{x}^1\dot{\eta}^2\} = 0 \tag{3.87}$$

which, upon using eqn (3.86), takes the form

$$X_{111}(\dot{x}^1)^3 + 3[X'_{111}(\dot{x}^1)^2\dot{\eta} - X_{1t1}a_t(\dot{x}^1)^3 + X''_{11}\dot{x}^1\dot{\eta}^2] = 0. \tag{3.88}$$

It is now observed that

$$\dot{x}^1 = 0 \tag{3.89}$$

is a solution; in fact, it represents the initial path and indicates that x^1 is not a suitable candidate for σ, unless it is deliberately decided to drop the initial path from the scope of the analysis. However, eqn (3.88) does not contain any indication that would prevent η from taking on the role of the parameter σ. Hence, setting $\sigma = \eta$, eqn (3.88) yields the derivatives of the post-critical paths as

$$x^{1,\prime} = -\frac{3}{2\bar{X}_{1111}}[X'_{111} \pm \surd\{(X'_{111})^2 - \tfrac{4}{3}X''_{11}\bar{X}_{1111}\}] \tag{3.90}$$

where

$$\bar{X}_{1111} = X_{1111} - 3X_{1t1}a_t \qquad (t = 2, 3, \ldots, n), \tag{3.91}$$

and $x^{1,\prime} = \partial x^1/\partial\eta|_c$.

Using eqn (3.90) yields (Huseyin 1984*a*)

$$x^1 = -\frac{3}{2\bar{X}_{1111}}[X'_{111} \pm \surd\{(X'_{111})^2 - \tfrac{4}{3}X''_{11}\bar{X}_{1111}\}]\mu, \tag{3.92}$$

which represents two intersecting post-critical equilibrium paths (Fig. 3.8). In view of the initial path and eqn (3.92), this phenomenon is

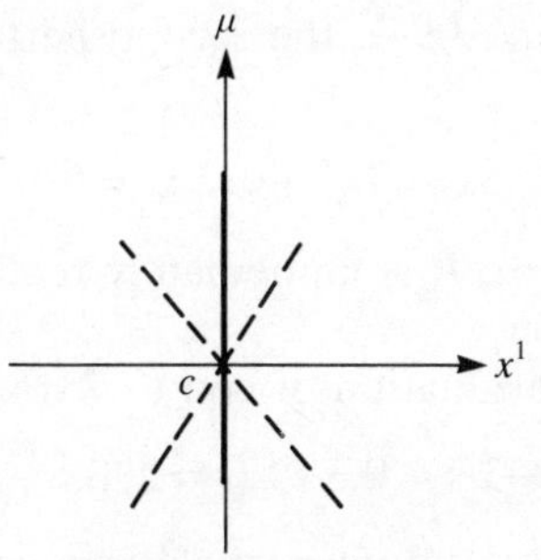

Fig. 3.8 Tri-furcation.

described as a *tri-furcation* in analogy with potential systems (Huseyin 1975).

After recognizing that $\dot{x}^1 = 0$ represents the initial path and dropping it from eqn (3.88), one may also proceed to obtain the post-critical paths by setting $\sigma = x^1$. It then follows from eqns (3.86) and (3.88) that

$$x^{t,11}(\equiv \ddot{x}^t) = -a_t$$

and

$$X_{1111} + 3\{X'_{111}\eta^{,1} - X_{1t1}a_t + X''_{11}(\eta^{,1})^2\} = 0 \tag{3.93}$$

which yields

$$\mu = -\frac{1}{2X''_{11}}[X'_{111} \pm \sqrt{\{(X'_{111})^2 - \tfrac{4}{3}X''_{11}\bar{X}_{1111}\}}]x^1, \tag{3.94}$$

and x^t given in eqn (3.83) remains unaffected.

It can readily be verified that eqn (3.94) is in compliance with eqn (3.92).

In this case the stable initial path passes through the critical point without losing its stability while two unstable post-critical paths intersect at the critical point. This suggests a high degree of imperfection sensitivity and will be studied later. Note, however, that post-critical paths may not exist, and this happens when the expression under the square-root is negative. The existence condition is

$$D \triangleq \{(X'_{111})^2 - \tfrac{4}{3}X''_{11}\bar{X}_{1111}\} > 0$$

Case 4. A tangential point of bifurcation. Consider next the situation when, instead of condition (3.84), the condition

$$X'_{11} = 0 \tag{3.95}$$

is satisfied at the critical point c.

As before, eqn (3.73) yields

$$\dot{x}^t = 0 \qquad (t = 2, 3, \ldots, n), \tag{3.96}$$

and, under the condition (3.95), the second perturbation equation (3.74) results in

$$\dot{x}^1 = 0 \quad \text{and} \quad \ddot{x}^t = 0. \tag{3.97}$$

By virtue of eqn (3.76), it is immediately realized that $\dot{\eta} = 1$, and η is an ideal candidate for σ.

Proceeding with the evaluation of eqn (3.75) yields

$$x^{t,'''}(\equiv \dddot{x}^t) = 0, \qquad (t = 2, 3, \ldots, n); \tag{3.98}$$

but no other information emerges, making it necessary to differentiate eqn (3.72) for a fourth time with respect to σ ($\equiv \eta$). Evaluation of this fourth perturbation equation at c leads to

$$3X_{111}(x^{1,''})^2 + 6X''_{11}x^{1,''} = 0 \quad \text{for } i = 1, \tag{3.99}$$

and

$$x^{t,''''} = -12a_t\left(\frac{X''_{11}}{X_{111}}\right)^2 \qquad \text{for } i \neq 1, \tag{3.100}$$

resulting in the initial path $x^i = 0$, and the post-critical path (Huseyin 1984*a*)

$$\begin{aligned} x^1 &= -\frac{X''_{11}}{X_{111}}(\mu)^2, \\ x^t &= -\tfrac{1}{2}a_t\left(\frac{X''_{11}}{X_{111}}\right)^2(\mu)^4, \end{aligned} \tag{3.101}$$

which indicates a *tangential bifurcation*, again in analogy with potential systems (Huseyin 1975). As depicted in Fig. 3.9, it can be shown that an initially stable path does not lose its stability as it passes through the critical point c, while a totally unstable equilibrium path also passes through c tangentially, indicating an even higher degree of imperfection sensitivity than the point of tri-furcation.

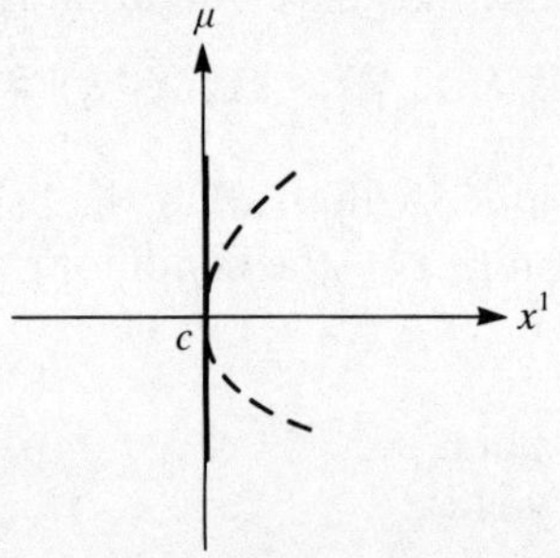

Fig. 3.9 Tangential bifurcation.

It is evident that under further assumptions with regard to vanishing system coefficients (derivatives of X evaluated at c), more degenerate bifurcation phenomena may be produced. Consider the following problem:

Problem 3.2. Assuming that

$$X_{111} = X'_{11} = X''_{11} = X'_{111} = X''_{111} = 0,$$

show that the post-critical equilibrium path takes the form of a cusp (Fig. 3.10) in the vicinity of the critical point c described by

$$(x^1)^2 = -\frac{X'''_{11}}{\bar{X}_{1111}}(\mu)^3 \tag{3.102}$$

and

$$x^t = -\tfrac{1}{2}a_t(x^1)^2, \qquad (t = 2, 3, \ldots, n)$$

where $\bar{X}_{1111}$ and a_t are given by eqns (3.91) and (3.80), respectively. [Hint: consider perturbations up to and including the ninth order (Yu 1984).]

So far, bifurcations arising due to a vanishing real eigenvalue of the Jacobian matrix at a critical value of the parameter η have been explored under certain additional conditions. A similar pattern of bifurcations emerges if a pair of eigenvalues vanish, as discussed in Section 3.1.1. A full analytical investigation of the phenomena arising in this case can be carried out in a parallel treatment; however, the analysis will now have to be based on the transformed system (3.62) since the transformation (3.65) cannot normally be employed due to continuity problems. The analysis will be left as an exercise for the reader, and the following

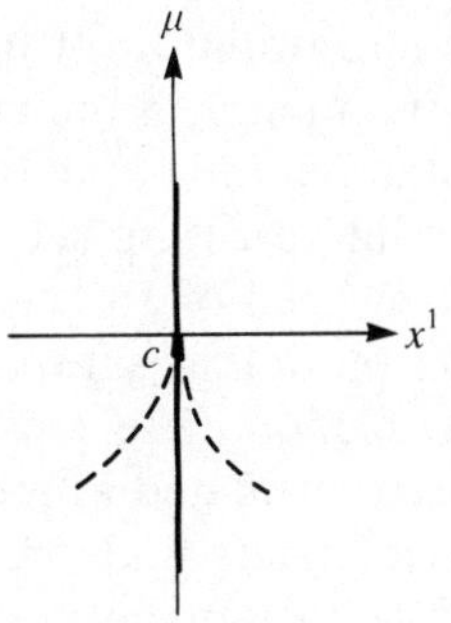

Fig. 3.10 A cusp.

problems are designed for this purpose:

Problem 3.3. Suppose that an initially stable equilibrium path of system (3.62) exhibits a critical point characterized by the Jacobian matrix (3.64) where

$$J = \begin{bmatrix} 0 & 1 \\ 0 & 0 \end{bmatrix}. \tag{3.103}$$

Determine the post-critical path via an appropriate perturbation scheme and show that the first-order equations of the path are given by

$$\mu = -\frac{W_{211}}{6W'_{21}} w^1, \qquad w^t = \tfrac{1}{2} g_t (w^1)^2, \qquad (t = 2, 3, \ldots, n), \tag{3.104}$$

where w^1 is the critical coordinate and g_t are certain expressions in terms of the derivatives of W, indicating an *asymmetric* point of bifurcation as depicted in Fig. 3.6.

Problem 3.4. Analyse the above problem under the additional condition

$$W_{211} = 0, \tag{3.105}$$

and show that the post-critical path is now described by

$$\mu = \frac{\bar{W}_{2111}}{2W'_{21}} (w^1)^2 \quad \text{and} \quad w^t = \tfrac{1}{2} h_t (w^1)^2, \tag{3.106}$$

where $\bar{W}_{2111} = W_{2111} - 3W_{2t1} h_t$, and h_t are certain constants. Demonstrate on a plot of μ versus w^1 that the critical point is a *symmetric* point of bifurcation as depicted in Fig. 3.7 (Mandadi and Huseyin 1980).

3.1.3 *Stability of equilibrium paths in the vicinity of simple bifurcation points*

The stability of an equilibrium state associated with a potential system can be studied conveniently on the basis of the extremum properties of a potential function as demonstrated in Chapter 2, and the stability distribution on the equilibrium paths in the vicinity of a critical point can also be established by evaluating the variations of this potential function at the equilibrium states of interest (also see Huseyin 1975).

In this section, a similar method of analysis will be developed to study the stability distribution on the initial and post-critical equilibrium paths explored in the preceding section. The method is based on the eigenvalues of the Jacobian matrix evaluated at arbitrary equilibrium points in the vicinity of a divergence point, and produces explicit stability and instability criteria through a systematic perturbation scheme (Huseyin 1981*a*).

In order to examine the stability of the equilibrium paths associated with system (3.66) under various critical conditions, consider the expansion of $X_i(x^j, \eta)$ into Taylor's series around a certain critical point $\eta = \eta_c$,

$$X_i(x^j, \mu) = \left.\frac{\partial X_i}{\partial x^j}\right|_c x^j + \left.\frac{\partial X_i}{\partial \eta}\right|_c (\eta - \eta_c) + \frac{1}{2}\left.\frac{\partial^2 X_i}{\partial x^j\,\partial x^k}\right|_c x^j x^k$$

$$+ \left.\frac{\partial^2 X_i}{\partial x^j\,\partial \eta}\right|_c x^j(\eta - \eta_c) + \ldots. \tag{3.107}$$

The Jacobian associated with a given equilibrium state in the vicinity of the critical point c can then be expressed as

$$X_{ij}(x^j, \mu) = X_{ij} + X_{ijk}x^k + X'_{ij}\mu + \ldots, \tag{3.108}$$

in the familiar notation. Evaluations of eqn (3.108) on the initial and post-critical paths intersecting at the critical point c will yield information about the stability properties of these paths.

Case 1. Stability distribution in the vicinity of an asymmetric point of bifurcation. By evaluating eqn (3.108) on the initial path $x^i = 0$, using the canonical form of X_{ij} (eqn 3.68) and keeping to a first-order approximation, one observes that the eigenvalues of the Jacobian can be determined from the uncoupled characteristic equations associated with each block of eqn (3.68). Clearly, complex conjugate eigenvalues associated with the blocks $\boldsymbol{K}_m$ continue to have negative real parts for sufficiently small μ; in fact, this was assumed in the beginning of Section 3.1.2. On the other hand, the eigenvalues associated with the diagonal block J follow from

$$|X_{ij}(0, \mu) - \lambda I| = 0, \qquad (i, j = 1, 2), \tag{3.109}$$

where

$$X_{ij}(0, \mu) = \begin{bmatrix} 0 & 0 \\ 0 & \alpha_2 \end{bmatrix} + \begin{bmatrix} X'_{11} & \\ & X'_{22} \end{bmatrix}\mu, \tag{3.110}$$

as $\lambda_1 = X'_{11}\mu$ and $\lambda_2 = \alpha_2 + X'_{22}\mu$. It is clear that λ_2 remains negative for sufficiently small μ since $\alpha_2 < 0$, and one has the following stability criterion for the initial path in the vicinity of the critical point $\eta = \eta_c$:

$$X'_{11}\mu \begin{Bmatrix} <0 \\ =0 \\ >0 \end{Bmatrix} \text{ for } \begin{Bmatrix} \text{stable} \\ \text{critical} \\ \text{unstable} \end{Bmatrix} \text{ equilibrium.} \tag{3.111}$$

In applications, as discussed earlier, $\alpha_1 < 0$ for $\eta < \eta_c$ and $\alpha_1 > 0$ for $\eta > \eta_c$, while $\alpha_1 = 0$ for $\eta = \eta_c$, implying that $X'_{11} > 0$, and the criterion (3.111) simply expresses the fact that the initial path is stable (unstable) for $\mu < 0$ ($\mu > 0$). Note, however, that the criterion (3.111) remains valid even if the reverse situation occurs; an unstable path gaining stability upon passing through the critical point c.

Next, evaluate the Jacobian (eqn 3.108) on the post-critical path (eqn 3.80); to this end, substitute for x^t and x^1 into eqn (3.108) to obtain

$$X_{ij}(\mu) = X_{ij} + X_{ij1}\left(-\frac{2X'_{11}}{X_{111}}\mu\right) + X'_{ij}\mu + 0(\mu^2) + \ldots \tag{3.112}$$

in which (2×2) blocks are no longer in an uncoupled form as in the case of the initial path. It is assumed that the eigenvalues α_2 and $(\alpha_m \pm \mathrm{i}\omega_m)$ stay away from the origin and their real parts continue to have negative real parts as $\alpha_1(\mu)$ crosses the origin, $\alpha_1(0) = 0$, when the post-critical path (eqn 3.80) passes through the critical point c. In order to examine the variation of $\alpha_1(\mu)$ along the post-critical path (eqn 3.80), let $\alpha_1(\mu)$ be expanded into Taylor's series

$$\alpha_1(\mu) = 0 + \alpha'_1\mu + \ldots \tag{3.113}$$

where $\alpha'_1 = \partial\alpha_1/\partial\mu|_{\mu=0}$.

The coefficient α'_1 can be obtained conveniently by recognizing that if $\alpha_1(\mu)$ is an eigenvalue of the Jacobian (3.112), it must satisfy the characteristic equation

$$|X_{ij}(\mu) - \alpha_1(\mu)\boldsymbol{I}| = 0 \tag{3.114}$$

identically. Hence, by differentiating eqn (3.114) by columns and evaluating at $\mu = 0$, one observes that all determinants except the leading one vanish, resulting in

$$\begin{vmatrix} X'_{11} + X_{111}(-2X'_{11}/X_{111}) - \alpha'_1 & 0 & 0 & 0 & \cdot \\ 0 + X_{211}(-2X'_{11}/X_{111}) & \alpha_2 - 0 & 0 & 0 & \cdot \\ 0 + \ldots & 0 & \alpha_3 - 0 & -\omega_3 & \cdot \\ \cdot & \cdot & \omega_3 & \alpha_3 - 0 & \cdot \\ \cdot & \cdot & \cdot & \cdot & \cdot \end{vmatrix} = 0 \tag{3.115}$$

which yields

$$\alpha'_1 = -X'_{11}. \tag{3.116}$$

It follows that the eigenvalue $\alpha_1(\mu)$ along the post-critical path may be expressed as

$$\alpha_1(\mu) = -X'_{11}\mu + 0(\mu^2) + \ldots. \tag{3.117}$$

In the vicinity of the critical point $\eta = \eta_c$, one then has the following criterion for the stability of the post-critical path:

$$X'_{11}\mu \begin{Bmatrix} >0 \\ =0 \\ <0 \end{Bmatrix} \text{ for } \begin{Bmatrix} \text{stable} \\ \text{critical} \\ \text{unstable} \end{Bmatrix} \text{ equilibrium.} \tag{3.118}$$

If $X'_{11} > 0$, the post-critical path is unstable for $\mu < 0$, and gains stability upon passing through the critical point $\eta = \eta_c$, in contrast with the initial path. More generally, the criteria (3.111) and (3.118) reveal that *an exchange of stabilities occurs at an asymmetric point of bifurcation*, in complete analogy with potential systems.

Case 2. Stability distribution in the vicinity of a symmetric point of bifurcation. Consider now the critical point at which the condition (3.81) is satisfied. Obviously, the arguments pertaining to the initial path in Case 1 are also valid here, and one proceeds to evaluate the Jacobian (3.108) on the post-critical path (3.83). Thus, substituting for μ and x^t yields

$$X_{ij}(x^1) = X_{ij} + X_{ij1}x^1 + \tfrac{1}{2}X_{ij11}(x^1)^2 - \tfrac{1}{2}X_{ijt}a_t(x^1)^2 - \tfrac{1}{2}X'_{ij}a_0(x^1)^2 + \dots . \tag{3.119}$$

Let the eigenvalue $\alpha_1(x^1)$ along the post-critical path (3.83) be again expressed as a Taylor's expansion around $\eta = \eta_c$,

$$\alpha_1(x^1) = 0 + \alpha_{1,1}x^1 + \tfrac{1}{2}\alpha_{1,11}(x^1)^2 + \dots , \tag{3.120}$$

where $\alpha_{1,1} = \partial\alpha_1/\partial x^1|_{x^1=0}$, etc.

If $\alpha_1(x^1)$ is an eigenvalue of the Jacobian (3.119), then it identically satisfies the characteristic equation

$$|X_{ij}(x^1) - \alpha_1(x^1)I| = 0. \tag{3.121}$$

It follows that eqn (3.121) may be differentiated as many times as required to generate a sequence of equations which will yield the coefficients $\alpha_{1,1}$, $\alpha_{1,11}$, etc. Thus, differentiating eqn (3.121) with respect to x^1 and evaluating at the critical point yields

$$\begin{vmatrix} 0-\alpha_{1,1} & 0 & 0 & \cdot & \cdot \\ X_{211} & \alpha_2 - 0 & 0 & \cdot & \cdot \\ X_{311} & 0 & \alpha_3 & -\omega_3 & \cdot \\ X_{411} & 0 & \omega_3 & \alpha_3 & \cdot \\ \cdot & \cdot & \cdot & \cdot & \cdot \\ \cdot & \cdot & \cdot & \cdot & \cdot \end{vmatrix} = 0, \tag{3.122}$$

resulting in

$$\alpha_{1,1} = 0. \tag{3.123}$$

A second differentiation of eqn (3.121) with respect to x^1, and evaluation at the critical point $x^1 = 0$ with the aid of eqn (3.123), yield an equation whose left-hand side consists of a sum of determinants. For clarity, this equation is given here fully:

$$\begin{vmatrix} X_{111} - X_{11t}a_t - X'_{11}a_0 - \alpha_{1,11} & 0 & 0 & 0 & \cdot \\ X_{2111} - X_{21t}a_t & \alpha_2 & 0 & 0 & \cdot \\ X_{3111} - X_{31t}a_t & 0 & \alpha_3 & -\omega_3 & \cdot \\ X_{4111} - X_{41t}a_t & 0 & \omega_3 & \alpha_3 & \cdot \\ \cdot & \cdot & \cdot & \cdot & \cdot \end{vmatrix}$$

$$+\begin{vmatrix} 0 & X_{121} & 0 & 0 & \cdot \\ X_{211} & X_{221} & 0 & 0 & \cdot \\ X_{311} & X_{321} & \alpha_3 & -\omega_3 & \cdot \\ X_{411} & X_{421} & \omega_3 & \alpha_3 & \cdot \\ \cdot & \cdot & \cdot & \cdot & \cdot \end{vmatrix} + \begin{vmatrix} 0 & 0 & X_{131} & 0 & \cdot \\ X_{211} & \alpha_2 & X_{231} & 0 & \cdot \\ X_{311} & 0 & X_{331} & -\omega_3 & \cdot \\ X_{411} & 0 & X_{431} & \alpha_3 & \cdot \\ \cdot & \cdot & \cdot & \cdot & \cdot \end{vmatrix}$$

$$+\begin{vmatrix} 0 & 0 & 0 & X_{141} & \cdot \\ X_{211} & \alpha_2 & 0 & X_{241} & \cdot \\ X_{311} & 0 & \alpha_3 & X_{341} & \cdot \\ X_{411} & 0 & \alpha_3 & X_{441} & \cdot \\ \cdot & \cdot & \cdot & \cdot & \cdot \end{vmatrix} + \begin{vmatrix} 0 & X_{121} & 0 & 0 & \cdot \\ X_{211} & X_{221} & 0 & 0 & \cdot \\ X_{311} & X_{321} & \alpha_3 & -\omega_3 & \cdot \\ X_{411} & X_{421} & \omega_3 & \alpha_3 & \cdot \\ \cdot & \cdot & \cdot & & \cdot \end{vmatrix}$$

$$+\begin{vmatrix} 0 & 0 & X_{131} & 0 & \cdot \\ X_{211} & \alpha_2 & X_{231} & 0 & \cdot \\ X_{311} & 0 & X_{331} & -\omega_3 & \cdot \\ X_{411} & 0 & X_{431} & \alpha_3 & \cdot \\ \cdot & \cdot & \cdot & \cdot & \cdot \end{vmatrix} + \begin{vmatrix} 0 & 0 & 0 & X_{141} & \cdot \\ X_{211} & \alpha_2 & 0 & X_{241} & \cdot \\ X_{311} & 0 & \alpha_3 & X_{341} & \cdot \\ X_{411} & 0 & \omega_3 & X_{441} & \cdot \\ \cdot & \cdot & \cdot & \cdot & \cdot \end{vmatrix} + \ldots = 0. \tag{3.124}$$

Expanding the determinants results in

$$\alpha_{1,11} = X_{1111} - X_{11t}a_t - X'_{11}a_0 - \frac{X_{121}X_{211}}{\alpha_2} - 2X_{131}\left(\frac{\alpha_3 X_{311} + \omega_3 X_{411}}{\alpha_3^2 + \omega_3^2}\right)$$
$$- 2X_{141}\left(\frac{\alpha_3 X_{411} - \omega_3 X_{311}}{\alpha_3^2 + \omega_3^2}\right) - \ldots. \tag{3.125}$$

Recalling the definitions of a_0 and a_t ($t = 2, 3, \ldots, n$), recognizing the fact that $X_{1t1} = X_{11t}$, and generalizing the summations in eqn (3.125) leads to

$$\alpha_{1,11} = X_{1111} - 3X_{11t}a_t - X'_{11}a_0 \tag{3.126}$$

where a_t and a_0 are defined by eqns (3.80) and (3.82), respectively.

In view of eqn (3.82), this derivative may also be expressed as

$$\alpha_{1,11} = 2X'_{11}a_0 \tag{3.127}$$

or

$$\alpha_{1,11} = \tfrac{2}{3}\bar{X}_{1111}, \tag{3.128}$$

where $\bar{X}_{1111} = X_{1111} - 3X_{11t}a_t$.

The Taylor's expansion (3.120) now yields

$$\alpha_1(x^1) = X'_{11}a_0(x^1)^2 + \dots \tag{3.129}$$

or equivalently

$$\alpha_1(x^1) = \tfrac{1}{3}\bar{X}_{1111}(x^1)^2 + \dots .$$

The stability criterion for the symmetric post-critical path (3.83) is then given by

$$X'_{11}a_0 \begin{Bmatrix} <0 \\ =0 \\ >0 \end{Bmatrix} \text{ for } \begin{Bmatrix} \text{stable} \\ \text{critical} \\ \text{unstable} \end{Bmatrix} \text{ equilibrium.} \tag{3.130a}$$

In other words, if one assumes $X'_{11} > 0$ as before and $a_0 < 0$, then the post-critical path (eqn 3.83) exists only for $\mu > 0$ and, according to eqn (3.130*a*), is totally stable, while the initial path is unstable for the same range of the parameter ($\mu > 0$) as indicated by eqn (3.111). On the other hand if $a_0 > 0$, post-critical equilibrium states (eqn 3.83) exist only for $\mu < 0$, and by virtue of eqn (3.130*a*) the path (eqn 3.83) is totally unstable while the initial path is stable. More generally, by comparing the criteria (3.111) and eqn (3.130*a*) in conjunction with eqn (3.83), one observes that *the post-critical path is totally stable* (*unstable*) *for* $\mu > 0$ *or* $\mu < 0$ *if the initial path is unstable* (*stable*) *for the same range of* μ (Fig. 3.7).

It is understood that the stability criterion (3.130*a*) is intended for a direct comparison with the criterion of (3.111); the stability criterion, however, can also be expressed as

$$(X_{1111} - 3X_{11t}a_t) \begin{Bmatrix} <0 \\ =0 \\ >0 \end{Bmatrix} \text{ for } \begin{Bmatrix} \text{stable} \\ \text{critical} \\ \text{unstable} \end{Bmatrix} \text{ equilibrium} \tag{3.130b}$$

which is independent of μ (Huseyin 1981*a*).

Problem 3.5. Following the procedure described in this section, examine the stability distribution on the initial and post-critical paths in the vicinity of a point of *tri-furcation* (Case 3, Section 3.1.2), and establish explicit stability criteria. Show that a stable (unstable) initial path remains stable (unstable) as it passes through the critical point c, while the post-critical paths are totally unstable (stable) in the vicinity of c if they exist.

Problem 3.6. Consider the phenomenon described as *tangential bifurcation* (Case 4, Section 3.1.2), and perform a similar stability analysis as in Problem 3.5 to establish stability criteria in general terms for the stability distribution on the initial and post-critical paths. Show that the initial path remains stable (unstable) as it passes through the critical point c, and the post-critical path (eqn (3.101) is totally unstable (stable) in this neighbourhood.

3.1.4 Compound branching

Let the critical point c on the initial path of system (3.62) be an r-fold compound critical point. According to definitions given in Section 3.1, an *r-fold compound* critical point is associated with a Jacobian matrix of *nullity* r. At an r-fold point, if the *multiplicity* of a zero eigenvalue is r, then its *index* is also r so that there exist r linearly independent eigenvectors. However, the multiplicity of a zero eigenvalue can exceed its index, and an *r-fold compound critical point* may be exhibited in various ways provided the index of the repeated zeros is r. The structure of the canonical form of the Jacobian matrix becomes known through an investigation of the eigenvectors corresponding to repeated roots or corresponding elementary divisors.

An investigation of a general nature covering all possible canonical structures of the Jacobian matrix is not practical, and attention here will be restricted to two representative situations:

1) *Multiplicity is equal to index*. Under this assumption, the Jordan canonical form corresponding to r repeated zeros is actually a diagonal matrix, and it is assumed that the vanishing eigenvalues become positive (real) upon crossing the origin. The formulation (3.66), with all its essential properties, may, therefore, be adopted for the analysis of this case.

Thus, the matrix $\boldsymbol{J}$ in eqn (3.68) is assumed to be an r-square null matrix,

$$\boldsymbol{J} = \begin{bmatrix} 0 & & & \\ & 0 & & \\ & & \ddots & \\ & & & 0 \end{bmatrix}, \tag{3.131}$$

and the blocks $\boldsymbol{K}_m$ are essentially the same as in eqn (3.69), with $m = (r+1), (r+3), \ldots$.

For simplicity, no real (negative) eigenvalues are considered here; it is understood, however, that incorporating such eigenvalues in the formulation poses no fundamental difficulties.

Under these assumptions, the modified properties (eqn 3.70) of system (3.66) read

$$X_{ij}(0, \eta) = X'_{ij}(0, \eta) = X''_{ij}(0, \eta) = \ldots = 0 \qquad (3.132)$$

for $i \neq j$ provided $j \neq i + 1$ when $i = r + 1, r + 3, \ldots$ and $j \neq i - 1$ when $i = r + 2, r + 4, \ldots$.

The first perturbation equation (3.73) now yields

$$\dot{x}^t = 0, \qquad \{t = (r+1), (r+2), \ldots, n\}. \qquad (3.133)$$

Using these derivatives, the second perturbation equation (3.74) leads to

$$X_{abc}\dot{x}^b\dot{x}^c + 2X'_{aa}\dot{x}^a\dot{\eta} = 0 \qquad (3.134)$$

and

$$\begin{bmatrix} \alpha_m & -\omega_m \\ \omega_m & \alpha_m \end{bmatrix}\begin{bmatrix} \ddot{x}^m \\ \ddot{x}^{m+1} \end{bmatrix} = -\begin{bmatrix} X_{mbc} \\ X_{(m+1)bc} \end{bmatrix}\dot{x}^b\dot{x}^c \qquad (3.135)$$

where $a, b, c = 1, 2, \ldots, r$ and $m = (r+1), (r+3), \ldots$.

Suppose now that η is capable of describing all possible equilibrium paths through c and set $\sigma = \eta$; it is then observed that eqn (3.134) represents a set of r second-order equations in r unknowns, $\dot{x}^a \equiv x^{a,\prime}$, which yield 2^r sets of *solution rays* in the real or complex field, provided the Jacobian of the eqns (3.134) does not vanish identically. This latter condition ensures that eqns (3.134) are independent; otherwise one solution follows from another and the system has infinitely many solutions.

The 2^r sets of solution rays include the initial path, and the number of post-critical paths is, therefore, $(2^r - 1)$. This result can also be deduced without setting $\sigma = \eta$. Indeed, instead of making this assumption, one considers eqn (3.76) in conjunction with eqn (3.134), which is now in the form

$$\sum_{a=1}^{r} (\dot{x}^a)^2 + (\dot{\eta})^2 = 1. \qquad (3.136)$$

The problem is now concerned with the solution of $(r+1)$ equations (eqns 3.134 and 3.136), in $(r+1)$ unknowns. Each equation has a polynomial of degree 2, and the number of solution rays, therefore, appears to be 2^{r+1} provided the associated Jacobian is not identically zero. It is observed, however, that if $(\dot{x}^a, \dot{\eta})$ is a solution ray, $(-\dot{x}^a, -\dot{\eta})$ represents the same ray, and the actual number of solution rays is, therefore, given by $\frac{1}{2}2^{r+1} = 2^r$ as predicted earlier.

Since complex roots occur in pairs, and due to the fact that a real path (initial) is already known, there must be at least one post-critical path passing through c. In the particular case of $r = 1$, for example, eqn

(3.134) yields the post-critical path (3.80) as expected. If $r = 2$, then, there is at least one and possibly three post-critical paths. For $r = 3$, a maximum of seven, and in general a maximum of $(2^r - 1)$ real equilibrium paths exist.

Also notice that the curvatures $\ddot{x}^t$ of non-critical coordinates are readily evaluated by substituting for $\dot{x}^a$ in eqn (3.135).

2) *Multiplicity is twice the index*. This condition implies that if the eigenvector space corresponding to the repeated zero roots of the Jacobian is r-dimensional, then the multiplicity of zeros is $l = 2r$. In other words, the Jordan canonical form J in eqn (3.64) has $l = 2r$ zero eigenvalues and r Jordan blocks; it is an $l \times l$ matrix of the form

$$\boldsymbol{J} = \begin{bmatrix} 0 & 1 & & & & & \\ & 0 & & & & & \\ & & 0 & 1 & & & \\ & & & 0 & & & \\ & & & & \cdot & \cdot & \\ & & & & & \cdot & \\ & & & & & & \cdot \;\; \cdot \\ & & & & & & \cdot \end{bmatrix}. \tag{3.137}$$

The blocks $\boldsymbol{K}_m$ of the Jacobian are as defined in eqn (3.66).

In this case the Jacobian has r elementary divisors of second degree associated with repeated zero roots.

The analysis of post-critical behaviour can be carried out on the basis of eqn (3.62). The structure of the Jacobian does not lend itself to a transformation of the type (3.65), and the formulation (3.66) is, therefore, not applicable to this case.

Assuming the solutions are again in the parametric form

$$\boldsymbol{w} = \boldsymbol{w}(\sigma), \qquad \eta = \eta(\sigma), \tag{3.138}$$

and substituting these assumed solutions back into the equilibrium equations, yields the identities

$$W_i[w^j(\sigma), \eta(\sigma)] \equiv 0 \tag{3.139}$$

which may be used to generate a sequence of perturbation equations.

Thus, differentiating eqn (3.139) with respect to σ and evaluating at the critical point c with the aid of the essential properties (3.63) results in

$$\dot{w}^e = 0, \qquad \dot{w}^t = 0, \tag{3.140}$$

where $e = 2, 4, \ldots, l$ and $t = (l + 1), \ldots, n$.

A second differentiation of eqn (3.139), and evaluation at c results in a split into three types of relations,

$$W_{eab}\dot{w}^a\dot{w}^b + 2W'_{ea}\dot{w}^a\dot{\eta} = 0, \tag{3.141}$$

$$W_{(e-1)ab}\dot{w}^a\dot{w}^b + 2W'_{(e-1)a}\dot{w}^a\dot{\eta} + \ddot{w}^e = 0, \tag{3.142}$$

$$\begin{bmatrix}\alpha_m & -\omega_m\\ \omega_m & \alpha_m\end{bmatrix}\begin{bmatrix}\ddot{w}^m\\ \ddot{w}^{m+1}\end{bmatrix} = -\begin{bmatrix}W_{mab}\dot{w}^a\dot{w}^b + 2W'_{ma}\dot{w}^a\dot{\eta}\\ W_{(m+1)ab}\dot{w}^a\dot{w}^b + 2W'_{(m+1)a}\dot{w}^a\dot{\eta}\end{bmatrix}, \tag{3.143}$$

where $a, b, c = 1, 3, \ldots, l-1$; $e, f, g = 2, 4, \ldots, l$ and $m = (l+1), (l+3), \ldots$.

In addition, one has the auxiliary equation

$$\sum_a (\dot{w}^a)^2 + (\dot{\eta})^2 = 1. \tag{3.144}$$

It is recognized that there are $(n+1)$ equations in $(n+1)$ unknowns. The procedure to be followed in determining these unknowns (path derivatives) is dictated by the structure of eqns (3.141) to (3.144). Indeed, eqns (3.141) and (3.144) are solved first to obtain $\dot{w}^a$ and $\dot{\eta}$; this can be done since there are $(r+1)$ equations in $(r+1)$ unknowns. One may then substitute for these first derivatives into eqns (3.142) and (3.143) to determine the second derivatives $\ddot{w}^e$ and $\ddot{w}^t$. It is noted that eqns (3.141) and (3.144) have quadratic polynomials, and if the associated Jacobian does not vanish identically, these equations yield 2^{r+1} solution rays. However, if $(\dot{w}^a, \dot{\eta})$ is a solution ray, so is $(-\dot{w}^a, -\dot{\eta})$, and the actual number of solution rays is 2^r as in the preceding case. Furthermore, if the initial path is excluded, the number of post-critical paths emerges as $2^r - 1$, as before.

Finally, consider the situations when certain key coefficients vanish at an r-fold compound critical point. If, for example, all $W_{eab} = 0$, it can be demonstrated that the system exhibits *symmetric* post-critical paths as in the case of simple critical points. Indeed, in this case, it follows from eqn (3.141) that

$$\dot{\eta} = 0, \tag{3.145}$$

and a third perturbation of eqn (3.139) under this condition yields

$$W_{eabc}\dot{w}^a\dot{w}^b\dot{w}^c + 3W_{efc}\ddot{w}^f\dot{w}^c + 3W_{etc}\ddot{w}^t\dot{w}^c + 3W'_{ec}\dot{w}^c\ddot{\eta} = 0, \tag{3.146}$$

$$\ddot{w}^e = -W_{(e-1)ab}\dot{w}^a\dot{w}^b \triangleq -p_{eab}\dot{w}^a\dot{w}^b, \tag{3.147}$$

and

$$\ddot{w}^t = -p_{tab}\dot{w}^a\dot{w}^b, \tag{3.148}$$

where p_{tab} are determined by solving eqn (3.143).

Introducing eqns (3.147) and (3.148) into eqn (3.146) results in

$$(W_{eabc} - 3W_{efc}p_{fab} - 3W_{etc}p_{tab})\dot{w}^a\dot{w}^b\dot{w}^c + 3W'_{ec}\dot{w}^c\ddot{\eta} = 0. \tag{3.149}$$

It is observed that eqn (3.149) represents r equations in $(r+1)$ derivatives, $\dot{w}^a$, and $\ddot{\eta}$. These equations, in conjunction with eqn (3.144) which is now in the form

$$\sum_a (\dot{w}^a)^2 = 1, \tag{3.150}$$

yield $(\frac{1}{2} \cdot 2 \cdot 3^r) = 3^r$ solution rays in the real or complex domain, in analogy with conservative systems (Huseyin 1975).

As stated earlier, compound critical points may arise in a variety of ways, depending on the properties of the Jacobian—associated with a critical point—which lend structure to a system, influencing its post-critical behaviour. Although the analyses presented in this section were concerned with two particular cases only, the underlying methodology may be adopted for the analysis of other situations involving compound critical points. It has to be noted once more that the formulation (eqn 3.66) is not always applicable because of the assumptions underlying the transformation (3.65); however, the restrictions associated with system (3.66) do not apply to the formulation (3.62), which may serve as a basis for exploring various branching phenomena.

Problem 3.7. Consider an r-fold critical point of system (3.66), and assume that the index of the repeated zero eigenvalue is also r as in the first case treated in this section. Assuming further that all the cubic coefficients X_{abc} vanish at the critical point, establish the equations—through the perturbation procedure—necessary to determine the path derivatives for a first-order estimation of all post-critical paths.

Problem 3.8. Consider a *3-fold compound* critical point c of system (3.62) characterized by the Jacobian matrix (3.64) where the Jordan canonical form $\boldsymbol{J}$ is given by

$$\boldsymbol{J} = \begin{bmatrix} 0 & & & \\ & 0 & & \\ & & 0 & 1 \\ & & & 0 \end{bmatrix}. \tag{3.151}$$

Here, one has a zero eigenvalue of multiplicity four but index three. Assuming further that no other coefficient vanishes at c, derive the equations which yield the path derivatives necessary for a first-order estimation of all post-critical paths. What is the number of post-critical paths passing through the critical point?

3.1.5 *Imperfection sensitivity of critical points*

As in the case of potential systems, critical bifurcation points can be very sensitive to imperfections—a phenomenon of extreme practical significance since it may imply drastic reductions in the critical value of a parameter. In this section, an imperfection parameter will be introduced into the formulation of systems treated in Section 3.1.2 in order to shed some light on the important question of imperfection-sensitivity of certain branching points. It is realized that, depending on the nature of the critical point (singularity), a single imperfection parameter may not be sufficient to produce a universal unfolding of the singularity; nevertheless, the results will serve as a prelude to a more comprehensive treatment of multiple-parameter systems in the following sections, as well as providing valuable information concerning the behaviour of many systems.

Consider once again the system characterized by eqn (3.66), and assume that there exist certain imperfections in the system which may be represented by a parameter ε so that the governing equations of the corresponding imperfect system are given by

$$\frac{\mathrm{d}\boldsymbol{x}}{\mathrm{d}t} = \boldsymbol{X}(\boldsymbol{x}, \eta, \varepsilon), \tag{3.152}$$

which possesses all the essential properties of eqn (3.66), described by eqns (3.67) to (3.70), for $\varepsilon = 0$. In other words, setting $\varepsilon = 0$ in eqn (3.152) gives eqn (3.66) with all its properties; thus, eqns (3.67), (3.68), and (3.70) are expressed as

$$\boldsymbol{X}(\mathbf{0}, \eta, 0) = \boldsymbol{X}'(\mathbf{0}, \eta, 0) = \ldots = \mathbf{0}, \tag{3.153}$$

$$\left[\frac{\partial \boldsymbol{X}}{\partial \boldsymbol{x}}\right]_{\substack{\boldsymbol{X}=\mathbf{0}\\ \varepsilon=0}} = \mathrm{diag}\,[\boldsymbol{J}, \boldsymbol{K}_3, \boldsymbol{K}_5, \ldots], \tag{3.154}$$

where the eigenvalues $\alpha_i(\eta, \varepsilon) \pm \omega_i(\eta, \varepsilon)$ are used to form

$$\boldsymbol{J} = \begin{bmatrix} \alpha_1(\eta, 0) & 0 \\ 0 & \alpha_2(\eta, 0) \end{bmatrix}, \qquad \boldsymbol{K}_m = \begin{bmatrix} \alpha_m(\eta, 0) & -\omega_m(\eta, 0) \\ \omega_m(\eta, 0) & \alpha_m(\eta, 0) \end{bmatrix},$$

$(m = 3, 5, \ldots)$, $\alpha_1(\eta_c, 0) = 0$ and $\alpha_t(\eta_c, 0) < 0$ for $t \neq 1$; and

$$X_{ij}(0, \eta, 0) = X'_{ij}(0, \eta, 0) = \ldots = 0 \tag{3.155}$$

for $i \neq j$ provided $j \neq i + 1$ when $i = 3, 5, \ldots$ and $j \neq i - 1$ when $i = 4, 6, \ldots$, respectively.

System (3.152) contains two independent parameters, η and ε, and the solutions of the equilibrium equations $X_i(x^j, \eta, \varepsilon) = 0$ in the vicinity of a bifurcation point c have to be expressed in terms of two perturbation parameters σ^α, $(\alpha = 1, 2)$. Thus, let

$$x^i = x^i(\sigma^\alpha), \qquad \eta = \eta(\sigma^\alpha) \quad \text{and} \quad \varepsilon = \varepsilon(\sigma^\alpha), \tag{3.156}$$

and substitute these assumed solutions back into the equilibrium equations to obtain the identities

$$X_i[x^j(\sigma^\alpha), \eta(\sigma^\alpha), \varepsilon(\sigma^\alpha)] \equiv 0 \tag{3.157}$$

which yield the perturbation equations

$$X_{ij}x^{j,\alpha} + X'_i\eta^{,\alpha} + X_{i\varepsilon}\varepsilon^{,\alpha} = 0, \tag{3.158}$$

$$\begin{aligned}(X_{ijk}x^{k,\beta} + X'_{ij}\eta^{,\beta} + X_{ij\varepsilon}\varepsilon^{,\beta})x^{j,\alpha} + X_{ij}x^{j,\alpha\beta}\\ + (X'_{ij}x^{j,\beta} + X''_i\eta^{,\beta} + X'_{i\varepsilon}\varepsilon^{,\beta})\eta^{,\alpha} + X'_i\eta^{,\alpha\beta}\\ + (X_{ij\varepsilon}x^{j,\beta} + X'_{i\varepsilon}\eta^{,\beta} + X_{i\varepsilon\varepsilon}\varepsilon^{,\beta})\varepsilon^{,\alpha} + X_{i\varepsilon}\varepsilon^{,\alpha\beta} = 0,\end{aligned} \tag{3.159}$$

etc.,

where the subscript ε on X's denotes differentiation with respect to ε, and $\alpha, \beta = 1, 2$.

Case 1. Imperfection sensitivity of an asymmetric point of bifurcation. Consider the asymmetric critical point which was studied in Section 3.1.2. This phenomenon arises when a real eigenvalue goes through zero at $\eta = \eta_c$, and $X_{11} \equiv \alpha_1(\eta_c, 0) = 0$ while other system coefficients remain non-zero. Evaluating eqn (3.158) at the critical point c yields

$$X_{1\varepsilon}\varepsilon^{,\alpha} = 0 \qquad \text{for } i = 1, \tag{3.160}$$

which results in

$$\varepsilon^{,\alpha} = 0, \qquad (\alpha = 1, 2), \tag{3.161}$$

under the assumption that $X_{1\varepsilon} \neq 0$.

By introducing this result into eqn (3.158), one obtains

$$x^{2,\alpha} = 0, \qquad \text{for } i = 2,$$

and

$$\begin{bmatrix}\alpha_m & -\omega_m\\ \omega_m & \alpha_m\end{bmatrix}\begin{bmatrix}x^{m,\alpha}\\ x^{(m+1),\alpha}\end{bmatrix} = \begin{bmatrix}0\\0\end{bmatrix}, \qquad \text{for } m = 3, 5, \ldots, \tag{3.162}$$

which may be expressed simply as

$$x^{t,\alpha} = 0, \qquad (\alpha = 1, 2;\ t = 2, 3, \ldots, n). \tag{3.163}$$

Guided by these derivatives, one may set $\sigma^1 = x^1$ and $\sigma^2 = \eta$ before proceeding to evaluate the second perturbation equation (3.159) which then yields

$$\varepsilon^{,11} = -\frac{X_{111}}{X_{1\varepsilon}}, \tag{3.164}$$

$$x^{2,11} = -\frac{1}{\alpha_2}\left(X_{211} - X_{111}\frac{X_{2\varepsilon}}{X_{1\varepsilon}}\right) \triangle -d_2, \tag{3.165}$$

$$\begin{aligned}x^{m,11} = -\frac{1}{\alpha_m^2 + \omega_m^2}\bigg\{\alpha_m\left(X_{m11} - X_{111}\frac{X_{m\varepsilon}}{X_{1\varepsilon}}\right)\\ + \omega_m\left(X_{(m+1)11} - X_{111}\frac{X_{(r+1)\varepsilon}}{X_{1\varepsilon}}\right)\bigg\} \triangle -d_m,\end{aligned} \tag{3.166}$$

$$x^{(m+1),11} = -\frac{1}{\alpha_m^2+\omega_m^2}\left\{\alpha_m\left(X_{(m+1)11} - X_{111}\frac{X_{(m+1)\varepsilon}}{X_{1\varepsilon}}\right)\right.$$

$$\left.+\,\omega_m\left(X_{m11} - X_{111}\frac{X_{m\varepsilon}}{X_{1\varepsilon}}\right)\right\} \triangleq -d_{m+1}, \tag{3.167}$$

and

$$\varepsilon^{,1\prime} = -\frac{X'_{11}}{X_{1\varepsilon}}, \tag{3.168}$$

$$x^{2,1\prime} = \frac{X'_{11}}{\alpha_2}\frac{X_{2\varepsilon}}{X_\varepsilon} \triangleq e_2, \tag{3.169}$$

$$x^{m,1\prime} = \frac{1}{\alpha_m^2+\omega_m^2}(\alpha_m X_{m\varepsilon} + \omega_m X_{(m+1)\varepsilon})\frac{X'_{11}}{X_{1\varepsilon}} \triangleq e_m, \tag{3.170}$$

$$x^{(m+1),1\prime} = \frac{1}{\alpha_m^2+\omega_m^2}(\alpha_m X_{(m+1)\varepsilon} - \omega_m X_{m\varepsilon})\frac{X'_{11}}{X_{1\varepsilon}} \triangleq e_{m+1}, \tag{3.171}$$

$$\varepsilon^{,\prime\prime} = 0, \qquad x^{t,\prime\prime} = 0, \qquad (t = 2, 3, \ldots, n), \tag{3.172}$$

where the superscripts 1 and $'$ following commas denote differentiation with respect to x^1 and η, respectively, as before.

Using these derivatives yields the first-order equations of the equilibrium surface in the vicinity of the asymmetric point of bifurcation c as

$$X_{1\varepsilon}\varepsilon + \tfrac{1}{2}X_{111}(x^1)^2 + X'_{11}x^1\mu = 0 \tag{3.173}$$

and

$$x^t + \tfrac{1}{2}d_t(x^1)^2 + e_t x^1\mu = 0. \tag{3.174}$$

Clearly, setting $\varepsilon = 0$ in eqn (3.173) gives the perfect system which exhibits the asymmetric point of bifurcation explored in Section 3.1.2. The surface described by eqn (3.173) is immediately recognized as an *anticlastic fold catastrophe* in complete analogy with potential systems (Fig. 3.11). Intersections of this surface with planes described by ε = constant yield a family of hyperbolae sharing the solutions of the

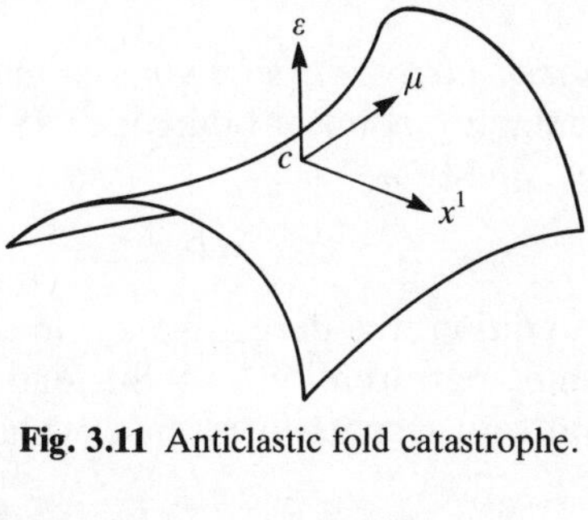

Fig. 3.11 Anticlastic fold catastrophe.

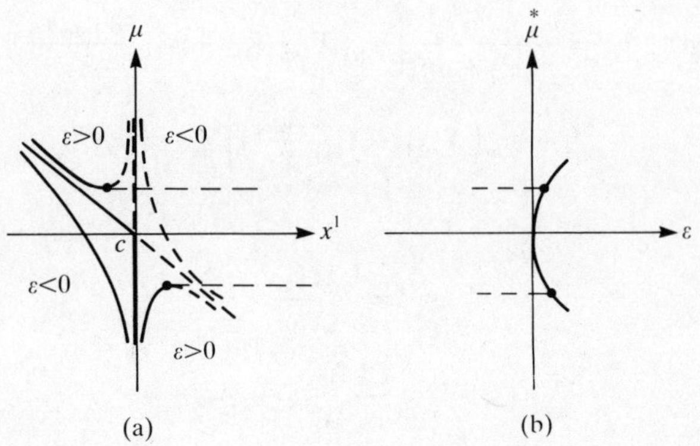

Fig. 3.12 (a) Equilibrium paths in the vicinity of the asymmetric point of bifurcation for $\varepsilon =$ constant and (b) parabolic relationship between critical peak values of parameter μ^* and imperfection parameter ε (drawn for $X'_{11} < 0$, $X_{111} < 0$, $X_{1\varepsilon} < 0$).

perfect system, $x^1 = 0$ and $\mu = -(X_{111}/2X'_{11})x^1$, as the asymptotes. Fig. 3.12*a* illustrates the equilibrium paths of the system in the vicinity of the asymmetric point of bifurcation c for $\varepsilon =$ constant. It is seen that for a given value of ε, the equilibrium paths of the imperfect system may either exhibit a limit point or remain totally stable as the parameter is increased from a value corresponding to a stable state. It is important to notice that the imperfect system may lose stability at a reduced value of the parameter since the limit points occur before the level of the bifurcation point is reached. The relationship between the critical peak values of the parameter ($\overset{*}{\mu}$) and the imperfection parameter ε is parabolic (Fig. 3.12*b*); it may be obtained by differentiating eqn (3.173) with respect to x^1, and then eliminating x^1 between the resulting criticality condition and eqn (3.173). Thus, one obtains the critical line

$$X_{111}\overset{*}{x}{}^1 + X'_{11}\overset{*}{\mu} = 0 \tag{3.175}$$

and the parameter-imperfection relation

$$X_{1\varepsilon}\varepsilon - \frac{1}{2}\frac{(X'_{11})^2}{X_{111}}(\overset{*}{\mu})^2 = 0. \tag{3.176}$$

Case 2. Imperfection sensitivity of a symmetric point of bifurcation. Next, Consider the symmetric point of bifurcation (Section 3.1.2, Case 2) which arises under the condition

$$X_{11} = X_{111} = 0. \tag{3.177}$$

As a result of this condition, the derivative $\varepsilon^{,11}$ in eqn (3.164) vanishes, thus necessitating a third perturbation. To this end, differentiating eqn (3.157) for a third time with respect to x^1, and evaluating at c under the condition (3.177), yields the relationship

$$X_{1111} + X_{1\varepsilon}\varepsilon^{,111} + 3X_{1t1}x^{t,11} = 0, \tag{3.178}$$

where

$$x^{t,11} = -a_t, \qquad (t = 2, 3, \ldots, n), \tag{3.179}$$

and the coefficients a_t are given in eqns (3.80).

It immediately follows that

$$\varepsilon^{,111} = -\frac{\bar{X}_{1111}}{X_{1\varepsilon}} \tag{3.180}$$

where $\bar{X}_{1111}$ is as defined by eqn (3.82).

Using the derivatives obtained earlier and eqn (3.180) leads to the first-order equation of the equilibrium surface,

$$X_{1\varepsilon}\varepsilon + X'_{11}x^1\mu + \frac{1}{3!}\bar{X}_{1111}(x^1)^3 = 0. \tag{3.181}$$

The relationship between the non-critical coordinates x^t and the perturbations parameters x^1 and μ remains identical to eqn (3.174) in form except for the coefficients d_t and l_t which have to be modified by simply setting $X_{111} = 0$ in the corresponding expressions.

It is recognized that eqn (3.181) describes a surface associated with the *cusp catastrophe* (Fig. 3.13), again in complete analogy with potential systems. Setting $\varepsilon = 0$ in eqn (3.181) results in the initial path $x^1 = 0$ and the post-critical path (eqns 3.83) of the perfect system, as expected. A

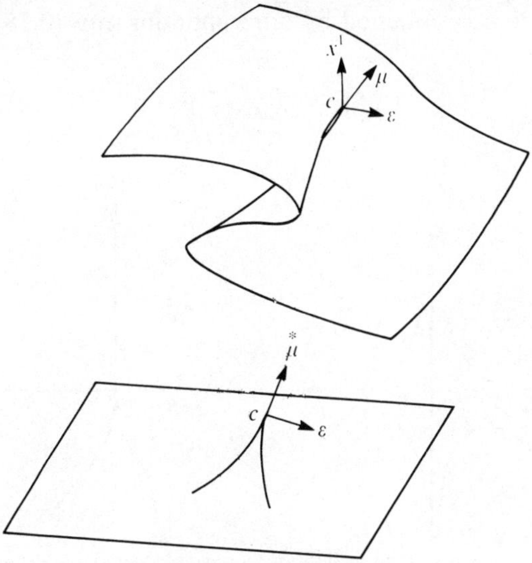

Fig. 3.13 Cusp catastrophe.

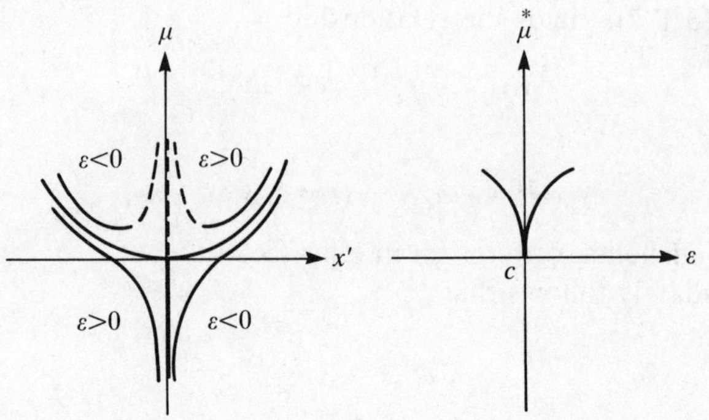

Fig. 3.14 Equilibrium paths in the vicinity of a stable symmetric point of bifurcation for $\varepsilon = \text{constant}$ in eqn (3.181).

family of imperfect systems is obtained by setting ε constant in eqn (3.181), and the equilibrium paths of such imperfect systems in the vicinity of a stable and unstable symmetric point of bifurcation are illustrated in Figs. 3.14 and 3.15, respectively. It is observed that, unlike the case of an asymmetric point of bifurcation, the sign of the imperfection parameter here does not have an effect on the nature of the response. In other words, the equilibrium paths of an imperfect system remain stable in the vicinity of a stable symmetric point, for both $\varepsilon > 0$ and $\varepsilon < 0$, when the behaviour of the system is traced by increasing η from an initial value $\eta_0 < \eta_c$. On the other hand, an imperfect system loses its stability at a limit point for both $\varepsilon > 0$ and $\varepsilon < 0$ if the symmetric point of bifurcation is unstable (Fig. 3.15).

The critical line is obtained by differentiating eqn (3.181) with respect to x^1 as

$$X'_{11}\overset{*}{\mu} + \tfrac{1}{2}\bar{X}_{1111}(\overset{*}{x}{}^1)^2 = 0 \qquad (3.182)$$

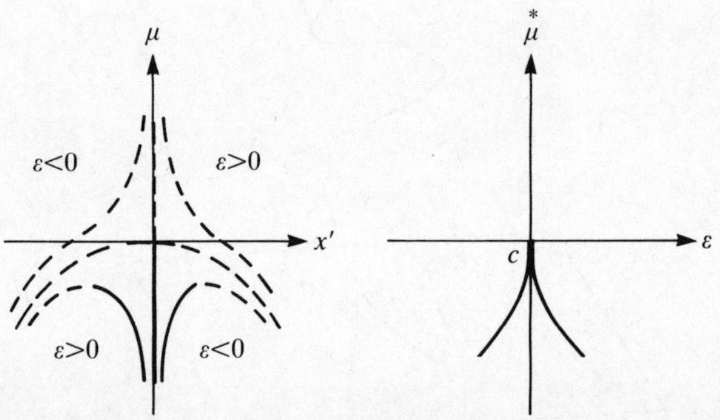

Fig. 3.15 Equilibrium paths in the vicinity of an unstable symmetric point of bifurcation for $\varepsilon = \text{constant}$ in eqn (3.181).

which, together with eqn (3.181), yields the stability boundary

$$X'_{11}\overset{*}{\bar{\mu}} + \tfrac{1}{2}\bar{X}_{1111}^{1/3}(3X_{1\varepsilon}\varepsilon)^{2/3} = 0. \tag{3.183}$$

This is the familiar cusp, indicating a high degree of imperfection-sensitivity. It is also noted that the intermediate sheet of the equilibrium surface shown in Fig. 3.13 is stable, while the two exterior sheets are unstable if c is an unstable point of bifurcation, as illustrated in Fig. 3.15. The stability properties of the equilibrium surface are reversed in the case of a stable symmetric point of bifurcation (Fig. 3.14).

Case 3. Imperfection sensitivity of a point of tri-furcation. Consider now a critical point of tri-furcation at which

$$X_{11} = X_{111} = X'_{11} = 0. \tag{3.184}$$

This situation was discussed (Case 3) in Section 3.1.2. The behaviour of the imperfect system (3.152) in the vicinity of such a critical point may be studied as before. It is first observed that in view of the conditions (3.184), one has

$$\varepsilon,^{1} = \varepsilon,' = \varepsilon,^{11} = \varepsilon,^{1\prime} = \varepsilon,'' = 0. \tag{3.185}$$

A third perturbation of eqn (3.157) yields, upon evaluation at the tri-furcation point c,

$$\varepsilon,^{11\prime} = -\frac{X'_{111}}{X_{1\varepsilon}}, \qquad \varepsilon,^{1\prime\prime} = -\frac{X''_{11}}{X_{1\varepsilon}} \quad \text{and} \quad \varepsilon,''' = 0, \tag{3.186}$$

in addition to the derivative (3.180) which remains unaffected.

Using these derivatives results in the first-order equation (Yu 1984)

$$X_{1\varepsilon}\varepsilon + \frac{1}{3!}\bar{X}_{1111}(x^1)^3 + \tfrac{1}{2}X'_{111}(x^1)^2\mu + \tfrac{1}{2}X''_{11}x^1(\mu)^2 = 0. \tag{3.187}$$

Setting $\varepsilon = 0$ in eqn (3.187) leads to the corresponding perfect system described by eqns (3.92) or (3.94), which are equivalent.

Introducing the transformation

$$\bar{\mu} = \mu + \frac{X'_{111}}{2X'_{11}}x^1 \tag{3.188}$$

into eqn (3.187) reduces this relationship to a simpler form given by

$$X_{1\varepsilon}\varepsilon - \frac{D}{8X''_{11}}(x^1)^3 + \tfrac{1}{2}X''_{11}x^1(\bar{\mu})^2 = 0, \tag{3.189}$$

where

$$D = (X'_{111})^2 - \tfrac{4}{3}X''_{11}\bar{X}_{1111}. \tag{3.190}$$

A similar form is also obtained in the special case of $X'_{111} = 0$ which has no influence on the topological properties of the equilibrium surface.

The critical line follows from eqn (3.189) as

$$-\frac{3D}{8X''_{11}}(\overset{*}{x}{}^1)^2 + \tfrac{1}{2}X''_{11}(\bar{\mu})^2 = 0, \tag{3.191}$$

and by eliminating x^1 between eqns (3.189) and (3.191) the stability boundary is established as

$$X_{1\varepsilon}\varepsilon = \pm\frac{2}{3\sqrt{3}}\frac{(X''_{11})^2}{D^{1/2}}\overset{*}{\bar{\mu}}{}^3. \tag{3.192}$$

It is recalled that the post-critical paths (3.94) of the perfect system may not exist unless the existence condition

$$D > 0 \tag{3.193}$$

is satisfied, and this condition is, of course, also required for the stability boundary (3.192) to be meaningful.

The behaviour of the system is illustrated in Fig. 3.16. For constant values of ε, the imperfect system exhibits limit points and a loss of stability that occurs at considerably reduced values of the parameter η compared to the critical tri-furcation value η_c. The locus of the limit points consists of two straight lines as may be inferred from the equation of the critical line (3.191). In fact, this relationship may also be expressed as

$$\overset{*}{\bar{\mu}} = \pm\frac{\sqrt{3}}{2}\frac{D^{1/2}}{X''_{11}}\overset{*}{x}{}^1, \tag{3.194}$$

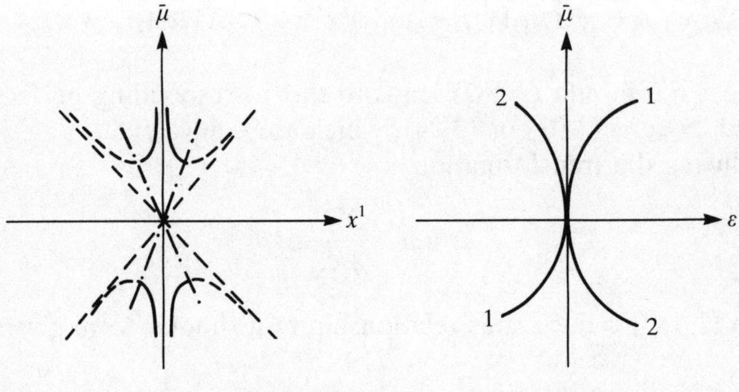

(a) (b)

Fig. 3.16 Behaviour of the imperfect system in the vicinity of a point of tri-furcation for ε = constant. In (a) the critical zone (eqn 3.194) is shown by dash-dotted lines. In (b) the stability boundary consists of two curves which share the critical point as a point of inflexion.

and the slopes of the critical lines are linked to the slopes of the post-critical paths (3.94) by

$$\overset{*}{\bar{\mu}}{}^{,1} = \sqrt{3}\,\bar{\mu}^{,1}. \tag{3.195}$$

This follows readily from a comparison of eqn (3.194) with eqn (3.94) after introducing the transformation (3.188) into the latter. The critical zone (3.194) is shown by dash-dotted lines in Fig. 3.16*a*.

The stability boundary (3.192) consists of two curves which share the critical point as a point of inflexion, as shown in Fig. 3.16*b*. This indicates that a point of tri-furcation is more imperfection-sensitive than symmetric and asymmetric points of bifurcations examined earlier.

Case 4. Imperfection sensitivity of a tangential bifurcation. The phenomenon of tangential bifurcation arises when

$$X_{11} = X'_{11} = 0 \tag{3.196}$$

at the critical point c, as has been demonstrated (Case 4) in Section 3.1.2.

The derivatives (3.161) to (3.167) remain valid for the imperfect system in this case as well. However, it follows from eqn (3.168) that $\varepsilon^{,1\prime} = 0$, due to condition (3.196), and one proceeds to a third perturbation to obtain

$$\varepsilon^{,1\prime\prime} = -\frac{X''_{11}}{X_{1\varepsilon}} \qquad \text{and} \qquad \varepsilon^{,\prime\prime\prime} = 0, \tag{3.197}$$

which lead to

$$X_{1\varepsilon}\varepsilon + \tfrac{1}{2}X_{111}(x^1)^2 + \tfrac{1}{2}X''_{11}x^1(\mu)^2 = 0. \tag{3.198}$$

This is the asymptotic equation of the equilibrium surface in the vicinity of the tangential point of bifurcation (Yu 1984).

Setting $\varepsilon = 0$ in eqn (3.198) yields the initial path $x^1 = 0$ and the post-critical path (eqn 3.101) of the perfect system, as expected.

The critical line is now described by

$$\overset{*}{x}{}^1 = -\frac{1}{2}\frac{X''_{11}}{X_{111}}(\overset{*}{\mu})^2, \tag{3.199}$$

which is in the form of a parabola, and the stability boundary is given by

$$X_{1\varepsilon}\varepsilon - \frac{1}{8}\frac{(X''_{11})^2}{X_{111}}(\overset{*}{\mu})^4 = 0. \tag{3.200}$$

This phenomenon is depicted in Fig. 3.17. For a given value of ε, an initially stable equilibrium path of the imperfect system reaches a maximum (a limit point) where it loses its stability. The locus of all these limit points is a parabola (3.199) in the state-parameter space, and takes

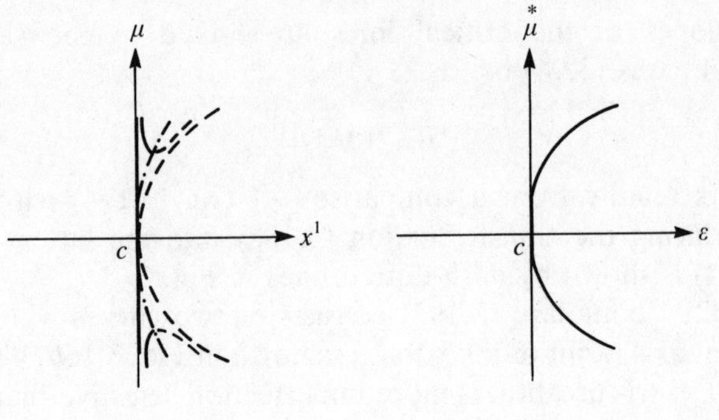

(a) (b)

Fig. 3.17 Behaviour of the imperfect system in the vicinity of a tangential bifurcation for $\varepsilon = \text{constant}$.

the form of eqn (3.200) when projected onto the parameter space (Fig. 3.17*b*). Clearly, the degree of imperfection sensitivity of a tangential point of bifurcation is more severe than in the cases discussed hitherto.

Problem 3.9. Consider the cusp bifurcation of Problem 3.2, and formulate the corresponding imperfect system following the procedure described in this section. Show that the first-order equation of the equilibrium surface under the conditions prescribed in Problem 3.2 is in the form

$$X_{1\varepsilon}\varepsilon + \frac{1}{3!}\bar{X}_{1111}(x^1)^3 + \frac{1}{3!}X'''_{11}x^1(\mu)^3 = 0. \tag{3.201}$$

The parameter-imperfection relationship is given by (Yu 1984)

$$243\bar{X}_{1111}(X_{1\varepsilon}\varepsilon)^2 + (X'''_{11})^3(\overset{*}{\mu})^9 = 0. \tag{3.202}$$

This phenomenon is depicted in Fig. 3.18. Discuss the behaviour

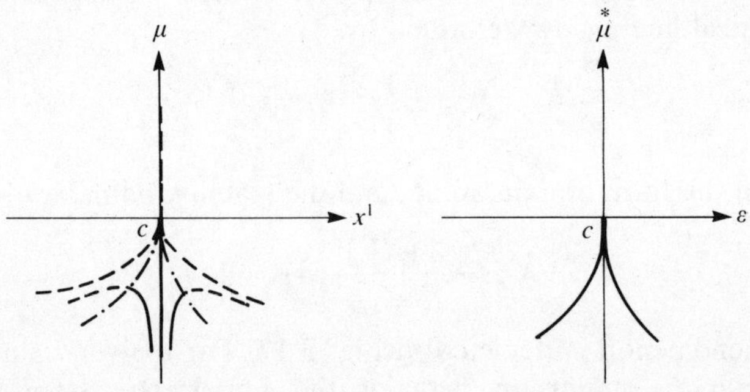

(a) (b)

Fig. 3.18 Behaviour of the imperfect system in the vicinity of a cusp bifurcation.

Table 3.1. Summary of results of Section 3.1

Type of bifurcation	Conditions	Stability boundary	
		Relationship	
Asymmetric point of bifurcation	$X_{11} = 0$	$\overset{*}{\mu} = A\varepsilon^{1/2}$	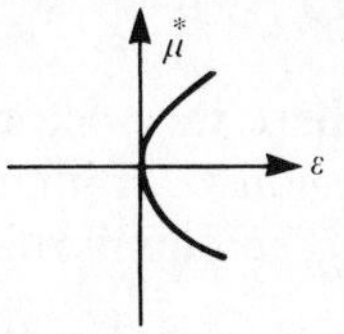
Symmetric point of bifurcation	$X_{11} = X_{111} = 0$	$\overset{*}{\mu} = B\varepsilon^{2/3}$	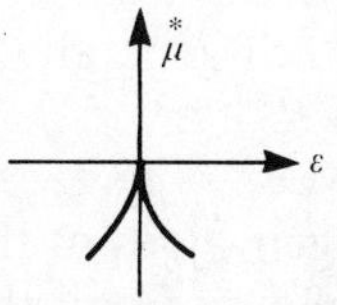
Tri-furcation	$X_{11} = X_{111} = X'_{11} = 0$	$\overset{*}{\mu} = C\varepsilon^{1/3}$	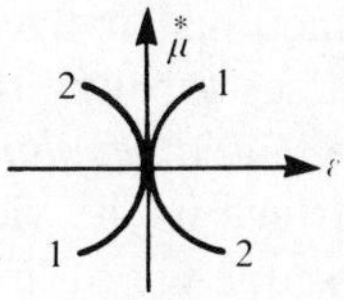
Tangential bifurcation	$X_{11} = X'_{11} = 0$	$\overset{*}{\mu} = D\varepsilon^{1/4}$	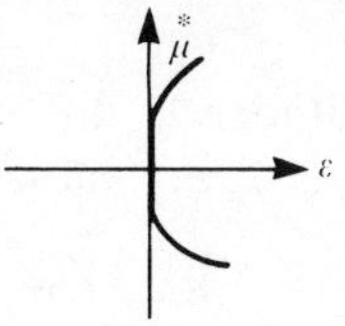
Cusp bifurcation	$X_{11} = X_{111} = 0$ $X'_{11} = X'_{111} = 0$ $X''_{11} = X''_{111} = 0$	$\overset{*}{\mu} = E\varepsilon^{2/9}$	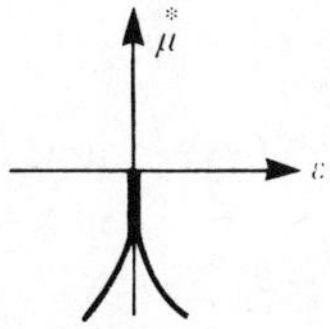

characteristics of the system, properties of the critical line and the degree of imperfection sensitivity.

The results of this section are summarized in Table 3.1. The relationships between the critical values of the parameter $\overset{*}{\mu}$ and the imperfection parameter ε are simply expressed as $\overset{*}{\mu} = (\text{constant})\varepsilon^p$, where p is the power corresponding to each case. The constants are denoted by A, B, C, D, and E.

3.2 Multiple-parameter systems

3.2.1 Classification of critical points

Consider an autonomous system described by

$$\frac{\mathrm{d}z}{\mathrm{d}t} = \mathbf{Z}(z, \boldsymbol{\eta}), \qquad z \in R^n, \qquad \boldsymbol{\eta} \in R^m \tag{3.203}$$

where the vector function $\mathbf{Z}(z, \boldsymbol{\eta})$ is assumed to be analytic—at least in a region G of interest.

The equilibrium states of system (3.203) satisfy

$$\mathbf{Z}_i(z^j, \eta^\alpha) = 0; \qquad (i, j = 1, 2, \ldots, n), \qquad (\alpha = 1, 2, \ldots, m), \tag{3.204}$$

which generally defines an m-dimensional *equilibrium surface* (manifold) in R^{n+m}.

Let c be a stable equilibrium point on this surface at which the eigenvalues of the corresponding Jacobian all have negative real parts. It follows from the fundamental existence theorem concerning implicit functions that a unique and single-valued equilibrium surface, $z = z(\boldsymbol{\eta})$, passes through this point so that $z_c = z(\boldsymbol{\eta}_c)$. Such a non-critical point will be called *regular* as before. The asymptotic equations of the equilibrium surface in the vicinity of a *regular* equilibrium point may be obtained via a sequence of linear perturbation equations generated from the identities

$$Z_i[z^j(\eta^\alpha), \eta^\alpha] \equiv 0. \tag{3.205}$$

Indeed, differentiating these functions with respect to η^α ($\alpha = 1, 2, \ldots, m$) and evaluating at c yields

$$\begin{aligned} &Z_{ij}z^{j,\alpha} + Z_{i\alpha} = 0, \\ &(Z_{ijk}z^{k,\beta} + Z_{ij\beta})z^{j,\alpha} + Z_{ij}z^{j,\alpha\beta} + Z_{ij\alpha}z^{j,\beta} + Z_{i\alpha\beta} = 0, \\ &\text{etc.} \end{aligned} \tag{3.206}$$

These equations may be solved successively for the surface derivatives $z^{j,\alpha}$, $z^{j,\alpha\beta}$, etc., which are then used to construct the surface through c as

$$z^j = z^{j,\alpha}\mu^\alpha + \frac{1}{2!} z^{j,\alpha\beta}\mu^\alpha\mu^\beta + \ldots \tag{3.207}$$

where $\eta^\alpha = \eta^\alpha_c + \mu^\alpha$.

As the parameters η^α are varied, static instabilities may occur at critical combinations of these parameters when the corresponding Jacobian matrices become *singular*. Definitions given in Section 3.1 concerning the types of critical points are valid here as well. Thus, a critical point is said to be *simple* if the nullity of the Jacobian matrix at that point is one. Similarly, a critical point will be called *r-fold compound* if the *nullity* of the Jacobian at that point is r.

In order to give a further characterization of the critical points associated with multiple parameters, introduce the transformations

$$z = z_c + Qy \quad \text{and} \quad \boldsymbol{\eta} = \boldsymbol{\eta}_c + P\boldsymbol{\varphi} \tag{3.208}$$

where c is a critical point corresponding to a critical combination of parameters, $\eta^{\alpha} = \eta_c^{\alpha}$, $\boldsymbol{Q}$ is the same transformation as in eqns (3.19), and the non-singular matrix $\boldsymbol{P}$ will be specified during the analysis according to the nature of the critical point under consideration such that the *significant* parameters may be identified and distinguished readily, resulting in a more revealing form for the equilibrium surface.

Introducing the transformation (3.208) into eqn (3.203) results in the system

$$\frac{\mathrm{d}\boldsymbol{y}}{\mathrm{d}t} = \boldsymbol{Y}(\boldsymbol{y}, \boldsymbol{\varphi}) \tag{3.209}$$

where

$$\left[\frac{\partial \boldsymbol{Y}}{\partial \boldsymbol{y}}\right]_c = \mathrm{diag}\,[J, \boldsymbol{K}_q, \boldsymbol{K}_{q+2}, \ldots], \tag{3.210}$$

$\boldsymbol{J}$ is the Jordan canonical form corresponding to the zero eigenvalues, and $\boldsymbol{K}_q$ are 2×2 blocks associated with complex conjugate eigenvalues with negative real parts. It is assumed that $\boldsymbol{J}$ has nullity r. Consider now the Taylor's expansion of eqn (3.209) around c,

$$\frac{\mathrm{d}y_i}{\mathrm{d}t} = Y_{ij}y^j + Y_{i\alpha}\varphi^{\alpha} + \ldots, \tag{3.211}$$

and assume that a sub-matrix $\boldsymbol{\Omega}$ of order $(r \times m)$ is formed from the rows of $Y_{i\alpha}$ directly corresponding to zero rows of $\boldsymbol{J}$. The following definitions were given by Mandadi and Huseyin (1977):

i) A critical point c is said to be *general* if the rank of the parameter matrix $\boldsymbol{\Omega}$ is equal to the nullity of the Jacobian matrix;

ii) A critical point c is said to be *special* if the rank of the parameter matrix $\boldsymbol{\Omega}$ is less than the nullity of the Jacobian matrix.

This broad classification of critical points is in complete analogy with that of potential systems, and *general* points may be associated with *generic* situations while *special* points arise in idealized mathematical models.

3.2.2 *Simple general points of order 2*

Case 1. A real eigenvalue vanishes. In this case the Jordan canonical form $\boldsymbol{J}$ in eqn (3.210) is given by

$$\boldsymbol{J} = \begin{bmatrix} 0 & 0 \\ 0 & \alpha_2 \end{bmatrix}, \tag{3.212}$$

and

$$\boldsymbol{K}_q = \begin{bmatrix} \alpha_q & -\omega_q \\ \omega_q & \alpha_q \end{bmatrix}, \qquad (\text{for } q = 3, 5, \ldots),$$

where $\alpha_t < 0$ $(t = 2, 3, 5, \ldots)$.

It is understood that the inclusion of only one non-zero real eigenvalue, α_2, in eqn (3.212) is for convenience and does not imply a loss of generality.

The columns of the matrix $\boldsymbol{P}$ are now formed from a set of orthogonal vectors which may be chosen such that (Huseyin and Mandadi 1980)

$$[Y_{1\alpha}] \triangleq \left[\frac{\partial Y_1}{\partial \varphi^\alpha}\right]_c = [-1 \quad 0 \quad 0 \quad \ldots \quad 0]. \tag{3.213}$$

Suppose the equilibrium equations $Y_i(y^j, \varphi^\alpha) = 0$ have an analytic solution in the neighbourhood of c that may be given the parametric representation

$$\boldsymbol{y} = \boldsymbol{y}(\boldsymbol{\sigma}), \qquad \boldsymbol{\varphi} = \boldsymbol{\varphi}(\boldsymbol{\sigma}), \tag{3.214}$$

where $\boldsymbol{\sigma}$ is an m-component real vector, and $\boldsymbol{\sigma} = \boldsymbol{0}$ gives the critical point c.

On the basis of the transformation (3.213) and previous experience, σ may be chosen as follows:

$$\sigma^1 = y^1, \qquad \sigma^\nu = \varphi^\nu, \qquad (\nu = 2, 3, \ldots, m). \tag{3.215}$$

The equilibrium surface (3.214) is then expressed as

$$y^t = y^t(y^1, \varphi^\nu), \qquad \varphi^1 = \varphi^1(y^1, \varphi^\nu), \tag{3.216}$$

where $t = 2, 3, \ldots, n$.

Substituting this assumed solution back into the equilibrium equations yields the identities

$$Y_i[y^j(y^1, \varphi^\nu), \varphi^\alpha(y^1, \varphi^\nu)] \equiv 0 \tag{3.217}$$

where $y^j(y^1, \varphi^\nu)$ and $\varphi^\alpha(y^1, \varphi^\nu)$ reduce to y^1 and φ^ν for $j = 1$ and $\alpha = \nu = 2, 3, \ldots, m$, respectively.

Successive differentiations of eqn (3.217) with respect to y^1 and φ^ν result in

$$Y_{ij} y^{j,1} + Y_{i\alpha} \varphi^{\alpha,1} = 0, \tag{3.218}$$

$$Y_{ij} y^{j,\nu} + Y_{i\alpha} \varphi^{\alpha,\nu} = 0, \tag{3.219}$$

$$(Y_{ijk} y^{k,1} + Y_{ij\alpha} \varphi^{\alpha,1}) y^{j,1} + Y_{ij} y^{j,11} + (Y_{ij\alpha} y^{j,1} + Y_{i\alpha\beta} \varphi^{\beta,1}) \varphi^{\alpha,1} + Y_{i\alpha} \varphi^{\alpha,11} = 0, \tag{3.220}$$

$$(Y_{ijk} y^{k,\nu} + Y_{ij\alpha} \varphi^{\alpha,\nu}) y^{j,1} + Y_{ij} y^{j,1\nu} + (Y_{ij\alpha} y^{j,\nu} + Y_{i\alpha\beta} \varphi^{\beta,\nu}) \varphi^{\alpha,1} + Y_{i\alpha} \varphi^{\alpha,1\nu} = 0, \tag{3.221}$$

and

$$(Y_{ijk}y^{k,\xi} + Y_{ij\alpha}\varphi^{\alpha,\xi})y^{j,\nu} + Y_{ij}y^{j,\nu\xi} + (Y_{ij\alpha}y^{j,\xi} + Y_{i\alpha\beta}\varphi^{\beta,\xi})\varphi^{\alpha,\nu} + Y_{i\alpha}\varphi^{\alpha,\nu\xi} = 0, \quad (3.222)$$

where the sets of indices $(\alpha, \beta, \gamma, \ldots)$, $(\nu, \xi, \ldots)$ range from 1 to m, and 2 to m, respectively.

Evaluating eqns (3.218) and (3.219) at c yields

$$\begin{aligned} &\varphi^{1,1} = 0, \qquad \varphi^{1,\nu} = 0, \qquad y^{t,1} = 0, \\ &y^{2,\nu} = -Y_{2\nu}/\alpha_2 \triangleq A_{2\nu}, \\ &y^{q,\nu} = -(\alpha_q Y_{q\nu} + \omega_q Y_{(q+1)\nu})/(\alpha_q^2 + \omega_q^2) \triangleq A_{q\nu}, \\ &y^{(q+1),\nu} = -(\alpha_q Y_{(q+1)\nu} - \omega_q Y_{q\nu})/(\alpha_q^2 + \omega_q^2) \triangleq A_{(q+1)\nu}, \end{aligned} \quad (3.223)$$

where $(t = 2, 3, \ldots)$ and $(q = 3, 5, \ldots)$.

Evaluating the second perturbation equations (3.220), (3.221), and (3.222) at c results in

$$\varphi^{1,11} = Y_{111}, \qquad \varphi^{1,1\nu} = Y_{11\nu} + Y_{11t}A_{t\nu} \triangleq \psi_\nu, \quad (3.224)$$

$$\varphi^{1,\nu\xi} = Y_{1\nu\xi} + Y_{1t\nu}A_{t\xi} + Y_{1t\xi}A_{t\nu} + Y_{1tp}A_{t\nu}A_{p\xi} \triangleq \psi_{\nu\xi}, \quad (3.225)$$

$$y^{2,11} = \frac{1}{\alpha_2}(Y_{111}Y_2^1 - Y_{211}) \triangleq B_2, \quad (3.226)$$

$$y^{q,11} = (\alpha_q k_q + \omega_q k_{q+1})/(\alpha_q^2 + \omega_q^2) \triangleq B_q, \quad (3.227)$$

$$y^{(q+1),11} = (\alpha_q k_{q+1} - \omega_q k_q)/(\alpha_q^2 + \omega_q^2) \triangleq B_{q+1}, \quad (3.228)$$

where

$$k_q = -Y_{q11} + Y_{111}Y_q^1, \qquad (Y_q^1 = -\partial Y_q/\partial\varphi^1|_c), \quad (3.229)$$

and $(t, p = 2, 3, \ldots)$.

The first-order equations of the equilibrium surface can now be constructed as

$$\varphi^1 = \tfrac{1}{2}Y_{111}(y^1)^2 + \psi_\nu y^1\varphi^\nu + \tfrac{1}{2}\psi_{\nu\xi}\varphi^\nu\varphi^\xi, \quad (3.230)$$

and

$$y^t = A_{t\nu}\varphi^\nu + \tfrac{1}{2}B_t(y^1)^2. \quad (3.231)$$

Topologically, this equilibrium surface is identical to that described by eqns (2.66) and (2.67). In other words, the critical point under consideration is a *general point of order 2*, and the phenomenon is recognized as the *fold catastrophe* (Fig. 2.1). Obviously, the discussion presented in relation to eqn (2.66) is valid for eqn (3.230) as well, and the complete analogy between potential and non-potential systems is observed. Thus, the surface (3.230) may be described as *synclastic*,

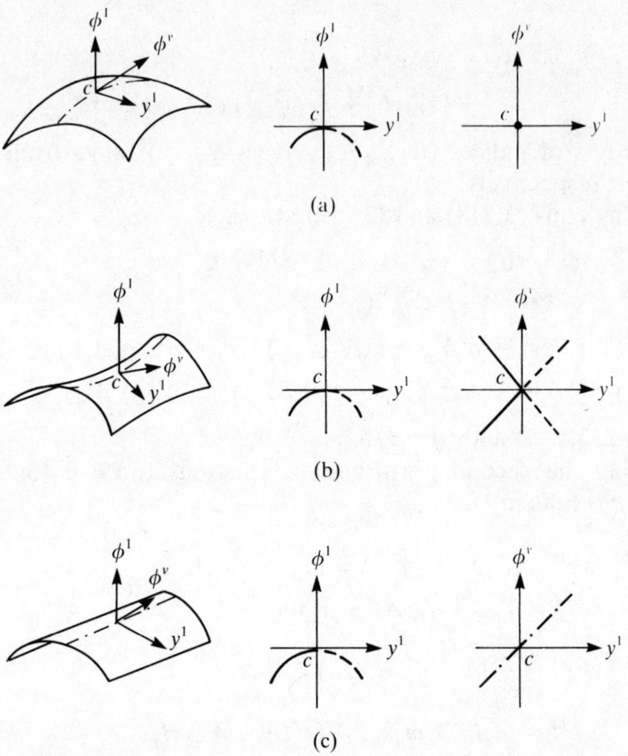

Fig. 3.19 (a) Synclastic surface, (b) anticlastic surface and (c) parabolic surface.

anticlastic, or *parabolic* according to whether the matrix

$$[Y_{111}\psi_{\nu\xi} - \psi_\nu \psi_\xi] \tag{3.232}$$

is positive definite, negative definite, or null, respectively (Fig. 3.19).

The matrix of the second fundamental tensor associated with the surface (3.230) is given by

$$\begin{bmatrix} Y_{111} & \psi_\nu \\ \psi_\nu & \psi_{\nu\xi} \end{bmatrix} \tag{3.233}$$

which yields the principal curvatures and principal directions. It can be shown, on the basis of the principal curvatures, that a synclastic surface, for example, is located entirely on one side of a plane tangent to the surface at c, and the intersections of the surface with planes $\varphi^1 =$ constant form a family of closed curves, shrinking to zero with φ^1 (Fig. 3.19*a*). This phenomenon was first observed in potential systems (Huseyin 1970*b*, 1975). In other words, if the surface is synclastic, a parameter ray in the φ^ν subspace yields an equilibrium state located at the origin which is *isolated* (Fig. 3.19*a*).

Stability distribution and stability boundary. It was assumed that a real eigenvalue of the Jacobian goes through zero at c while all the other eigenvalues remain in the left half-plane in the vicinity of c. The expansion of the Jacobian around the critical point c,

$$Y_{ij}(y^k, \varphi^\alpha) = Y_{ij} + Y_{ijk}y^k + Y_{ij\alpha}\varphi^\alpha + \ldots, \tag{3.234}$$

may be used to examine the stability distribution on the equilibrium surface as in the case of one-parameter systems. To this end, consider the equilibrium paths (sub manifolds) corresponding to arbitrary control rays in parameter space described by

$$\varphi^\alpha = \varphi^{\alpha,\rho}\rho \qquad (\varphi^{1,\rho} \neq 0), \tag{3.235}$$

where the $\varphi^{\alpha,\rho}$ are constants. Introduction of eqn (3.235) into eqn (3.230) yields the path

$$\varphi^{1,\rho}\rho = \tfrac{1}{2}Y_{111}(y^1)^2. \tag{3.236}$$

Let the variation of α_1 along this path be expressed as

$$\alpha_1(y^1) = \alpha_{1,1}y^1 + \tfrac{1}{2}\alpha_{1,11}(y^1)^2 + \ldots \tag{3.237}$$

where $\alpha_1(0) = 0$. Using eqns (3.230), (3.231), and (3.236) in the expansion (3.234) yields

$$Y_{ij}(y^1) = Y_{ij} + Y_{ij1}y^1 + 0\{(y^1)^2\} + \ldots \tag{3.238}$$

Since α_1 is an eigenvalue of the Jacobian, it satisfies

$$|Y_{ij}(y^1) - \alpha_1(y^1)I| = 0 \tag{3.239}$$

identically, and differentiating eqn (3.239) with respect to y^1 by columns and evaluating at c results in

$$\begin{vmatrix} Y_{111} - \alpha_{1,1} & 0 & 0 & 0 & \cdot \\ Y_{211} & \alpha_2 & 0 & 0 & \cdot \\ Y_{311} & 0 & \alpha_3 & -\omega_3 & \cdot \\ Y_{411} & 0 & \omega_3 & \alpha_3 & \cdot \\ \cdot & \cdot & \cdot & \cdot & \cdot \end{vmatrix} = 0 \tag{3.240}$$

which yields $\alpha_{1,1} = Y_{111}$.

It follows from eqn (3.232) that the eigenvalue

$$\alpha_1 = Y_{111}y^1, \qquad (Y_{111} \neq 0),$$

is negative on one side of c and positive on the other side; i.e. *if an equilibrium path is stable on one side of a limit point it has to be unstable on the other side*.

If the surface is synclastic, all parameter rays (3.235) lead to limit points which possess the above stability property. On the other hand, if the surface is anticlastic, the critical point c appears as an asymmetric point of bifurcation if the rays are restricted to the sub-space spanned by φ^ν (i.e., rays with $\varphi^{1,\rho}=0$). In such cases an exchange of stabilities occurs between the paths (Fig. 3.19b), as demonstrated earlier in the case of one-parameter systems. Moving on the entire surface in the vicinity of c, the eigenvalue α_1 varies as a function of m independent parameters and for certain critical combinations of these parameters it remains zero, tracing the *critical zone*. It is inferred from the above discussion that the critical zone and the stability boundary may be obtained by differentiating eqn (3.230) with respect to y^1 and introducing the resulting relation into eqns (3.230) and (3.231). Thus,

$$Y_{111}\overset{*}{y}{}^1+\psi_\nu\overset{*}{\varphi}{}^\nu=0, \qquad \overset{*}{y}{}^t=A_{t\nu}\overset{*}{\varphi}{}^\nu, \tag{3.241}$$

$$\overset{*}{\varphi}{}^1=\frac{1}{2Y_{111}}(Y_{111}\psi_{\nu\xi}-\psi_\nu\psi_\xi)\overset{*}{\varphi}{}^\nu\overset{*}{\varphi}{}^\xi. \tag{3.242}$$

The latter equation defines the stability boundary, and the convexity properties discussed in Section 2.4.1 are valid here as well (also see Huseyin 1975, pp. 132–141), although the system under consideration is non-potential.

Case 2. A Jordan block of order 2. Consider again system (3.209) with the Jacobian (3.210), and let the Jordan canonical form be in the form of

$$\boldsymbol{J}=\begin{bmatrix}0 & 1\\ 0 & 0\end{bmatrix}, \tag{3.243}$$

while the blocks $\boldsymbol{K}_q$ ($q=3,5,\ldots$) remain unchanged.

In this case, the matrix $\boldsymbol{P}$ may be chosen such that

$$[Y_{2\alpha}]=\left[\frac{\partial Y_2}{\partial\varphi^\alpha}\right]_c=[-1 \quad 0 \quad 0 \quad \ldots \quad 0]. \tag{3.244}$$

Assuming that the equilibrium surface in the vicinity of the critical point c—which is now characterized by eqns (3.243) and (3.244)—may again be expressed by eqns (3.216), the identities (3.217) yield the perturbation equations (3.218) to (3.222). Evaluating the first perturbation equations at c results in

$$\begin{gathered}\varphi^{1,1}=0, \qquad \varphi^{1,\nu}=0, \qquad y^{t,1}=0,\\ y^{2,\nu}=-Y_{1\nu}\triangle A_{2\nu}, \qquad y^{q,\nu}=A_{q\nu}, \qquad y^{(q+1),\nu}=A_{(q+1)\nu}.\end{gathered} \tag{3.245}$$

The second perturbation equations yield

$$\begin{aligned}
\varphi^{1,11} &= Y_{211}, \qquad \varphi^{1,1v} = (Y_{21v} + Y_{21t}A_{tv}) \triangleq \chi_v, \\
\varphi^{1,v\xi} &= Y_{2v\xi} + Y_{2tv}A_{t\xi} + Y_{2t\xi}A_{tv} + Y_{2tp}A_{tv}A_{p\xi} \triangleq \chi_{v\xi}, \\
y^{2,11} &= -Y_{111} + Y_1^1 Y_{211} \triangleq C_2, \\
y^{q,11} &= (\alpha_q d_q + \omega_q d_{q+1})/(\alpha_q^2 + \omega_q^2) \triangleq C_q, \\
y^{(q+1)11} &= (\alpha_q d_{q+1} - \omega_q d_q)/(\alpha_q^2 + \omega_q^2) \triangleq C_{q+1}.
\end{aligned} \tag{3.246}$$

where

$$d_q = -Y_{q11} + Y_{211}Y_q^1, \qquad (Y_q^1 \triangleq -\partial Y_q/\partial \varphi^1|_c).$$

The equilibrium surface takes the form

$$\varphi^1 = \tfrac{1}{2}Y_{211}(y^1)^2 + \chi_v y^1 \varphi^v + \tfrac{1}{2}\chi_{v\xi}\varphi^v\varphi^\xi$$

and (3.247)

$$y^t = A_{tv}\varphi^v + \tfrac{1}{2}C_t(y^1)^2,$$

which is topologically identical to the surface described by eqns (3.230) and (3.231). The second fundamental tensor is now represented by the matrix

$$\begin{bmatrix} Y_{211} & \chi_v \\ \chi_v & \chi_{v\xi} \end{bmatrix}, \tag{3.248}$$

and the surface is called synclastic, anticlastic, or parabolic according to whether the matrix

$$[Y_{211}\chi_{v\xi} - \chi_v\chi_\xi] \tag{3.249}$$

is positive definite, negative definite, or null, respectively.

An important difference between the behaviour of the system in the vicinity of a critical point characterized by eqn (3.243) and that of Case 1 associated with the critical point (3.212) has to be recognized. Indeed, it can be demonstrated that *a loss of stability in the vicinity of the critical point* (3.243) *is not always associated with static instability* (*divergence*). On the contrary, the system under consideration here may exhibit *dynamic instability* under some circumstances. This situation arises due to the presence of more than one independent parameter, widening the possibilities concerning the manner in which the vanishing eigenvalues of the Jacobian may vary in the vicinity of c.

In order to demonstrate this interesting phenomenon, consider a general parameter path of the form

$$\varphi^\alpha = \varphi^{\alpha,\rho}\rho + \tfrac{1}{2}\varphi^{\alpha,\rho\rho}\rho^2 + \dots. \tag{3.250}$$

The corresponding first-order equations of the equilibrium path (sub-manifold) on the equilibrium surface (3.247) are then given by

$$\begin{aligned} \varphi^{1,\rho}\rho &= \tfrac{1}{2}Y_{211}(y')^2, \\ y^t &= A_{tv}\varphi^{v,\rho}\rho + \tfrac{1}{2}C_t(y^1)^2. \end{aligned} \tag{3.251}$$

Evaluating the expansion of the Jacobian (eqn 3.234) first on the surface (3.247) and then on the specific path (3.251) yields

$$Y_{ij}(y^1, \varphi^v) = Y_{ij} + Y_{ij1}y^1 + Y_{ijt}A_{tv}\varphi^v + Y_{ijv}\varphi^v + \ldots \tag{3.252}$$

and

$$Y_{ij}(y^1) = Y_{ij} + Y_{ij1}y^1 + 0\{(y^1)^2\} + \ldots. \tag{3.253}$$

The latter expression is in the same form as that associated with one-parameter systems (eqn 3.54), discussed at the end of Section 3.1.1. Following precisely the same pattern of arguments leads to the conclusion that a stable path (3.251) can lose stability only by divergence upon passing through the critical point (eqn 3.243).

However, this is not so for all sub-manifolds on the surface. Consider, for example, a special equilibrium path on the surface by choosing the independent variables φ^v and y^1 appropriately. To this end, let

$$\varphi^v = l^v\rho, \qquad (v = 2, 3, \ldots, n),$$

and

$$y^1 = l_1\rho^2 \tag{3.254}$$

where l_1 and l^v are certain constants.

The first-order equations of the corresponding one-dimensional sub-manifold (equilibrium path) follow from eqn (3.247) as

$$\begin{aligned} \varphi^1 &= \tfrac{1}{2}\chi_{v\xi}l^v l^\xi\rho^2, \\ y^t &= A_{tv}l^v\rho. \end{aligned} \tag{3.255}$$

The Jacobian (3.229) evaluated on this path takes the form

$$Y_{ij}(\rho) = Y_{ij} + Y_{ij1}l_1\rho^2 + Y_{ijt}A_{tv}l^v\rho + \tfrac{1}{2}Y^1_{ij}\chi_{v\xi}l^v l^\xi\rho^2 + Y_{ijv}l^v\rho + \ldots \tag{3.256}$$

where $Y^1_{ij} = \partial Y_{ij}/\partial\varphi^1|_c$.

For simplicity, let the system be two-dimensional; i.e. let $i, j = 1, 2$ (and $t = 2$). Assume further that l^v are chosen such that (see Huseyin 1975)

$$(Y_{212}A_{2v} + Y_{21v})l^v = 0. \tag{3.257}$$

The elements of the Jacobian (3.256) are then expressed as

$$\begin{aligned}
Y_{11}(\rho) &= (Y_{112}A_{2\nu}l^{\nu} + Y_{11\nu}l^{\nu})\rho + 0(\rho^2) \triangleq d_{11}\rho + 0(\rho^2),\\
Y_{12}(\rho) &= 1 + (Y_{122}A_{2\nu}l^{\nu} + Y_{12\nu}l^{\nu})\rho + 0(\rho^2) \triangleq 1 + d_{12}\rho + 0(\rho^2),\\
Y_{21}(\rho) &= (Y_{211}l_1 + \tfrac{1}{2}Y^1_{21}\chi_{\nu\xi}l^{\nu}l^{\xi})\rho^2 \triangleq d_{21}\rho^2,\\
Y_{22}(\rho) &= (Y_{222}A_{2\nu}l^{\nu} + Y_{22\nu}l^{\nu})\rho + 0(\rho^2) \triangleq d_{22}\rho + 0(\rho^2).
\end{aligned} \tag{3.258}$$

The eigenvalues of the Jacobian $Y_{ij}(\rho)$ in the vicinity of the critical point c follow from

$$\begin{vmatrix} d_{11}\rho - \lambda & 1 \\ d_{21}\rho^2 & d_{22}\rho - \lambda \end{vmatrix} = 0, \tag{3.259}$$

which yields

$$\lambda^2 - (d_{11} + d_{22})\rho\lambda + (d_{11}d_{22} - d_{21})\rho^2 = 0, \tag{3.260}$$

resulting in

$$\lambda_{1,2} = \tfrac{1}{2}(d_{11} + d_{22})\rho \pm \tfrac{1}{2}\rho\surd\{(d_{11} - d_{22})^2 + 4d_{21}\}. \tag{3.261}$$

It follows from eqn (3.261) that if

$$(d_{11} + d_{22}) > 0 \quad \text{and} \quad \{(d_{11} - d_{22})^2 + 4d_{21}\} < 0, \tag{3.262}$$

the equilibrium path (3.255) is stable for $\rho < 0$ and flutter-unstable for $\rho > 0$ in the vicinity of c, while $\rho = 0$ gives the general critical point characterized by eqn (3.243). This observation has significant implications; indeed, it demonstrates that while a *general* critical point described by a Jordan block of order 2 appears as a divergence point—at which static instabilities occur—on most equilibrium paths, it may also seem to be the threshold of flutter instability on certain special paths. Recalling that a general point, by definition, may embrace bifurcation and limit points, this new phenomenon adds another dimension to an already complex situation. Indeed, a family of limit cycles may branch off from the equilibrium surface (3.247) in the vicinity of such a critical point, and it is likely that critical points with these complex properties occur at the intersections of flutter and divergence boundaries of the system (Huseyin and Plaut 1973). Flutter instability and limit cycles will be explored in the following chapter, but one can already deduce from topological considerations that the entire critical divergence zone on the equilibrium surface (3.247) may no longer represent the stability boundary of the system, since it may partially lie in an already unstable region, due to the presence of a flutter zone on the surface. The critical divergence zone is again obtained as in Case 1 by differentiating eqn (3.247) with respect to y^1. This procedure yields

$$\begin{aligned}
&Y_{211}\overset{*}{y}{}^1 + \chi_\nu\overset{*}{\varphi}{}^\nu = 0, \qquad \overset{*}{y}{}^t = A_{t\nu}\overset{*}{\varphi}{}^\nu,\\
&\overset{*}{\varphi}{}^1 = \frac{1}{2Y_{211}}(Y_{211}\chi_{\nu\xi} - \chi_\nu\chi_\xi)\overset{*}{\varphi}{}^\nu\overset{*}{\varphi}{}^\xi.
\end{aligned} \tag{3.263}$$

Also note that applying the Routh–Hurwitz criterion to eqn (3.260) results in the conditions

$$(d_{11}+d_{22})\rho<0 \quad \text{and} \quad (d_{11}d_{22}-d_{21})>0,$$

which are necessary and sufficient for stability, and may be used in the vicinity of c to examine the stability properties of neighbouring states.

The interesting features of the system discussed above will further be amplified on a simpler model analysed by Takens (1974) and Guckenheimer and Holmes (1983).

Consider a two-dimensional, two-parameter system described by

$$\begin{bmatrix}\dot{y}^1\\ \dot{y}^2\end{bmatrix}=\begin{bmatrix}0 & 1\\ 0 & 0\end{bmatrix}\begin{bmatrix}y^1\\ y^2\end{bmatrix}+\begin{bmatrix}0\\ \varphi^1+y^2\varphi^2+(y^1)^2+y^1y^2\end{bmatrix}. \tag{3.264}$$

The equations of the equilibrium surface are immediately obtained as

$$y^2=0, \qquad \varphi^1=-(y^1)^2, \tag{3.265}$$

which indicate a *parabolic* surface (Fig. 3.20) associated with a general critical point (fold catastrophe). The critical zone on this surface consists of the straight line defined by $\overset{*}{y}{}^1=0$, $\overset{*}{\varphi}{}^1=0$; i.e. the φ^2 axis itself is the critical divergence line (Fig. 3.20). It is not difficult to demonstrate that only a part of this line ($\varphi^2<0$) is associated with an initial loss of stability, while the rest of the line ($\varphi^2>0$) lies in an unstable region.

To this end, evaluating the Jacobian of the system on the equilibrium surface (3.265) yields

$$[Y_{ij}(y^1,\varphi^2)]=\begin{bmatrix}0 & 1\\ 2y^1 & y^1+\varphi^2\end{bmatrix}, \tag{3.266}$$

which has the eigenvalues

$$\lambda_{1,2}=\tfrac{1}{2}(y^1+\varphi^2)\pm\tfrac{1}{2}\sqrt{\{(y^1+\varphi^2)^2+8y^1\}}. \tag{3.267}$$

The necessary and sufficient conditions for stability can readily be established as

$$y^1<0 \quad \text{and} \quad y^1+\varphi^2<0, \tag{3.268}$$

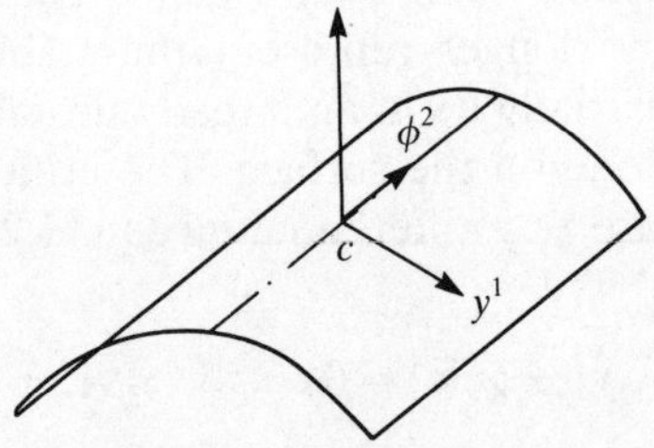

Fig. 3.20 A parabolic surface.

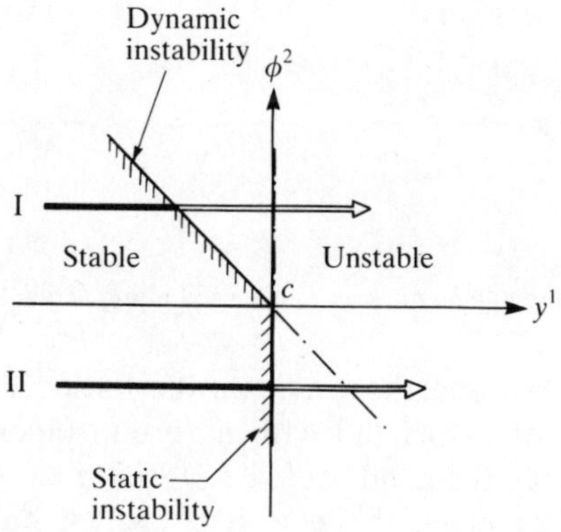

Fig. 3.21 Divergence and flutter lines in the (y^1-ϕ^2) plane.

which indicate that $y^1 = 0$ is indeed the critical (divergence) line and $y^1 + \varphi^2 = 0$ consists of flutter points (for $\varphi^2 > 0$). At the intersection of these two straight lines is a general critical point c associated with a Jordan block of order 2.

Figure 3.21 shows both divergence and flutter lines in the (y^1-φ^2) plane. It is observed that the divergence line as a whole is not associated with an initial loss of stability and, indeed, for $\varphi^2 > 0$ it lies in an unstable region. On the other hand, the portion of the line $y^1 + \varphi^2 = 0$ lying in the region $y^1 > 0$ does not represent flutter points at all. The portions of both lines associated with the stability boundary of the system are traced in Fig. 3.21.

It is instructive to observe various instability mechanisms and the movement of the eigenvalues of the Jacobian along a prescribed line in (y^1-φ^2) space. Consider first the straight line parallel to the y^1 axis and defined by $\varphi^2 = \text{constant} > 0$; the progress is described by y^1, and indicated by arrow I in Fig. 3.21. The positions of the eigenvalues, as y^1 is varied along arrow I, are shown in Fig. 3.22. Similarly, arrow II is along a straight line parallel to y^1 but defined by $\varphi^2 = \text{constant} < 0$, and Fig. 3.23 shows the positions of the eigenvalues along arrow II. It is noticed that a real eigenvalue crosses the imaginary axis as arrows cross the φ^2 axis (i.e. the divergence line) while a pair of complex conjugate

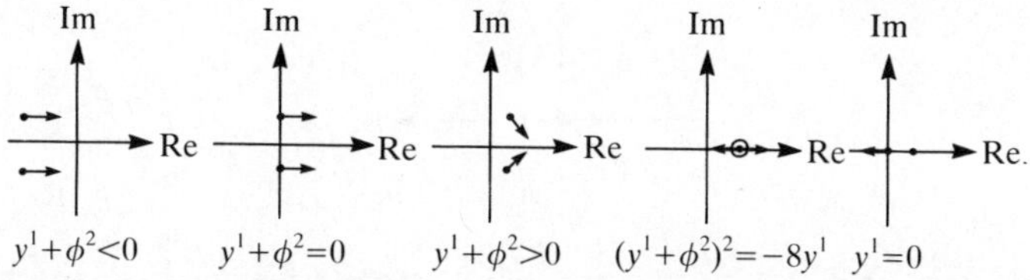

Fig. 3.22 Position of the eigenvalues as y^1 is varied along arrow I in Fig. 3.21 ($\phi^2 > 0$).

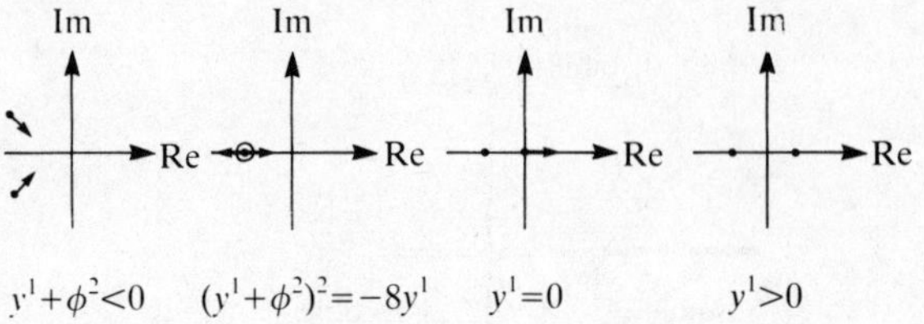

Fig. 3.23 Positions of the eigenvalues as y^1 is varied along arrow II in Fig. 3.21 ($\phi^2 < 0$).

eigenvalues crosses the imaginary axis on the flutter line. In other words, the only critical point associated with a Jordan block of order 2 is the general point c, occupying an *isolated position* at the intersection of divergence and flutter lines. If the y^1 axis itself is considered as a third arrow, instability at c will obviously be by divergence (Fig. 3.24). However, dynamic instability is also exhibited, as predicted earlier, if a special path is followed. To this end, suppose y^1 varies with φ^2 according to the relation $y^1 = -(\varphi^2)^2$ as shown in Fig. 3.25*a*; then the eigenvalues (eqn 3.267) are expressed by

$$\lambda_{1,2} = \tfrac{1}{2}\varphi^2 \pm \tfrac{1}{2}\varphi^2\sqrt{(-7)} \tag{3.269}$$

in the vicinity of the critical point c.

It is observed that here φ^2 takes on the role of the parameter ρ of eqn (3.261), and a stable system ($\varphi^2 < 0$) loses its stability by flutter ($\varphi^2 > 0$) as φ^2 passes through zero (Fig. 3.25*b*). This result also follows from the stability conditions (3.268) which reduce to $\varphi^2 < 0$ in the vicinity of the critical point c.

Finally, one may determine the stability boundary of the system in the parameter space (consisting of the flutter and divergence boundaries) by substituting for y^1 from eqns (3.265) into the flutter line ($y^1 + \varphi^2 = 0$) and divergence line ($y^1 = 0$), and observing the regions where an initial loss of stability takes place. Thus, the flutter boundary is given by

$$\overset{*}{\varphi}{}^1 = -(\overset{*}{\varphi}{}^2)^2, \qquad \varphi^2 > 0, \tag{3.270}$$

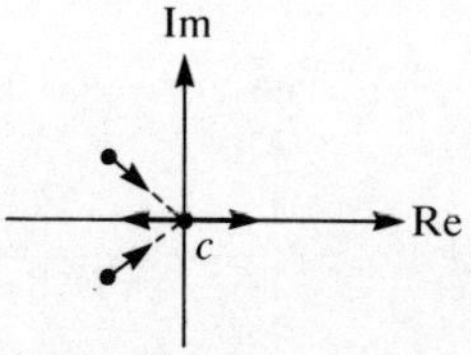

Fig. 3.24 Position of the eigenvalues if y^1 axis is considered as third arrow in Fig. 3.21. Instability at c is by divergence.

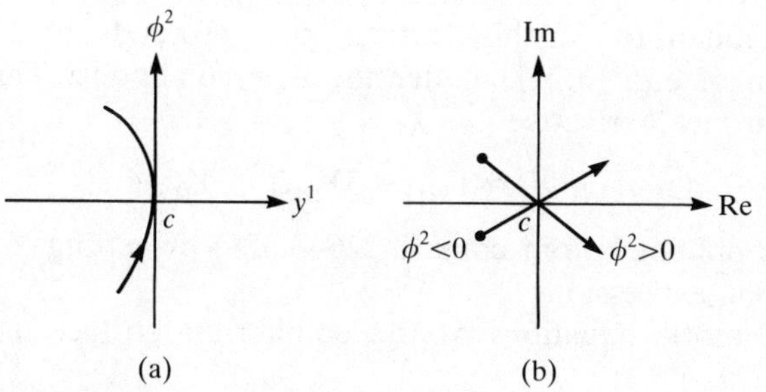

Fig. 3.25 Dynamic instability: (a) y^1 varies with ϕ^2 according to $y^1 = -(\phi^2)^2$, (b) a stable system ($\phi^2 < 0$) loses its stability by flutter ($\phi^2 > 0$) as ϕ^2 passes through zero.

and the divergence boundary takes the simple form

$$\overset{*}{\varphi}{}^1 = 0, \qquad \varphi^2 < 0, \tag{3.271}$$

together forming the stability boundary as depicted in Fig. 3.26.

3.2.3 Simple general points of order 3 (singular general points)

Simple general points of order 3 may arise in both cases of the preceding section (i.e. when a real eigenvalue or a pair of complex conjugate eigenvalues of the Jacobian vanishes) provided a certain additional system coefficient vanishes at the critical point.

These two situations will again be analysed separately.

Case 1. A real eigenvalue vanishes. Consider once more the system (3.209) with the properties described by eqns (3.212) and (3.213). In this case, however, let the coefficient $Y_{111} = 0$. As a result of this assumption, one of the key surface derivatives, namely $\varphi^{1,11}$, vanishes, rendering eqn (3.230) incomplete and pointing to a topologically different equilibrium surface in the vicinity of the critical point c. In order to obtain the

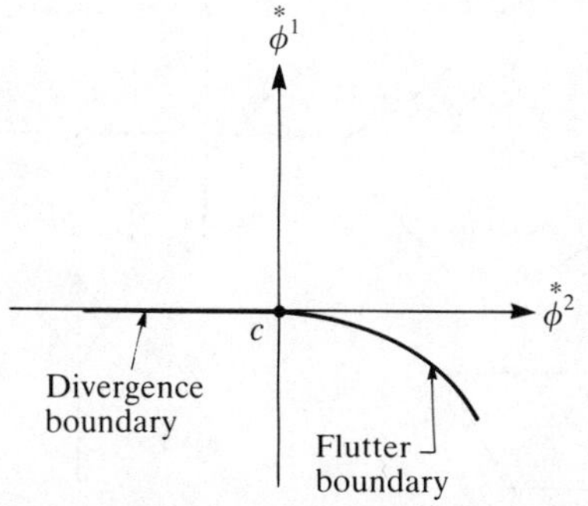

Fig. 3.26 Stability boundary of the system.

first-order equations of this surface, one proceeds to a third-order perturbation of eqn (3.217) under the new conditions. This procedure yields the surface derivative

$$\varphi^{1,111} = Y_{1111} + 3Y_{1t1}B_t \triangle \bar{Y}_{1111} \tag{3.272}$$

where B_t is obtained from eqns (3.226–3.228) by setting $Y_{111} = 0$ in the corresponding expressions.

The asymptotic equations of the equilibrium surface are then constructed as

$$\varphi^1 = \frac{1}{3!}\bar{Y}_{1111}(y^1)^3 + \psi_\nu y^1\varphi^\nu + \tfrac{1}{2}\psi_{\nu\xi}\varphi^\nu\varphi^\xi$$

and (3.273)

$$y^t = A_{t\nu}\varphi^\nu + \tfrac{1}{2}B_t(y^1)^2.$$

Topologically, this surface is identical to that examined in Section 2.4.2 for potential systems, and the critical point under consideration is a *general point of order 3* or a *singular general point*. One may reduce eqns (3.273) to a simpler form in various ways as discussed in Section 2.4.2 and in earlier paper (Huseyin 1977). One of these approaches consists of introducing $\varphi^\nu = l^\nu\rho$ and recognizing that the topological features of the surface remain qualitatively invariant for all such rays. Thus, one has

$$\varphi^1 = \frac{1}{3!}\bar{Y}_{1111}(y^1)^3 + \psi_\nu l^\nu y^1\rho + \tfrac{1}{2}\psi_{\nu\xi}l^\nu l^\xi\rho^2,$$

which yields

$$\varepsilon = \frac{1}{3!}\bar{Y}_{1111}(y^1)^3 + \psi_\nu l^\nu y^1\rho \tag{3.274}$$

upon defining a new parameter, $\varepsilon = \varphi^1 - \tfrac{1}{2}\psi_{\nu\xi}l^\nu l^\xi\rho^2$.

This is a two-parameter surface which is readily identified as the *cusp catastrophe* associated with potential systems (Fig. 3.27). It has already

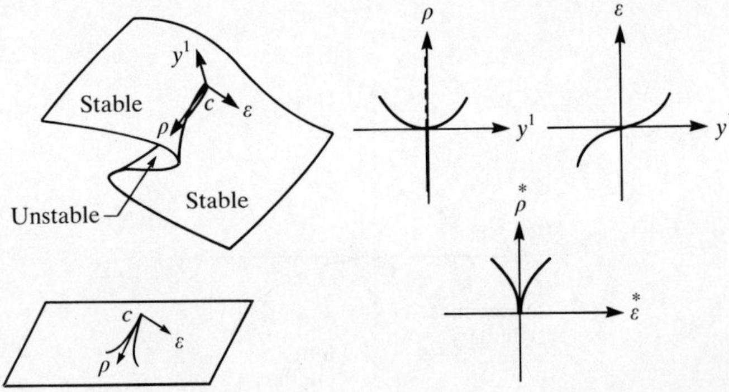

Fig. 3.27 Cusp catastrophe with a stable critical point c.

emerged in the imperfection analysis of symmetric points of bifurcation (Case 2, Section 3.1.5), under a different formulation. It is noted that on a plot of y^1 versus ρ, the critical point c appears as a symmetric point of bifurcation. It can be shown, by adopting the approach discussed in Section 3.1.3, that the post-critical path

$$\psi_v l^v \rho = -\tfrac{1}{6}\bar{Y}_{1111}(y^1)^2$$

is totally stable (unstable) if the trivial path $y^1 = 0$ is unstable (stable) for the same range of ρ. Combining this property with the stability distribution in the vicinity of folds (or limit points), as discussed in the preceding section (Case 1), it may be concluded that if the exterior sheets of the equilibrium surface (Fig. 3.27) are stable (unstable) the interior sheet is unstable (stable). Also note that the exterior sheets are stable if the coefficient $\bar{Y}_{1111} > 0$, i.e. if the critical point c is stable. The critical line, demarcating the regions of stability and instability, and the stability boundary are given by

$$\tfrac{1}{2}\bar{Y}_{1111}(\overset{*}{y}{}^1)^2 + \psi_v l^v \overset{*}{\rho} = 0 \tag{3.275}$$

and

$$\overset{*}{\varepsilon}{}^2 = -\frac{8}{9\bar{Y}_{1111}}(\psi_v l^v \overset{*}{\rho})^3. \tag{3.276}$$

Fig. 3.27 shows a stable critical point c, and it is understood that the stability distribution on the surface is reversed if c is unstable ($\bar{Y}_{1111} < 0$).

Case 2. A Jordan block of order 2. System (3.209), with the properties (3.243) and (3.244), will be treated here under the additional assumption that $Y_{211} = 0$.

It follows from the first derivative in eqn (3.246) that $\varphi^{1,11} = 0$, and a third perturbation of eqn (3.217) under this condition becomes necessary in order to construct the equilibrium surface. Following the familiar procedure yields

$$\varphi^{1,111} = Y_{2111} + 3Y_{21t}C_t \triangleq \bar{Y}_{2111}, \tag{3.277}$$

and the first-order equations of the equilibrium surface are obtained as

$$\begin{aligned} \varphi^1 &= \frac{1}{3!}\bar{Y}_{2111}(y^1)^3 + \chi_v y^1 \varphi^v + \tfrac{1}{2}\chi_{v\xi}\varphi^v\varphi^\xi, \\ y^t &= A_{tv}\varphi^v + \tfrac{1}{2}C_t(y^1)^2, \end{aligned} \tag{3.278}$$

where the coefficients follow from eqn (3.246) by setting $Y_{211} = 0$ in the corresponding expressions. Introducing the new parameters ρ and ε, as in the preceding case, eqns (3.278) may be expressed as

$$\varepsilon = \frac{1}{3!}\bar{Y}_{2111}(y^1)^3 + \chi_\nu l^\nu y^1 \rho$$

and (3.279)

$$y^t = A_{t\nu} l^\nu \rho + \tfrac{1}{2}C_t (y^1)^2,$$

where $\varepsilon = \varphi^1 - \frac{1}{2}\chi_{\nu\xi} l^\nu l^\xi \rho^2$ and $\varphi^\nu = l^\nu \rho$.

The critical divergence zone is described by

$$\tfrac{1}{2}\bar{Y}_{2111}(\overset{*}{y}{}^1)^2 + \chi_\nu l^\nu \overset{*}{\rho} = 0,$$
$$\overset{*}{\varepsilon}{}^2 = -\frac{8}{9\bar{Y}_{1111}}(\chi_\nu l^\nu \overset{*}{\rho})^3. \tag{3.280}$$

The question again arises as to the possibility of a dynamic instability and the presence of a flutter boundary in the vicinity of c which might become a part of the stability boundary. To examine the instability behaviour of the system, the expansion of the Jacobian (3.234) is evaluated on the equilibrium surface (3.279), resulting in

$$Y_{ij}(y^1, \rho) = Y_{ij} + Y_{ij1} y^1 + Y_{ijt}\{A_{t\nu} l^\nu \rho + \tfrac{1}{2}C_t (y^1)^2\}$$
$$+ Y_{ij\nu} l^\nu \rho + \tfrac{1}{2}Y_{2111}(y^1)^2 + \dots. \tag{3.281}$$

For simplicity, let $i, j = 1, 2$ (and $t = 2$); then

$$[Y_{ij}] = \begin{bmatrix} 0 & 1 \\ 0 & 0 \end{bmatrix}, \qquad C_2 = -Y_{111},$$

and

$$Y_{11}(y^1, \rho) = 0 + Y_{111} y^1 + (Y_{112} A_{2\nu} + Y_{11\nu}) l^\nu \rho + \dots, \tag{3.282}$$
$$Y_{12}(y^1, \rho) = 1 + Y_{121} y^1 + (Y_{122} A_{2\nu} + Y_{12\nu}) l^\nu \rho + \dots,$$
$$Y_{21}(y^1, \rho) = 0 + \tfrac{1}{2}(Y_{2111} - Y_{212} Y_{111})(y^1)^2 + \chi_\nu l^\nu \rho + \dots,$$
$$Y_{22}(y^1, \rho) = 0 + Y_{221} y^1 + (Y_{222} A_{2\nu} + Y_{22\nu}) l^\nu \rho + \dots,$$

where χ_ν is defined in eqn (3.246).

The characteristic equation

$$|Y_{ij}(y^1, \rho) - \lambda \boldsymbol{I}| = 0$$

yields

$$\lambda^2 - (\mathrm{Tr}\, \boldsymbol{Y})\lambda + \Delta = 0 \tag{3.283}$$

where $(\mathrm{Tr}\, \boldsymbol{Y})$ is the trace of $Y_{ij}(y^1, \rho)$, and $\Delta = \det|Y_{ij}(y^1, \rho)|$.

Using eqns (3.282) results in

$$\mathrm{Tr}\, \boldsymbol{Y} = (Y_{111} + Y_{221}) y^1 + \rho \sum_{i=1}^{2} (Y_{ii2} A_{2\nu} + A_{22\nu}) l^\nu + \dots$$
$$\Delta = -\{\tfrac{1}{2}\bar{Y}_{2111}(y^1)^2 + \chi_\nu l^\nu \rho\} + \dots. \tag{3.284}$$

For stability, the conditions

$$\operatorname{Tr} \boldsymbol{Y} < 0 \quad \text{and} \quad \Delta > 0 \tag{3.285}$$

must be satisfied.

It is first observed that, by setting $\Delta = 0$, one recovers the critical divergence zone (3.280) as should be expected. Similarly, $\operatorname{Tr} \boldsymbol{Y} = 0$ describes a flutter line if, in general,

$$(\operatorname{Tr} \boldsymbol{Y})^2 - 4\Delta < 0 \tag{3.286}$$

for both $\operatorname{Tr} \boldsymbol{Y} < 0$ and $\operatorname{Tr} \boldsymbol{Y} > 0$ in the vicinity of the boundary.

It seems that the critical point c may again appear as a divergence point on some paths and a flutter point on others. Indeed, if one assumes that y^1 and ρ are of the same order, i.e. if $y^1 = l_1\sigma$ and $\rho = l\sigma$ (where σ is a parameter and l_1 and l are non-zero constants), *a stable equilibrium path described by σ can lose stability by divergence only*. This can be verified by following the same arguments advanced in the case of eqn (3.59). In fact, the eigenvalues in the vicinity of c may be expressed as $\lambda_{1,2} = \frac{1}{2}M\sigma \pm \sqrt{(N\sigma)}$, where M and N are certain finite constants, and the above result follows for all combinations of positive and negative M and N, upon recognizing that $|\sigma|^{1/2} > |\sigma|$ if σ is very small.

On the other hand, consider a special path (sub-manifold) on the equilibrium surface (3.279) characterized by $\rho = l_0(y^1)^2$ where l_0 is a constant. Then

$$\operatorname{Tr} \boldsymbol{Y} = (Y_{111} + Y_{221})y^1 + 0\{(y^1)^2\} \ldots,$$
$$\Delta = -\{\tfrac{1}{2}\bar{Y}_{2111} + \chi_v l^v l_0\}(y^1)^2 + 0\{(y^1)^2\} \ldots,$$

and the eigenvalues, to a first-order approximation, are expressed as

$$\lambda_{1,2} = \tfrac{1}{2}(Y_{111} + Y_{221})y^1 \pm \tfrac{1}{2}y^1\sqrt{E} \tag{3.287}$$

where

$$E = (Y_{111} + Y_{221})^2 + 4(\tfrac{1}{2}\bar{Y}_{2111} + \chi_v l^v l_0).$$

If $E < 0$, then the eigenvalues cross the origin of the complex λ-plane from left to right (Fig. 3.28) as y^1 varies either from negative to positive

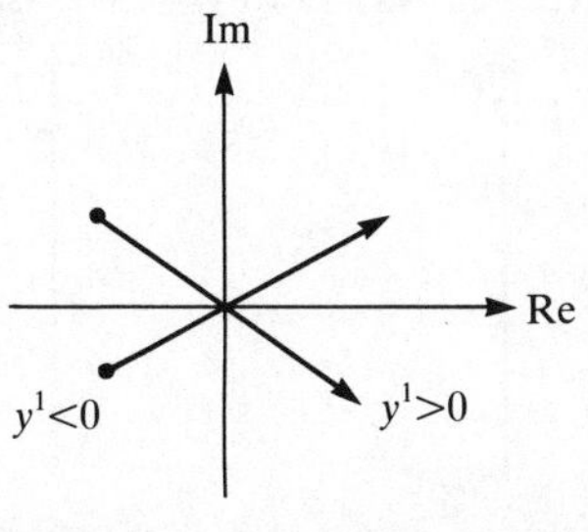

Fig. 3.28 Position of eigenvalues as y^1 varies from negative to positive and vice versa, $E < 0$, $(Y_{111} + Y_{22}) > 0$.

or vice versa—depending on the sign of $(Y_{111} + Y_{221})$—indicating dynamic instability. It is interesting to notice that $E < 0$ implies $\Delta > 0$ on the path under consideration.

It is thus observed that while the critical point c may, in general, be regarded as a divergence point on one-dimensional sub-manifolds (equilibrium paths) of the surface (3.279), it may also appear as a threshold of flutter instability on certain special paths. As in the case of general critical points of order 2 (fold catastrophe) associated with a Jordan block of order 2, *singular general points* here may, indeed, be expected to represent singular points, lying on intersections of divergence and flutter lines. The critical divergence zone (eqns 3.280) may not, therefore, represent the stability boundary of the system in its entirety.

Finally, it is noted that the equilibrium surface (3.273) associated with Case 1 has to be modified if $\bar{Y}_{1111} = 0$. Similarly, the surface (3.278) will not reflect the true picture if $\bar{Y}_{2111}$ vanishes. In both cases, higher order perturbations are required to construct the first-order equations of the equilibrium surfaces. Following the procedure and transformations outlined for potential systems in Section 2.4.3, it can be demonstrated that a fourth-order surface emerges in each case, and furthermore these surfaces correspond to the swallow tail associated with potential systems. In fact, one can keep on generating higher order structurally stable surfaces if no limit is imposed on the number of parameters.

Problem 3.10. Derive the critical zone associated with eqn (3.273) by considering the equilibrium equations $Y_i(y^j, \varphi^\alpha) = 0$ and the criticality condition $|Y_{ij}(y^k, \varphi^\alpha)| = 0$ (Hint: Assume the critical zone in the parametric form $y^t = y^t(y^1, \varphi^\nu)$ and $\varphi^1 = \varphi^1(y^1, \varphi^\nu)$). Verify eqns (3.275) and (3.276) by introducing $\varphi^\nu = l^\nu \rho$ and reducing the number of independent parameters to two.

Problem 3.11. Consider system (3.209). Simple general points of order 2 and 3 can also arise at a critical point characterized by a zero eigenvalue of multiplicity r if the index of the eigenvalue is one, and the Jordan canonical form $\boldsymbol{J}$ in eqn (3.210) is given by the r-square matrix

$$\boldsymbol{J} = \begin{bmatrix} 0 & 1 & & & \\ & 0 & 1 & & \\ & & 0 & \cdot & \\ & & & \cdot & \cdot \\ & & & & \cdot\ 1 \\ & & & & 0 \end{bmatrix}.$$

Formulate the problems of Sections 3.2.2 and 3.2.3 under this assumption and derive the asymptotic equation of the equilibrium surface in each case.

3.2.4 *Compound general points*

By definition, an r-fold compound critical point is associated with a Jacobian matrix of nullity r. Depending on the structure of the Jacobian matrix, several cases may arise. The first case (generic) that comes to mind is concerned with a critical point at which r real eigenvalues vanish and become positive upon crossing the origin of the complex λ-plane.

Consider system (3.209) with the block diagonal Jacobian matrix (3.210) where the Jordan canonical form $\boldsymbol{J}$ is represented by an $r \times r$ diagonal matrix whose elements are all zero. The transformation matrices in eqn (3.208) can then be chosen such that

$$[Y_{a\alpha}] = [Y_{a\varepsilon} \quad Y_{av}] = [\boldsymbol{I} \quad \boldsymbol{0}], \qquad (m > r), \tag{3.288}$$

where $Y_{a\alpha} \equiv \partial Y_a / \partial \varphi^\alpha |_c$; $a = 1, 2, \ldots, r$; $\varepsilon = 1, 2, \ldots, r$ and $v = r + 1, \ldots, m$. This transformation is based on the definition of general points, ensuring that the rank of the parameter matrix $Y_{a\alpha}$ is r.

The solution of the equilibrium equations $Y_i(y^j, \varphi^\alpha) = 0$ in the vicinity of the critical point c, identified by eqns (3.210) and (3.288), is normally described in a parametric form $y = y(\sigma)$ and $\varphi = \varphi(\sigma)$; however, the transformed parameter matrix (3.288) enables one to select the perturbation parameters here in advance. Indeed, the critical coordinates y^a $(a = 1, 2, \ldots, r)$ and φ^v $(v = r + 1, \ldots, m)$ are suitable for this purpose, and the equilibrium surface in the vicinity of c is expressed by

$$y^t = y^t(y^a, \varphi^v) \quad \text{and} \quad \varphi^\varepsilon = \varphi^\varepsilon(y^a, \varphi^v). \tag{3.289}$$

Substituting eqn (3.289) into the equilibrium equations yields the identities

$$Y_i[y^j(y^a, \varphi^v), \varphi^\alpha(y^a, \varphi^v)] \equiv 0 \tag{3.290}$$

where $y^j(y^a, \varphi^v)$ and $\varphi^\alpha(y^a, \varphi^v)$ reduce to y^a and φ^v whenever $j = a$ and $\alpha = v$, respectively.

Differentiating eqn (3.290) with respect to y^a and φ^v yields

$$Y_{ij} y^{j,a} + Y_{i\alpha} \varphi^{\alpha,a} = 0 \tag{3.291}$$

and

$$Y_{ij} y^{j,v} + Y_{i\alpha} \varphi^{\alpha,v} = 0. \tag{3.292}$$

Evaluating eqn (3.291) at c leads to

$$0 + \delta_{a\varepsilon} \varphi^{\varepsilon,a} = 0 \qquad \text{for } i = a,$$

resulting in

$$\varphi^{\varepsilon,a} = 0. \tag{3.293}$$

Introducing this result into eqn (3.291) leads to

$$\begin{bmatrix} \alpha_q & -\omega_q \\ \omega_q & \alpha_q \end{bmatrix} \begin{bmatrix} y^{q,a} \\ y^{(q+1),a} \end{bmatrix} = 0 \qquad \text{for } q = r+1, r+3, \ldots,$$

yielding

$$y^{t,a} = 0 \qquad (t = r+1, \ldots, n). \tag{3.294}$$

Similarly, evaluating eqn (3.292) at c yields

$$\varphi^{\varepsilon,v} = 0 \quad \text{and} \quad y^{t,v} = A_{tv} \tag{3.295}$$

where A_{tv} is exactly the same as given in eqn (3.223).

A second perturbation of eqn (3.290) yields the derivatives

$$\begin{aligned}
\varphi^{\varepsilon,ab} &= -\delta^{\varepsilon c} Y_{cab} \triangle \psi_{\varepsilon ab}, \\
\varphi^{\varepsilon,av} &= -\delta^{\varepsilon b}\{Y_{bav} + (Y_{ba} + A_{tv})\} \triangle \psi_{\varepsilon av}, \\
\varphi^{\varepsilon,v\xi} &= -\delta^{\varepsilon a}(Y_{av\xi} + Y_{atv}A_{t\xi} + Y_{at\xi}A_{tv} + Y_{atp}A_{tv}A_{p\xi}) \triangle \psi_{\varepsilon v\xi}, \\
y^{q,ab} &= (\alpha_q k_{qab} + \omega_q k_{(q+1)ab})/(\alpha_q^2 + \omega_q^2) \triangle \psi_{qab}, \\
y^{(q+1),ab} &= (\alpha_q k_{(q+1)ab} - \omega_q k_{qab})/(\alpha_q^2 + \omega_q^2) \triangle \psi_{(q+1)ab},
\end{aligned} \tag{3.296}$$

where $k_{qab} = -(Y_{qab} + \psi_{\varepsilon ab} Y_{q\varepsilon})$.

Using these derivatives, the first-order equations of the equilibrium surface in the vicinity of the compound general point c are constructed as

$$\varphi^{\varepsilon} = \tfrac{1}{2}\psi_{\varepsilon ab} y^a y^b + \psi_{\varepsilon av} y^a \varphi^v + \tfrac{1}{2}\psi_{\varepsilon v\xi}\varphi^v \varphi^\xi \tag{3.297}$$

and

$$y^t = A_{tv}\varphi^v + \tfrac{1}{2}\psi_{tab} y^a y^b. \tag{3.298}$$

A comparison of these equations with eqns (2.60) and (2.61) for potential systems reveals the analogy in the behaviour characteristics.

The second fundamental tensors associated with the surface (3.297) are represented by

$$\boldsymbol{\psi}^{(\varepsilon)} = \begin{bmatrix} \psi_{(\varepsilon)ab} & \psi_{(\varepsilon)av} \\ \psi_{(\varepsilon)va} & \psi_{(\varepsilon)v\xi} \end{bmatrix}, \qquad (\varepsilon = 1, 2, \ldots, r), \tag{3.299}$$

which characterize the local shape of the equilibrium surface. The primary parameters φ^ε span the normal space, while the secondary parameters φ^v, together with y^a, constitute a basis for the tangent space.

In the special case of $r = 2$, the surface (3.297) may be related to elliptic or hyperbolic umbilics, depending on the coefficients. In order to demonstrate this, introduce $\varphi^v = l^v \rho$ into eqn (3.297) to obtain

$$\begin{aligned}
\varphi^1 &= \tfrac{1}{2}\psi_{1ab} y^a y^b + \psi_{1av} l^v y^a \rho + \tfrac{1}{2}\psi_{1v\xi} l^v l^\xi \rho^2, \\
\varphi^2 &= \tfrac{1}{2}\psi_{2ab} y^a y^b + \psi_{2av} l^v y^a \rho + \tfrac{1}{2}\psi_{2v\xi} l^v l^\xi \rho^2.
\end{aligned} \tag{3.300}$$

Evidently, topological properties of the surface remains essentially unaltered for different rays. Introduce further two new parameters such that

$$\begin{aligned}
\varepsilon^1 &= \tfrac{1}{2}\psi_{1ab} y^a y^b + \psi_{1a} y^a \rho \\
\varepsilon^2 &= \tfrac{1}{2}\psi_{2ab} y^a y^b + \psi_{2a} y^a \rho,
\end{aligned} \tag{3.301}$$

where
$$\varepsilon^1 = \varphi^1 - \tfrac{1}{2}\psi_{1\nu\xi}l^\nu l^\xi \rho^2,$$
$$\varepsilon^2 = \varphi^2 - \tfrac{1}{2}\psi_{2\nu\xi}l^\nu l^\xi \rho^2,$$
$$\psi_{1a} = \psi_{1a\nu}l^\nu \quad \text{and} \quad \psi_{2a} = \psi_{2a\nu}l^\nu.$$

These equations are similar to eqns (2.100) and (2.101), except that the latter equations follow directly from a potential. If $\psi_{1ab}y^a y^b$ and $\psi_{2ab}y^a y^b$ are derivable from a homogeneous cubic polynomial in y^a which can be factorized into three linear forms such that the associated coefficient vectors are pairwise linearly independent and real (two of them complex conjugate), then the equilibrium surface (3.301) is associated with an elliptic (hyperbolic) umbilic. As described in Section 2.5.1, in applications the features of the surface (3.301) and the connection with the umbilics may also be examined on the basis of the principal curvatures which are the eigenvalues of the second fundamental tensors.

As far as the critical zone and the stability boundary are concerned, it is not possible to obtain the latter in general explicit terms. However, the critical zone may be expressed in terms of the critical coordinates y^a ($a = 1, 2$) by adopting an approach analogous to that described in Section 2.5.2. In this case, the critical extremum points on the surface (3.301) are obtained with the aid of the Jacobian determinant

$$\begin{bmatrix} \partial\varepsilon^1/\partial y^1 & \partial\varepsilon^1/\partial y^2 \\ \partial\varepsilon^2/\partial y^1 & \partial\varepsilon^2/\partial y^2 \end{bmatrix} = 0, \tag{3.302}$$

which yields a criticality condition similar to condition (2.118) associated with potential systems. Substituting for ρ in eqns (3.301) results in the critical zone in the form of

$$\varepsilon^1 = \varepsilon^1(y^a), \qquad \varepsilon^2 = \varepsilon^2(y^a), \qquad \rho = \rho(y^a) \tag{3.303}$$

which has computational advantages in numerical calculations. Critical surfaces constructed this way may then be compared with elliptic and hyperbolic umbilics (Figs. 2.7 and 2.8).

Due to symmetry properties that may prevail in a system, the coefficients Y_{cab} may all vanish, resulting in $\varphi^{\varepsilon,ab} = 0$. In this case a third-order perturbation of eqn (3.290) becomes essential to modify the equations of the equilibrium surface appropriately. This procedure yields the derivative

$$\varphi^{\varepsilon,abc} = -\delta^{\varepsilon d}(Y_{dcab} + 3Y_{tcd}\psi_{tab}) \triangle \psi_{\varepsilon abc}$$

which is used to construct the first-order equations of the equilibrium surface:

$$\varphi^\varepsilon = \frac{1}{3!}\psi_{\varepsilon abc}y^a y^b y^c + \psi_{\varepsilon a\nu}y^a \varphi^\nu + \tfrac{1}{2}\psi_{\varepsilon\nu\xi}\varphi^\nu \varphi^\xi \tag{3.304}$$

Clearly, eqn (3.298) remains unaffected.

Finally, as noted at the beginning of this section, an r-fold compound general point may be manifested in different ways. In other words, the Jacobian at the critical point may have a Jordan canonical form of a variety of structures with nullity r. Here, only the generic case, in which r real eigenvalues pass through zero, has been considered. Two more cases will be given as problems:

Problem 3.12. Consider system (3.209), and assume that r pairs of complex conjugate eigenvalues vanish at c. Assume further that the Jacobian (3.210) is in the form

$$\left[\frac{\partial \boldsymbol{Y}}{\partial \boldsymbol{y}}\right]_c = \operatorname{diag}\left[\begin{bmatrix}0 & 1\\ 0 & 0\end{bmatrix}, \begin{bmatrix}0 & 1\\ 0 & 0\end{bmatrix}, \ldots, \begin{bmatrix}\alpha_{l+1} & -\omega_{l+1}\\ \omega_{l+1} & \alpha_{l+1}\end{bmatrix}\right],$$

where $l = 2r$, and the parameter matrix has been canonized so that

$$[Y_{e\alpha}] = [Y_{e\varepsilon} \quad Y_{ev}] = [\boldsymbol{I} \quad \boldsymbol{0}]$$

(where $e = 2, 4, \ldots, l$; $\varepsilon = 1, 3, \ldots, l-1$; $v = 2, 4, \ldots, l, l+1, \ldots, m$). Derive the asymptotic equations of the equilibrium surface (see Huseyin and Mandadi 1980).

Problem 3.13. Suppose system (3.209) is a three-dimensional system again with m parameters. Assuming that the Jacobian at the critical point c has nullity 2, such that

$$\left[\frac{\partial Y}{\partial \boldsymbol{y}}\right]_c = \begin{bmatrix}0 & & \\ & 0 & 1\\ & 0 & 0\end{bmatrix},$$

formulate the problem along the lines of the aforegoing analysis to obtain the first-order equations of the equilibrium surface in the vicinity of c.

3.2.5 *Special critical points*

As in the case of potential systems, special points may arise in many mathematical models in a variety of ways. The relatively simple phenomena of one-parameter bifurcations, studied in Sections 3.1.2 to 3.1.5, are associated with special critical points according to definitions given earlier. The important imperfection sensitivities of bifurcation points were also explored in Section 3.1.5. In this section certain selected cases will be treated from a more general point of view.

Consider once more the autonomous system (3.203), and assume that the equilibrium equations, $\boldsymbol{Z}(z, \boldsymbol{\eta}) = \boldsymbol{0}$, possess a particular solution of the form $z = \boldsymbol{f}(\boldsymbol{\eta})$ which is single-valued in a region of interest—usually the neighbourhood of a certain critical point. The objective is to develop a method of analysis that is capable of yielding other equilibrium states that may exist in the neighbourhood.

Since the fundamental surface $\boldsymbol{z}=\boldsymbol{f}(\boldsymbol{\eta})$ is assumed to be single-valued in the region of interest, one may introduce a transformation of the form

$$\boldsymbol{z}=\boldsymbol{f}(\boldsymbol{\eta})+\boldsymbol{Q}\boldsymbol{w}, \qquad |\boldsymbol{Q}|\neq 0, \tag{3.305}$$

which is similar to eqn (3.61), except that here $\boldsymbol{f}(\boldsymbol{\eta})$ is a function of several variables, and the new coordinates w may be envisaged as sliding on a surface rather than a path (Huseyin 1971*a*). The transformation matrix $\boldsymbol{Q}$ is chosen such that the Jacobian at the point of interest is in a block diagonal form.

The first point of interest is a simple critical point $\boldsymbol{\eta}_c$ on the fundamental surface $f(\eta)$ at which a pair of eigenvalues vanishes such that the corresponding Jordan canonical form is given by

$$\boldsymbol{J}=\begin{bmatrix}0 & 1\\ 0 & 0\end{bmatrix}. \tag{3.306}$$

A non-singular transformation of parameters,

$$\boldsymbol{\mu}=\boldsymbol{P}\boldsymbol{\varphi}, \qquad \boldsymbol{\mu}=\boldsymbol{\eta}-\boldsymbol{\eta}_c, \tag{3.307}$$

is also introduced such that as a result of eqns (3.305) and (3.307), the transformed system

$$\frac{\mathrm{d}\boldsymbol{w}}{\mathrm{d}t}=\boldsymbol{W}(\boldsymbol{w},\boldsymbol{\varphi}) \tag{3.308}$$

has the following properties:

$$\boldsymbol{W}(\boldsymbol{0},\boldsymbol{\varphi})=\boldsymbol{W}_{\alpha}(\boldsymbol{0},\boldsymbol{\varphi})=\boldsymbol{W}_{\alpha\beta}(\boldsymbol{0},\boldsymbol{\varphi})=\ldots=\boldsymbol{0}, \tag{3.309}$$

$$\left[\frac{\partial \boldsymbol{W}}{\partial \boldsymbol{w}}\right]_c=\operatorname{diag}\,[\boldsymbol{J},\boldsymbol{K}_3,\boldsymbol{K}_5,\ldots], \tag{3.310}$$

and

$$W_{21\alpha}=[-1,0,\ldots,0], \tag{3.311}$$

where $\boldsymbol{J}$ is given by eqn (3.306) and

$$\boldsymbol{K}_q=\begin{bmatrix}\alpha_q & -\omega_q\\ \omega_q & \alpha_q\end{bmatrix}, \qquad \alpha_q<0, \qquad (q=3,5,\ldots).$$

The fundamental surface, $\boldsymbol{z}=\boldsymbol{f}(\boldsymbol{\eta})$, is now represented by $\boldsymbol{w}=\boldsymbol{0}$, and the post-critical surface in the vicinity of c is formally assumed to be in the form of $\boldsymbol{w}=\boldsymbol{w}(\boldsymbol{\sigma})$ and $\boldsymbol{\varphi}=\boldsymbol{\varphi}(\boldsymbol{\sigma})$ where $\boldsymbol{\sigma}$ is a vector parameter. On the basis of the transformations introduced above, however, one may identify the basic variables w^1 and φ^v $(v=2,3,\ldots,m)$ as suitable parameters for the role of this vector parameter. Thus, introducing the assumed solutions into the equilibrium equations yields

$$W_i[w^j(w^1, \varphi^\nu), \varphi^\alpha(w^1, \varphi^\nu)] \equiv 0, \tag{3.312}$$

where $w^j(w^1, \varphi^\nu)$ and $\varphi^\alpha(w^1, \varphi^\nu)$ reduce to w^1 and φ^ν for $j=1$ and $\alpha = \nu$, respectively.

A sequence of perturbation equations is now generated by successive differentiations of W_i with respect to the independent variables w^1 and φ^ν. Evaluating the first perturbation equations at c, with the aid of eqns (3.309) to (3.311), results in the derivatives

$$w^{t,1} = 0, \qquad w^{t,\nu} = 0. \tag{3.313}$$

The second perturbation equations yield

$$\begin{aligned}
w^{t,\nu\xi} &= 0, \qquad \varphi^{1,1} = \tfrac{1}{2}W_{211}, \qquad \varphi^{1,\nu} = 0,\\
w^{2,11} &= -W_{111} + W^1_{11}W_{211} \triangleq L_2,\\
w^{q,11} &= (\alpha_q e_q + \omega_q e_{q+1})/(\alpha_q^2 + \omega_q^2) \triangleq L_q,\\
w^{(q+1)11} &= (\alpha_q e_{q+1} - \omega_q e_q)/(\alpha_q^2 + \omega_q^2) \triangleq L_{q+1},\\
w^{2,1\nu} &= -W_{11\nu} \triangleq L_{2\nu},\\
w^{q,1\nu} &= -(W_{q1\nu}\alpha_q + W_{(q+1)1\nu}\omega_q)/(\alpha_q^2 + \omega_q^2) \triangleq L_{q\nu},\\
w^{(q+1),1\nu} &= (W_{q1\nu}\omega_q - W_{(q+1)1\nu}\alpha_q)/(\alpha_q^2 + \omega_q^2) \triangleq L_{(q+1)\nu},
\end{aligned} \tag{3.314}$$

where

$$e_q = -W_{q11} + W^1_{q1}W_{211}, \qquad (W^1_{11} \triangleq -\partial W_{11}/\partial\varphi^1|_c).$$

A third perturbation results in

$$\varphi^{1,\nu\xi} = W_{21\nu\xi} + W_{2t\nu}L_{t\xi} + W_{2t\xi}L_{t\nu} \triangleq L_{\nu\xi}. \tag{3.315}$$

The post-critical equilibrium surface in the vicinity of c is then described by

$$\varphi^1 = \tfrac{1}{2}W_{211}w^1 + \tfrac{1}{2}L_{\nu\xi}\varphi^\nu\varphi^\xi$$

and (3.316)

$$w^t = L_{t\nu}w^1\varphi^\nu + \tfrac{1}{2}L_t(w^1)^2.$$

In view of the fact that $w^i = 0$ represents the fundamental surface, one observes that any parameter ray of the form $\varphi^\alpha = l^\alpha\rho$ $(l^1 \neq 0)$ yields an *asymmetric point of bifurcation* on a plot of ρ versus w^1 (Fig. 3.29*a*). On the other hand, a ray with $l^1 = 0$ leads to a *tangential bifurcation* described by

$$\tfrac{1}{2}W_{211}w^1 + \tfrac{1}{2}L_{\nu\xi}l^\nu l^\xi\rho^2 = 0 \tag{3.317}$$

which is illustrated in Fig. 3.29*b*. In other words, the special critical point under consideration combines in itself asymmetric and tangential bifurcations of one-parameter systems which were explored in Section 3.1.

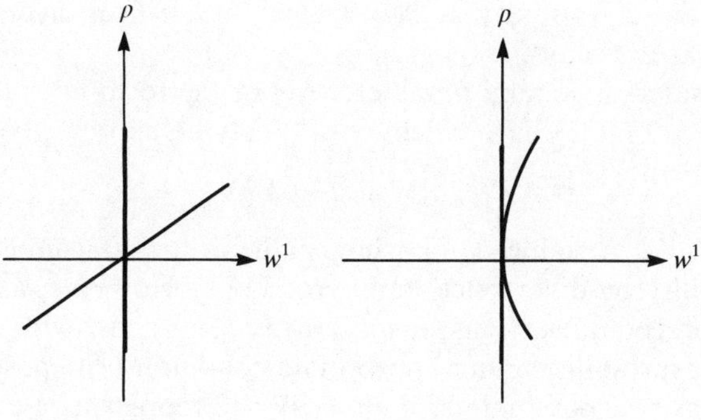

Fig. 3.29 (a) Asymmetric point of bifurcation and (b) tangential bifurcation.

The critical divergence zone consists of

$$\begin{aligned} \overset{*}{w}^1 &= 0, \\ \overset{*}{\varphi}^1 &= \tfrac{1}{2} L_{\nu\xi} \overset{*}{\varphi}{}^\nu \overset{*}{\varphi}{}^\xi, \end{aligned} \tag{3.318}$$

which represents the intersection of the fundamental and post-critical surfaces. This may be demonstrated by examining the Jacobian in the vicinity of the critical point. To this end, consider the expansion of the Jacobian matrix around c,

$$W_{ij}(w^k, \varphi^\alpha) = W_{ij} + W_{ijk} w^k + W_{ij\alpha} \varphi^\alpha + \tfrac{1}{2} W_{ij\alpha\beta} \varphi^\alpha \varphi^\beta + \ldots \tag{3.319}$$

and assume for simplicity that $i, j = 1, 2$.

Evaluating eqn (3.319) first on the fundamental surface—i.e. setting $w^i = 0$ in this expression—yields

$$\begin{aligned} W_{11}(\varphi^\alpha) &= 0 + W_{11\alpha} \varphi^\alpha + \ldots, \\ W_{12}(\varphi^\alpha) &= 1 + \ldots, \\ W_{21}(\varphi^\alpha) &= 0 - \varphi^1 + \tfrac{1}{2} W_{21\nu\xi} \varphi^\nu \varphi^\xi + \ldots, \\ W_{22}(\varphi^\alpha) &= 0 + W_{22\alpha} \varphi^\alpha + \ldots. \end{aligned} \tag{3.320}$$

The eigenvalues of the Jacobian are then obtained from

$$\lambda^2 - (\mathrm{Tr}\, \boldsymbol{W})\lambda + \Delta = 0 \tag{3.321}$$

where the trace of $W_{ij}(\varphi^\alpha)$ is

$$\mathrm{Tr}\, \boldsymbol{W} = (W_{11\alpha} + W_{22\alpha})\varphi^\alpha + \ldots, \tag{3.322}$$

and the determinant, $\Delta = |W_{ij}(\varphi^\alpha)|$, takes the form

$$\Delta = \varphi^1 - \tfrac{1}{2} L_{\nu\xi} \varphi^\nu \varphi^\xi + \ldots, \tag{3.323}$$

upon using eqns (3.314) and (3.315).

It is immediately noted that $\Delta = 0$ gives the critical divergence zone (eqns 3.318) as expected.

If one considers a general parameter ray of the form $\varphi^{\alpha} = l^{\alpha}\rho$ ($l^1 \neq 0$), the eigenvalues (3.321) in the vicinity of c may be expressed as

$$\lambda_{1,2} = \tfrac{1}{2}(W_{11\alpha} + W_{22\alpha})l^{\alpha}\rho \pm \tfrac{1}{2}\sqrt{(-4l^1\rho)}, \tag{3.324}$$

which shows that a stable equilibrium path on the fundamental surface loses its stability by divergence upon crossing c. However, as demonstrated in the preceding sections, a special ray (e.g. a ray with $l^1 = 0$) may lead to flutter instability under appropriate conditions. Suppose $v = 2$ and the system has two parameters; then $\mathrm{Tr}\,\boldsymbol{W} = 0$ represents a flutter line if $\mathrm{Tr}\,\boldsymbol{W}$ changes from negative to positive upon crossing the line $\mathrm{Tr}\,\boldsymbol{W} = 0$, while $\Delta > 0$. The latter condition indicates stability with regard to divergence, and flutter may, therefore, occur before the divergence line is reached (on certain equilibrium paths). On the other hand, this phenomenon does not occur in many mathematical models which remains stable—with imaginary eigenvalues—until the divergence boundary is reached, where an initial loss of stability takes place.

Consider next the post-critical equilibrium surface. Evaluating eqn (3.319) on the surface (3.316) yields

$$\begin{aligned}
W_{11}(w^1, \varphi^{v}) &= 0 + W_{111}w^1 + W^1_{11}(\tfrac{1}{2}W_{211}w^1) + W_{11v}\varphi^{v} + \ldots, \\
W_{12}(w^1, \varphi^{v}) &= 1 + \ldots, \\
W_{21}(w^1, \varphi^{v}) &= 0 + W_{211}w^1 - \tfrac{1}{2}(W_{211}w^1 + L_{v\xi}\varphi^{v}\varphi^{\xi}) + \tfrac{1}{2}W_{21v\xi}\varphi^{v}\varphi^{\xi} + \ldots, \\
W_{22}(w^1, \varphi^{v}) &= 0 + W_{221}w^1 + W^1_{22}(\tfrac{1}{2}W_{221}w^1) + W_{22v}\varphi^{v} + \ldots,
\end{aligned} \tag{3.325}$$

which result in

$$\begin{aligned}
\mathrm{Tr}\,\boldsymbol{W} = \{&W_{111} + W_{221} + \tfrac{1}{2}(W_{11\alpha} + W^1_{22})W_{211}\}w^1 \\
&+ (W_{11v} + W_{22v})\varphi^{v} + \ldots,
\end{aligned} \tag{3.326}$$

$$\Delta = -\tfrac{1}{2}W_{211}w^1 + \ldots. \tag{3.327}$$

The first-order expression (3.327) for Δ follows from the fact that the quadratic form in $\varphi^{v}\varphi^{\xi}$ vanishes when eqn (3.315) and L_{2v} are introduced into Δ.

It is noted that $w^1 = 0$ yields $\Delta = 0$ and the divergence zone (3.318). Also, substituting for w^1 in Δ gives

$$\Delta = -(\varphi^1 - \tfrac{1}{2}L_{v\xi}\varphi^{v}\varphi^{\xi}),$$

which, when compared with eqn (3.323), shows reversed stability properties with regard to divergence on the fundamental and post-critical surfaces. Indeed, at least for $\varphi^{v} = 0$, an *exchange of stabilities* occurs at the critical point c between the fundamental and post-critical paths associated with the parameter φ^1.

An exchange of stabilities would take place between the entire fundamental and post-critical surfaces in the vicinity of the divergence boundary if the critical point c were associated with the generic case in which a real eigenvalue crosses the origin of the λ-plane, rather than a pair described by eqn (3.306). The following problem is designed to illustrate this phenomenon:

Problem 3.14. Reformulate the entire problem of this section under the assumption that a real eigenvalue of the Jacobian vanishes at the critical point c while the remaining eigenvalues have negative real parts. Derive the equations of the post-critical equilibrium surface, and the critical divergence boundary in the vicinity of c. Demonstrate that the divergence boundary is indeed the stability boundary of the system, and an *exchange of stabilities* occurs between the fundamental and post-critical surfaces in the vicinity of this boundary. Explore the analogy between this behaviour and that of a potential system (Huseyin 1971*a*).

Effect of symmetry. A glance at the first equation in (3.316) reveals that the coefficient W_{211} plays a significant role in characterizing the post-critical surface. This coefficient may vanish due to symmetry properties and, if this happens, a third perturbation is required to determine the first-order equations of the surface. Thus, evaluating the third order perturbation equations at c yields

$$\begin{aligned}\varphi^{1,11} &= \tfrac{1}{3}(W_{2111} + 3W_{2t1}L_t) = \bar{W}_{2111},\\ \varphi^{1,1\nu} &= \tfrac{1}{2}(W_{211\nu} + 2W_{21t}L_{t\nu} + W_{2t\nu}L_t) \triangle L_\nu,\end{aligned} \tag{3.328}$$

where L_t and $L_{t\nu}$ are given by eqn (3.314) with $W_{211} = 0$.

The first-order equations of the equilibrium surface are then constructed as

$$\varphi^1 = \tfrac{1}{2}\bar{W}_{2111}(w^1)^2 + L_\nu w^1 \varphi^\nu + \tfrac{1}{2}L_{\nu\xi}\varphi^\nu\varphi^\xi$$

and

$$w^t = L_{t\nu}w^1\varphi^\nu + \tfrac{1}{2}L_t(w^1)^2. \tag{3.329}$$

Topologically, this surface is analogous to surface (3.247); however, the surface (3.329) is a post-critical surface, and a fundamental surface—described here by $w^i = 0$—intersects it along a critical divergence zone. The second fundamental tensor associated with eqn (3.329) is represented by the matrix

$$\begin{bmatrix} \bar{W}_{2111} & L_\nu \\ L_\nu & L_{\nu\xi} \end{bmatrix},$$

and the surface is called synclastic, anticlastic, or parabolic according to whether

$$[\bar{W}_{2111}L_{\nu\xi} - L_\nu L_\xi]$$

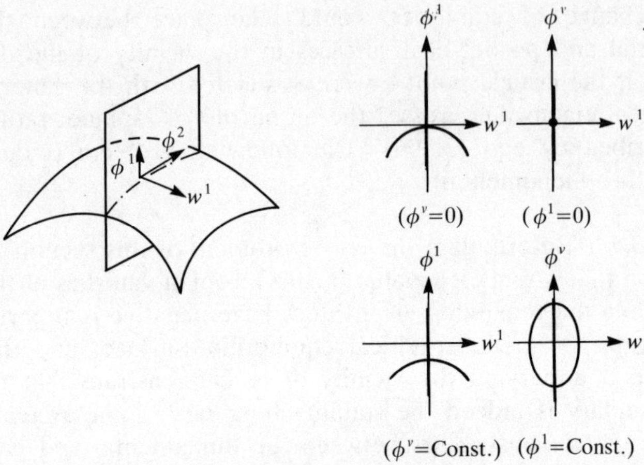

Fig. 3.30 Synclastic surface.

is positive definite, negative definite, or null, respectively. Synclastic and anticlastic surfaces are illustrated in Figs. 3.30 and 3.31, respectively, where the post-critical surfaces have negative curvatures $\varphi^{1,11}$. It is understood, however, that the curvatures of the post-critical surfaces can also be reversed.

The critical point c is *a symmetric point of bifurcation* on a general parameter ray described by $\varphi^\alpha = l^\alpha \rho$ ($l^1 \neq 0$), while it appears as a point of *tri-furcation* on a special ray with $l^1 = 0$.

The critical divergence zone consists of points of intersection of the fundamental and post-critical surfaces, and is given by

$$\begin{aligned} \overset{*}{w}{}^1 &= 0, \\ \overset{*}{\varphi}{}^1 &= \tfrac{1}{2} L_{\nu\xi} \overset{*}{\varphi}{}^\nu \overset{*}{\varphi}{}^\xi, \end{aligned} \tag{3.330}$$

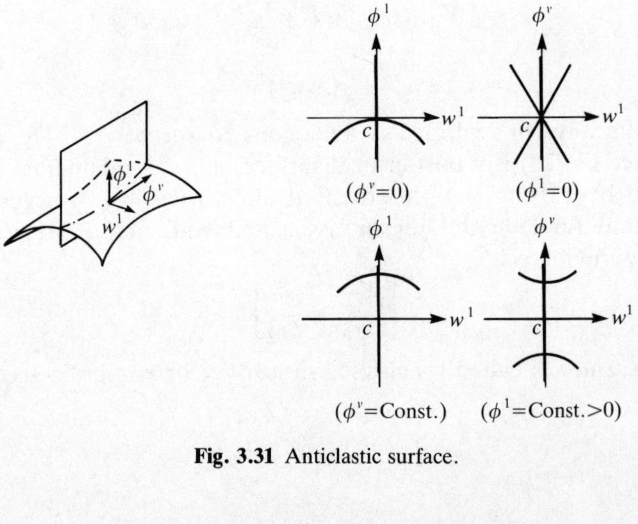

Fig. 3.31 Anticlastic surface.

which may be verified by considering the expansion of the Jacobian around c, as in the preceding asymmetric case.

Indeed, evaluating the Jacobian matrix on the fundamental surface $w^i = 0$ leads to conclusions that are formally identical to those pertaining to the asymmetric case, since $W_{211} = 0$ does not alter eqn (3.321). It is clear, therefore, that the critical point c may again be the intersection of *divergence* and *flutter* boundaries.

Problem 3.15. The equations of the post-critical surface obtained in Problem 3.14 are rendered inadequate in the case $W_{111} = 0$ at c. Reformulate Problem 3.14 under this symmetry condition, and derive the first-order equation of the equilibrium surface and the critical zone. Show that the critical zone is associated with the stability boundary of the system, and that the fundamental surface becomes unstable upon crossing this boundary. Also, demonstrate that the post-critical surface is totally stable or unstable, depending on whether a certain expression is positive or negative. Discuss the analogy between this system and symmetric potential systems (Huseyin 1971*b*).

Effect of imperfections. The influence of imperfections on one-parameter bifurcation points was studied in Section 3.1.5. The presence of several parameters may somewhat complicate the analysis. Nevertheless, it can be shown, through the systematic perturbation procedure, that a *special* critical point transforms into a *general* critical point if an appropriate number of imperfection parameters is introduced. Also, if a special critical point is characterized by a vanishing real eigenvalue which becomes positive upon crossing the origin, a new set of generalized coordinates may be employed which slide along the fundamental surface—rather than a path—and rotate such that the Jacobian matrix is kept in canonical form in the vicinity of the critical point. This formulation should be used in Problems 3.14 and 3.15. However, if the critical point is characterized by eqn (3.306), only the sliding system of variables characterized by eqn (3.309) may be employed.

Consider a class of imperfect systems characterized by $z_t = \boldsymbol{Z}(z, \boldsymbol{\eta})$ where one of the parameters, say $\eta^1 \equiv \varepsilon$, represents imperfections and η^ν $(\nu = 2, 3)$ are the independent parameters. Suppose the perfect system, $\varepsilon = 0$, exhibits a single-valued fundamental equilibrium surface, $z = \boldsymbol{f}(\eta^\nu)$, and a certain critical combination of the parameters is associated with a special critical point on this surface characterized by a Jordan block of order 2. Then, introducing a transformation of the form

$$z = \boldsymbol{f}(\eta^\nu) + \boldsymbol{Q}\boldsymbol{w}, \qquad |\boldsymbol{Q}| \neq 0, \tag{3.331}$$

results in a transformed system of equations,

$$\frac{d\boldsymbol{w}}{dt} = \boldsymbol{W}(\boldsymbol{w}, \varepsilon, \eta^{\nu}), \tag{3.332}$$

such that

$$\boldsymbol{W}(\boldsymbol{0}, 0, \eta^{\nu}) = \boldsymbol{W}_{\nu}(\boldsymbol{0}, 0, \eta^{\nu}) = \boldsymbol{W}_{\nu\xi}(\boldsymbol{0}, 0, \eta^{\nu}) = \ldots = 0,$$

and

$$\left[\frac{\partial \boldsymbol{W}}{\partial \boldsymbol{w}}\right]_c = \text{diag}\,[\boldsymbol{J}, \boldsymbol{K}_3, \boldsymbol{K}_5, \ldots], \tag{3.333}$$

where

$$\boldsymbol{J} = \begin{bmatrix} 0 & 1 \\ 0 & 0 \end{bmatrix}, \qquad \boldsymbol{K}_q = \begin{bmatrix} \alpha_q & -\omega_q \\ \omega_q & \alpha_q \end{bmatrix}, \qquad \alpha_q < 0 \qquad (q = 3, 5, \ldots).$$

It is assumed that the imperfection parameter ε is primary such that

$$W_{2\varepsilon} = \left[\frac{\partial W_2}{\partial \varepsilon}\right]_c \neq 0. \tag{3.334}$$

Let the equilibrium surface of the imperfect system be in the form of $\varepsilon = \varepsilon(w^1, \eta^{\nu})$, $w^t = w^t(w^1, \eta^{\nu})$, where $t = 2, 3, \ldots, n$. Following the familiar perturbation procedure finally leads to the asymptotic equations

$$\begin{aligned} \varepsilon &= \frac{1}{2W_{2\varepsilon}} W_{211}(w^1)^2 + \frac{1}{W_{2\varepsilon}} W_{21\nu} w^1 \mu^{\nu}, \\ w^t &= \tfrac{1}{2}\bar{W}_t (w^1)^2 + \bar{W}_{t\nu} w^1 \mu^{\nu}, \end{aligned} \tag{3.335}$$

where $\boldsymbol{\mu} = \boldsymbol{\eta} - \boldsymbol{\eta}_c$, and $\bar{W}_t$ and $\bar{W}_{t\nu}$ are certain combinations of system coefficients.

It is now observed that the equilibrium surface (3.335) is topologically equivalent to the surface (3.247)—the surface associated with a *general* critical point of order 2. In fact, surface (3.335) is identified as the *fold catastrophe* as in the case of surface (3.247), and the behaviour characteristics associated with the latter may be exhibited by the system here as well. The reader is referred, therefore, to Case 2, Section 3.2.2, for a complete discussion.

If $W_{211} = 0$, a further perturbation yields the surface

$$\varepsilon = \frac{1}{6W_{2\varepsilon}} \bar{W}_{2111}(w^1)^3 + \frac{1}{W_{2\varepsilon}} W_{21\nu} w^1 \mu^{\nu} \tag{3.336}$$

where

$$\bar{W}_{2111} = W_{2111} + 3W_{21t}\bar{w}_t.$$

It is now noted that the *special* critical point c has been transformed to a *general* critical point of order 3 (*singular point*) by the addition of an imperfection parameter. Indeed, the surface becomes topologically identical to surface (3.279) if one introduces $\varphi^\nu = l^\nu \rho$, and it may readily be identified as the *cusp catastrophe*. All the properties and behaviour characteristics related to *singular general points* (Section 3.2.3, Case 2) are also applicable to surface (3.336), and will not be repeated here. An important feature associated with the critical points under consideration, however, is worth noting once more: unlike the generic case, in which a real eigenvalue crosses the origin, both fold and cusp catastrophes in the vicinity of a critical point characterized by a Jordan block of order 2 may not fully describe the behaviour of the system. Indeed, it is possible that a family of limit cycles branches off the equilibrium surface, and the critical point becomes the intersection of divergence and flutter boundaries as discussed in Section 3.2.2.

Problem 3.16. Verify the first-order equation (3.335) of the equilibrium surface (fold).

Problem 3.17. Verify the first-order equation (3.336) of the equilibrium surface (cusp).

Problem 3.18. Derive the first-order equations of the equilibrium surface of the imperfect system (3.332) in the generic case when a real eigenvalue crosses the origin at the special critical point. Establish the conditions and equations of both fold and cusp catastrophes and associated stability boundaries. Discuss the imperfection sensitivities in both cases.

Problem 3.19. Consider the system

$$z_{,t} = \boldsymbol{Z}(z, \boldsymbol{\eta}), \qquad z \in R^n, \qquad \boldsymbol{\eta} \in R^m, \qquad ({}_{,t} = \mathrm{d}/\mathrm{d}t).$$

Assume that r $(r < m)$ parameters η^ε $(\varepsilon = 1, 2, \ldots, r)$ represent imperfections, and for $\eta^\varepsilon = 0$ the (perfect) system exhibits an r-fold special critical point on a single-valued fundamental surface, $z = \boldsymbol{f}(\eta^\nu)$, $(\nu = r+1, \ldots, m)$. Assuming further that the r-fold critical point is characterized by r vanishing real eigenvalues of the Jacobian, introduce the sliding and rotating coordinates by the transformation

$$z = \boldsymbol{f}(\eta^\nu) + \boldsymbol{Q}(\eta^\nu)\boldsymbol{x}$$

to obtain a new set of governing equations $\dot{\boldsymbol{x}} = \boldsymbol{X}(\boldsymbol{x}, \eta^\varepsilon, \eta^\nu)$ with the properties that the Jacobian is in an appropriate canonical form at all points of the fundamental surface.

Show that the equilibrium surface of the imperfect system in the sub-space of critical coordinates is in the general form of

$$\varphi^{\varepsilon} = \tfrac{1}{2}\bar{X}_{\varepsilon ab}x^a x^b + \bar{X}_{\varepsilon av}x^a \varphi^v$$

where $\bar{X}_{\varepsilon ab}$ and $\bar{X}_{\varepsilon av}$ are constants, $\varphi^{\alpha} = \eta^{\alpha} - \eta^{\alpha}_{c}$, and x^a are the critical coordinates $(a, b = 1, 2, \ldots, r)$.

This equation corresponds to eqn (3.297); identify the differences in formulation and the final forms of the equations. Discuss the behaviour characteristics of the system.

3.3 Concluding remarks

The formulation, analysis, and results concerning the static instabilities of non-potential systems treated in this chapter have been designed to bring to light the analogies as well as differences between the behaviour characteristics of potential and non-potential systems. Even within the chapter, the treatment is on a comparative basis so that the common features and phenomena associated with seemingly different systems may be recognized. Attention has also been drawn to distinct properties of the transformations employed to facilitate the analytical treatment of various phenomena efficiently. In this regard, a brief summary of notation and transformations as well as the contexts in which they have been used may provide a convenient reference for the reader:

1) System $\boldsymbol{z}_{,t} = \boldsymbol{Z}(\boldsymbol{z}, \eta)$: Initial formulation. No special properties $({}_{,t} = \mathrm{d}/\mathrm{d}t)$.

2) System $\boldsymbol{y}_{,t} = \boldsymbol{Y}(\boldsymbol{y}, \mu)$: The Jacobian evaluated at a point of interest is in a block-diagonal form.

The transformation involved is in the form of $\boldsymbol{z} = \boldsymbol{z}_c + \boldsymbol{Q}\boldsymbol{y}$, $\eta = \eta_c + \mu$.

It facilitates the analysis of *general* critical points.

In the case of multiple parameter systems, an additional transformation of parameters, $\boldsymbol{\eta} = \boldsymbol{\eta}_c + \boldsymbol{P}\boldsymbol{\varphi}$, is introduced for convenience. The system is then expressed as $\boldsymbol{y}_{,t} = \boldsymbol{Y}(\boldsymbol{y}, \boldsymbol{\varphi})$, where the parameter matrix is also in a canonical form.

3) System $\boldsymbol{w}_{,t} = \boldsymbol{W}(\boldsymbol{w}, \eta)$: This system makes use of *sliding coordinates* through the transformation $\boldsymbol{z} = \boldsymbol{f}(\eta) + \boldsymbol{Q}\boldsymbol{w}$.

The Jacobian *at the point of interest* is again in block-diagonal form.

It facilitates the analysis of *special* critical points like bifurcations, but excludes extrema and hence *general* points.

In the case of multiple-parameter systems, an additional transformation of parameters is introduced to canonize the parameter matrix, and the system takes the form $\boldsymbol{w}_{,t} = \boldsymbol{W}(\boldsymbol{w}, \boldsymbol{\varphi})$.

4) System $\boldsymbol{x}_{,t} = \boldsymbol{X}(\boldsymbol{x}, \eta)$: This system involves both sliding and rotating coordinates through the transformation $\boldsymbol{z} = \boldsymbol{f}(\eta) + \boldsymbol{Q}(\eta)\boldsymbol{x}$.

The Jacobian is now in a block-diagonal form along $\boldsymbol{f}(\eta)$.

It facilitates the analysis of special critical points like bifurcations in the generic cases. Because of continuity problems, however, it cannot be introduced in the case of critical points associated with Jordan blocks of order 2 or higher.

It may also involve an additional transformation of parameters if the system contains several parameters.

With the aid of these transformations, it has been possible to explore static instabilities of autonomous systems analytically and conveniently. The study of dynamic instabilities will also be based on one of these formulations.

There exist a number of other approaches concerning divergence instability of non-potential systems. Plaut (1976, 1977) examined the branching phenomena associated with non-conservative systems by setting the frequency in the Lagrange's equations equal to zero and assuming that the system exhibits a trivial initial path. Post-buckling behaviour of continuous non-conservative elastic systems has also been studied (Plaut, 1978). The monographs by Ioos and Joseph (1980), and Chow and Hale (1982) contain general theories concerning static instabilities. Haken (1983), Potier-Ferry (1979, 1981*a*, *b*), Nicolis and Prigogine (1977), Wilson (1981), Poore (1976), Kolkka (1984), and Troger (1981) are among several other authors who studied related problems.

4

DYNAMIC INSTABILITY OF AUTONOMOUS SYSTEMS

4.1 Introductory remarks

As noted in the preceding chapters, non-linear autonomous systems may lose or gain stability at a critical equilibrium point from which a family of limit cycles (periodic orbits) bifurcates. Historically, the theory of such dynamic—as well as static—bifurcations originated in the works of Poincaré (1892). Following these pioneering studies, dynamic bifurcations were treated by Andronov and Witt (1930), and later by Hopf (1942). The latter author extended the analysis of a certain dynamic bifurcation phenomenon—commonly known as a Hopf bifurcation—from two dimensions to higher dimensions. He showed that if a pair of complex conjugate eigenvalues of the Jacobian crosses the imaginary axis with non-zero velocity, as a parameter varies through a critical equilibrium point—while the remaining eigenvalues have negative real parts—there exists a family of limit cycles in the vicinity of the critical point which actually bifurcates from this point. This is the simplest structurally stable dynamic bifurcation phenomenon associated with autonomous systems, and is exhibited by a variety of real systems. Indeed, the Hopf bifurcation theory has wide-ranging applications in physics, mechanics, electrical networks, chemical processes, biology, ecology and many other disciplines. It is not, therefore, surprising that the subject has attracted the attention of a substantial number of scientists, engineers and mathematicians alike. These activities have led to further developments concerning *degenerate Hopf bifurcations,* which arise when the basic conditions underlying Hopf's theorem are violated, and *generalized Hopf bifurcations* which are associated with multiple parameters as well as multiple eigenvalues.

The methodology underlying these dynamic bifurcation phenomena embraces a wide range of approaches, including the center manifold theory, Lyapunov–Schmidt reduction scheme, factorization and implicit function theorems, Poincaré normal form, averaging techniques, multiple time scaling and harmonic balancing. Some of these methods are mathematically more powerful and perhaps more elegant than others; however, it has to be borne in mind that in applications conceptual simplicity and applicability are virtues that cannot be ignored as a particular method is adopted. Indeed, motivated by this fact and a desire to tackle a whole range of problems concerning oscillatory bifurcations through a unified perturbation approach that parallels the multiple-parameter perturbation technique employed in the analyses of static bifurcations (see the preceding chapter), an *intrinsic method of harmonic balancing* has been developed in recent years (Huseyin and Atadan 1981, 1983; Atadan and Huseyin 1982*a, b,* 1984*a*). This method is essentially a variation on the classical method of harmonic balancing, and is designed to eliminate the drawbacks and shortcomings associated with the latter. As a matter of fact, it has been demonstrated that the new approach does, indeed, yield consistent approximations for a variety of non-linear dynamic bifurcation problems in a systematic way, and it will form the basis of the treatment throughout this chapter. An additional advantage of the method—apart from being conceptually simple and systematic—is that it is capable of producing *explicit* results in general terms, as in the case of the static bifurcations, thus eliminating the need for extensive analysis in each specific application. Indeed, the explicit formulas generated in the course of the general analyses may be used directly and conveniently for the solution of specific problems in mechanics, electrical networks, chemical processes, and other areas, provided the governing equations of a specific problem fall within the scope of the general formulation or they are brought within that scope by appropriate transformations. The task of the analyst is then reduced to evaluating certain derivatives and substituting them into the appropriate formulas to construct ordered approximations for the limit cycles, bifurcation paths, frequency-amplitude relationships, and other relevant information concerning the behaviour of a system.

The next two sections are concerned with an introduction of this methodology.

4.2 Non-linear oscillations and the method of harmonic balancing

Essentially, the phenomena of oscillatory bifurcations are related to non-linear oscillations, and the methods available in that context—such as averaging, multiple time scaling, and the Lindstedt–Poincaré method—may be expected to be potential candidates for adoption in the analyses of dynamic bifurcations as well. This is also true for the well-known harmonic balancing technique which is relatively simple and leads to a set of algebraic equations rather than differential or integral-differential equations. As mentioned earlier, however, the method may also lead to inaccurate and/or inconsistent results in certain applications (Mahaffey 1976), and this drawback has recently led to justifiable criticism of the method (Nayfeh and Mook 1979, Mees and Chua 1979). As a matter of fact, Nayfeh and Mook (1979) abandon this technique altogether after demonstrating its shortcomings in their comprehensive book concerning non-linear oscillations. Nevertheless, in view of the relative conceptual as well as practical simplicity of this classical method, one may attempt to introduce appropriate procedural modifications in its application such that the consistency of the results is ensured.

Small-amplitude oscillations of a non-linear system can be analysed by perturbation methods. It is well known in the theory of non-linear oscillations, however, that one of the features distinguishing non-linear oscillations from linear systems is the frequency-amplitude relationship; and a standard perturbation technique, which does not allow for such an interaction, is not appropriate for the analysis of non-linear systems. In fact, assuming that the solution can be represented by a straight-forward expansion in terms of a perturbation parameter leads to *secular terms* which are not periodic. In order to circumvent the difficulties associated with secular terms, several *singular perturbation techniques* have been developed over the years. The Linstedt–Poincaré method, the method of multiple scales, and a variety of averaging methods, for example, are designed for this purpose. The methods based on averaging include the Krylov–Bogoliubov and Krylov–Bogoliubov–Mitropolsky techniques, the generalized method of averaging, and averaging using Lagrangians. These methods and many other techniques are discussed in several excellent books on the subject (e.g. Minorsky 1962, Hayashi 1964, Nayfeh and Mook 1979, Hagedorn 1981).

Illustrative example No. 1. Consider now a one-degree-of-freedom, conservative system described by the second-order, non-linear differential equation

$$x_{tt} + \alpha_1 x + \alpha_2 x^2 + \alpha_3 x^3 = 0, \tag{4.1}$$

where $x_{tt} = \mathrm{d}^2x/\mathrm{d}t^2$ and $x = 0$ is the origin of the non-linear system.

It is first noted that the system here is conservative and the periodic motions in the vicinity of the origin are associated with a *centre*. In other words, unlike self-excited oscillations and the corresponding limit cycles, the response of the system here depends on the initial conditions which are normally given by specifying the initial position and velocity.

In a perturbation approach, one aims at predicting the response of the non-linear system by perturbing the response of the corresponding linear system

$$x_{tt} + \alpha_1 x = 0, \tag{4.2}$$

which has a natural frequency $\omega_c = \sqrt{\alpha_1}$. An ordinary perturbation technique consists of expressing the solution in terms of a perturbation parameter ε as

$$x(t; \varepsilon) = \varepsilon x_1(t) + \varepsilon^2 x_2(t) + \ldots, \tag{4.3}$$

and setting the coefficient of each power of ε equal to zero after substituting eqn (4.3) into eqn (4.1). The underlying assumptions in this approach are that ε is independent of $x_1(t)$, $x_2(t)$, etc. This procedure results in a set of differential equations which are solved sequentially with the aid of the initial conditions. It can be demonstrated, however, that at one stage of this procedure secular terms, which are not periodic, like $t^n \cos \omega t$, $t^n \sin \omega t$ appear in the solutions. It follows, therefore, that the assumed solution (4.3) does not represent a *uniformly valid* asymptotic solution for the non-linear system, in the sense that x_m may not provide a small correction for x_{m-1}, x_{m-2}, etc. as t increases.

As noted earlier, one of the singular perturbation techniques designed to overcome this difficulty is the Linstedt–Poincaré method. In this approach, the above perturbation technique is modified appropriately so that the secular terms are eliminated. To accomplish this, introduce

$$\tau = \omega(\varepsilon)t, \tag{4.4}$$

and let $\omega(\varepsilon)$ and $x(\tau; \varepsilon)$ be expressed as

$$\omega(\varepsilon) = \omega_c + \varepsilon\omega_1 + \varepsilon^2\omega_2 + \ldots, \tag{4.5}$$

and

$$x(\tau; \varepsilon) = \varepsilon x_1(\tau) + \varepsilon^2 x_2(\tau) + \ldots. \tag{4.6}$$

Substituting these expansions back into the original differential equation (4.1) yields a sequence of perturbation equations. In solving these equations one then chooses ω_1, ω_2, etc., such that secular terms are eliminated.

Following the steps described above, Nayfeh and Mook (1979) obtained the periodic solution and frequency-amplitude relationship of the non-linear system (4.1) as

$$x(t) = q_1 \cos \varphi - q_1^2(\alpha_2/2\alpha_1)(1 - \tfrac{1}{3}\cos 2\varphi) + \ldots \tag{4.7}$$

and

$$\omega = \sqrt{\alpha_1}\,[1 + \{(9\alpha_3\alpha_1 - 10\alpha_2^2)/24\alpha_1^2\}q_1^2] + \ldots, \tag{4.8}$$

where $\varphi = \omega t + \beta$ and q_1 is the amplitude of the first harmonic. It was also demonstrated that the method of multiple scales leads to precisely the same solution. The amplitude q_1 and the phase angle β can be determined from the two initial conditions which have to be specified.

Consider now the conventional method of harmonic balancing which is not a perturbation technique in the sense of the aforegoing discussion. In this approach, one assumes a periodic solution of the form

$$x(t) = \sum_{m=0}^{M} (p_m \cos m\omega t + r_m \sin m\omega t) \tag{4.9}$$

or

$$x(t) = \sum_{m=0}^{M} q_m \cos m\varphi, \quad \text{where} \quad \varphi = \omega t + \beta. \tag{4.10}$$

The assumed Fourier series is then introduced into the original differential equation, and the coefficient of each harmonic is equated to zero, yielding a set of algebraic equations—in contrast with the sets of differential equations in many other methods. These equations relate the frequency and amplitudes, and are often solved in terms of the amplitude of the first harmonic. Using eqn (4.10), for example, one observes that the resulting $M+1$ equations contain $M+2$ unknowns (ω, q_0, $q_1, \ldots,$ q_M), and may be solved for ω, q_2, $q_3, \ldots, q_M$ in terms of q_1.

Consider first the simple one-term solution

$$x = q_1 \cos \varphi. \tag{4.11}$$

Introducing this assumed solution into (4.1) yields

$$-(\omega^2 - \alpha_1)q_1 \cos \varphi + \tfrac{1}{2}\alpha_2 q_1^2(1 + \cos 2\varphi) + \tfrac{1}{4}\alpha_3 q_1^3(3 \cos \varphi + \cos 3\varphi) = 0, \tag{4.12}$$

which results in

$$\omega = \sqrt{\alpha_1}(1 + \tfrac{3}{8}\alpha_3\alpha_1^{-1}q_1^2), \tag{4.13}$$

if q_1 is assumed to be small.

Clearly, eqn (4.13) does not fully agree with eqn (4.8), and the reason for this inaccurate result is rather evident. In fact, as a consequence of the assumption (4.11), which was quite arbitrary, the terms with a potential to generate further second-order contributions (i.e. terms $O(q_1^2)$) to the result (eqn 4.13) were left out at the outset. However, it is often impossible to identify *a priori* the harmonics which would contribute to a certain order of approximation, and one may simply have to rely on intuition in selecting the harmonics to be retained in an assumed solution. Such an approach is likely to lead to inaccurate results. Consider, for example, the next step (Mahaffey 1976), and assume a solution of the form

$$x = q_0 + q_1 \cos \varphi, \tag{4.14}$$

which follows from eqn (4.10) for $M = 1$.

Substituting eqn (4.14) into eqn (4.1), and setting the constant term and the coefficient of $\cos \varphi$ to zero result in

$$\alpha_1 q_0 + \alpha_2 q_0^2 + \tfrac{1}{2}\alpha_2 q_1^2 + \alpha_3 q_0^3 + \tfrac{3}{2}\alpha_3 q_0 q_1^2 = 0,$$

and (4.15)

$$-(\omega^2 - \alpha_1) + 2\alpha_2 q_0 + 3\alpha_3 q_0^2 + \tfrac{3}{4}\alpha_3 q_1^2 = 0,$$

which can be solved to yield

$$q_0 = -\tfrac{1}{2}\alpha_2\alpha_1^{-1}q_1^2 + O(q_1^4) + \ldots, \tag{4.16}$$

and

$$\omega = \sqrt{\alpha_1}\{1 + (3\alpha_3\alpha_1 - 4\alpha_2^2)q_1^2/8\alpha_1^2\}. \tag{4.17}$$

Using eqn (4.16), the solution (4.14) may be expressed as

$$x = q_1 \cos \varphi - \alpha_2 q_1^2/2\alpha_1 + \dots. \tag{4.18}$$

Comparing now eqn (4.7) with eqn (4.18), and eqn (4.8) with eqn (4.17) leads to the conclusion that the new results again fail to embrace all the second-order terms in q_1.

Next, consider

$$x = q_0 + q_1 \cos \varphi + q_2 \cos 2\varphi \tag{4.19}$$

as an assumed solution. Following the procedure described above finally yields precisely the same results as obtained via other methods (eqns 4.7 and 4.8).

In the light of this analysis, Nayfeh and Mook (1979) conclude that 'to obtain a consistent solution by using the method of harmonic balance, one needs either to know a great deal about the solution *a priori* or to carry enough terms in the solution to check the order of the coefficients of all the neglected harmonics . . . Therefore we prefer not to use this technique'.

If one, nevertheless, perceives certain advantages in this method and wishes to be able to use it by somehow rectifying its deficiencies, the basic question that poses itself is as follows: is it possible to introduce a systematic procedure whereby the method of harmonic balancing will yield an ordered form of approximations accurately and consistently? An affirmative response to this question is provided in the next section within the spirit of the static perturbation technique used in the asymptotic analysis of static bifurcations.

4.3 An intrinsic method of harmonic analysis

The new approach will be illustrated on the same system (4.1) for comparison. It is first realized that any procedure with the above stated goals should preferably be capable of engaging all the harmonics which will contribute to a particular order of approximation and discard the higher order harmonics *automatically,* thus eliminating speculations on the part of the analyst. It is clear that the harmonics contributing to a particular order of approximation may vary from one specific problem to another, and it is essential that such variations are systematically accounted for.

Suppose now that the solution of eqn (4.1) is expressed in the parametric form

$$x(t; \varepsilon) = \sum_{m=0}^{M} q_m(\varepsilon) \cos m\{\omega(\varepsilon)t + \beta\} \tag{4.20}$$

where ε is a small parameter, and ω is a function of ε (Atadan and Huseyin 1984*a*).

By identifying the origin with $\varepsilon = 0$, so that $x(t; 0) = 0$, one has $q_m(0) \triangleq q_{m0} = 0$ in the Fourier series (4.20). The series is kept open-ended; in other words, it is not confined to one, two or any particular number of harmonics, since it is not generally possible to predict *a priori* the contributions of various harmonics.

If $x(t; \varepsilon)$ is a solution of eqn (4.1), it satisfies this equation identically; that is,

$$x_{tt}(t; \varepsilon) + f\{x(t; \varepsilon)\} \equiv 0, \tag{4.21}$$

where

$$f\{x(t; \varepsilon)\} = \alpha_1 x(t; \varepsilon) + \alpha_2 x^2(t; \varepsilon) + \alpha_3 x^3(t; \varepsilon). \tag{4.22}$$

In analogy with the static perturbation procedure, a sequence of perturbation equations can now be generated by differentiating the identity (4.21) with respect to ε, and evaluating the derivatives at the origin. Thus, one obtains

$$x'_{tt} + f_x x' = 0, \tag{4.23}$$

$$x''_{tt} + f_{xx}(x')^2 + f_x x'' = 0, \tag{4.24}$$

$$x'''_{tt} + f_{xxx}(x')^3 + 3f_{xx}x'x'' + f_x x''' = 0, \tag{4.25}$$

etc.,

where the primes and subscript x's denote differentiations with respect to ε and x, respectively, and all equations are evaluated at the origin.

Substituting eqn (4.20) into the first-order perturbation equation (4.23) yields

$$\sum_{m=0}^{M} (m^2 - 1)q'_m \cos m(\omega_c t + \beta) = 0, \tag{4.26}$$

upon evaluation at $\varepsilon = 0$ and defining

$$\omega(0) = \omega_c. \tag{4.27}$$

Now, setting the constant term $(m = 0)$ and the coefficients of $\cos m(\omega_c t + \beta)$ equal to zero for each m results in the derivatives

$$q'_0 = q'_m = 0 \quad \text{for} \quad m \geq 2. \tag{4.28}$$

Here, it is immediately observed that q_1 has a unique feature, compared to other amplitudes, and it is, indeed, a potentially suitable candidate for replacing the unidentified perturbation parameter ε, in analogy with the *critical coordinate* of the static bifurcation problems treated in the preceding chapters.

Setting $\varepsilon = q_1$ yields

$$\dot{q}_1 = 1, \qquad \ddot{q}_1 = \dddot{q}_1 = \ldots = 0, \tag{4.29}$$

where dots replace primes, indicating differentiation with respect to q_1. It is noted that the first-order term of the solution, $q_1 \cos(\omega t + \beta)$, is already discernable as expected.

Next, introduce eqn (4.20) into the second-order perturbation equation (4.24) to obtain

$$\sum_{m=1}^{M} (m^2 - 1)\ddot{q}_m \omega_c^2 \cos m(\omega_c t + \beta) + 4\dot{\omega}\omega_c \cos(\omega_c t + \beta) - \alpha_2\{1 + \cos 2(\omega_c t + \beta)\} = 0 \quad (4.30)$$

which yields

$$\ddot{q}_0 = -3\ddot{q}_2 = -\alpha_2/\omega_c^2, \quad (4.31)$$

$$\ddot{q}_m = 0 \quad \text{for} \quad m \geq 3, \quad (4.32)$$

and

$$\dot{\omega} = 0. \quad (4.33)$$

Note that eqn (4.32) does not require the solution of $(M - 2)$ equations as it might first seem; in fact, this result follows from

$$(m^2 - 1)\ddot{q}_m \cos m\beta = 0 \quad \text{and} \quad (m^2 - 1)\ddot{q}_m \sin m\beta = 0, \quad (4.34)$$

where $m \geq 3$. In general, similar situations arise with regard to other derivatives of amplitudes which can be treated in the same manner. This is a key feature of the procedure and enables one not to limit M to a certain value at the outset.

Expanding the coefficients of the harmonics into Taylor's series with the aid of the derivatives (4.28), (4.29), (4.31), and (4.32), the periodic solution (4.20) may now be constructed up to (and including) second-order terms as

$$x(t; q_1) = q_1 \cos(\omega t + \beta) + \alpha_2 q_1^2/2\omega_c^2\{-1 + \tfrac{1}{3}\cos 2(\omega t + \beta)\} + O(q_1^3) + \ldots \quad (4.35)$$

which is in full agreement with eqn (4.7).

It is important to observe that the first-order term (in q_1) emerges from the first-order perturbation equation, and the second-order terms (associated with q_1^2) are obtained from the second-order perturbation equations—including what may be called the *drift* term $\alpha_2 q_1^2/2\omega_c^2$. In general, the kth order perturbation equation produces the kth order terms in the periodic solution but $(k - 1)$th order terms in the expansion of $\omega(q_1)$. In analogy with the static perturbation procedure, the results of the previous $(k - 1)$ perturbations are utilized in solving the kth order problem.

As eqn (4.33) indicates, the first derivative of ω is zero, and further perturbations are needed to generate the *second-order terms* for $\omega(q_1)$. Following the procedure described above, the third-order perturbation equation (4.25) yields

$$\ddot{\omega} = (9\alpha_3\omega_c^2 - 10\alpha_2^2)/12\omega_c^3. \tag{4.36}$$

Using eqns (4.33) and (4.36), the Taylor's expansion of $\omega(q_1)$ yields

$$\omega(q_1) = \omega_c + q_1^2(9\alpha_3\omega_c^2 - 10\alpha_2^2)/24\omega_c^2 + O(q_1^3) + \ldots \tag{4.37}$$

which is in full agreement with eqn (4.8).

The process leading to the derivative (4.36) also produces the third-order derivatives of amplitudes which may be used to construct the third-order terms of the periodic solution.

It is evident from the foregoing analysis that once a particular *order of approximation* is generated for the periodic solution, there is no danger of inadvertently excluding any potential contributions from higher harmonics to that approximation. For instance, if a term of $O(\varepsilon^2)$ does not involve the second harmonic ($\cos 2\varphi$), then, the latter can only appear in a term at least of $O(\varepsilon^3)$. The conventional harmonic balancing technique does not possess this important property, as eqn (4.18) clearly manifests. This deficiency is of course due to the fact that fixing the number of harmonics at the outset amounts to pre-empting a certain solution without justification. One may suggest the substitution of the general Fourier series (4.10) into eqn (4.1) directly, without specifying M *a priori*. If this route is chosen, however, one immediately encounters serious difficulties that help explain why the number of harmonics in applications is always limited.

The new approach, on the other hand, eliminates such basic difficulties and provides a systematic and reliable procedure leading to an *ordered form of consistent approximations*. Algebraic manipulations become comparatively simple if ε is replaced by an appropriate basic variable of the system at an early stage of the analysis. In the example analysed here, for instance, it became clear at the end of the first perturbation that the amplitude of the first harmonic is suitable for this purpose. Indeed, eqn (4.28) reveals that while the derivatives $q_0', q_2', q_3', \ldots, q_M'$ vanish, indicating that none of these amplitudes are suitable for the role of ε, the amplitude of the first harmonic, q_1, is not in this category. In general, the functions $q_m(\varepsilon)$ should have a non-zero projection in the direction of the perturbation parameter. In some problems, the first perturbation may not reveal an appropriate candidate for ε, and one may have to proceed with ε unidentified until a clear indication emerges. In other words, the choice of a basic variable of ε depends on the specific problem in hand, and *the selection process is guided by the analysis itself.* It should be noted that the replacement of ε by a basic variable is not a mandatory step—although desirable for the sake of simplicity. Indeed, the analysis can be carried out with ε; in this case, all the amplitudes and the frequency are expressed in terms of ε.

Finally, it should be noted that, in general, one may have to treat the phase angles β_m ($m = 1, 2, \ldots$) as independent variables. In fact, each β_m may be expressed as a function of ε as well, in which case the analysis becomes somewhat lengthier. It can be shown, however, that if this formulation is used in the analysis of system (4.1), the results obtained above remain unaffected.

Illustrative example No. 2. Consider now the non-linear differential equation

$$\frac{d^2x}{dt^2} + x = a + \varepsilon x^2, \qquad |\varepsilon| \ll 1, \tag{4.38}$$

subject to the initial conditions

$$x(0) = A \quad \text{and} \quad dx(0)/dt = 0. \tag{4.39}$$

In recognition of the fact that the system (4.38) is autonomous, one may introduce the time scaling

$$\tau = \omega t \tag{4.40}$$

to facilitate the analysis; here, ω is the frequency of the periodic solution of eqn (4.38).

Introducing eqn (4.40) into eqn (4.38) results in the transformed system

$$\omega^2 x_{\tau\tau} + x = a + \varepsilon x^2 \tag{4.41}$$

where the subscript τ indicates differentiation with respect to τ. The periodic solution may be expressed in a parametric form in terms of ε,

$$x = x(\tau; \varepsilon), \qquad \omega = \omega(\varepsilon), \tag{4.42}$$

which is now 2π-periodic in τ, with $\omega(0) = \omega_c = 1$.

The initial conditions take the form

$$x(0; \varepsilon) = A \quad \text{and} \quad x_\tau(0; \varepsilon) = 0. \tag{4.43}$$

Instead of eqn (4.10), let the periodic solution now be in the form of eqn (4.9) which is expressed as (Atadan and Huseyin 1982*a*)

$$x(\tau; \varepsilon) = p_0(\varepsilon) + \sum_{m=1}^{M} \{p_m(\varepsilon)\cos m\tau + r_m(\varepsilon)\sin m\tau\}. \tag{4.44}$$

The linearized equation, defined by $\varepsilon = 0$, is in the form of

$$x_{\tau\tau} + x = a, \tag{4.45}$$

and substituting eqn (4.44) into eqn (4.45), setting the coefficient of $\cos m\tau$ and $\sin m\tau$ equal to zero, and evaluating at $\varepsilon = 0$ yields

$$p_0(0) \triangle p_{0c} = a \tag{4.46}$$

and

$$\left.\begin{aligned} p_m(0) \triangleq p_{mc} = 0 \\ r_m(0) \triangleq r_{mc} = 0 \end{aligned}\right\} \quad \text{for} \quad m \geq 2. \tag{4.47}$$

It follows that

$$x(\tau; \varepsilon) = a + p_{1c} \cos \tau + r_{1c} \sin \tau + O(\varepsilon) + \ldots \tag{4.48}$$

as expected. The initial conditions (4.43), on the other hand, yield

$$x(0; 0) = a + p_{1c} = A \quad \text{and} \quad x_\tau(0; 0) = r_{1c} = 0, \tag{4.49}$$

which are introduced into eqn (4.48) to obtain

$$x(\tau; \varepsilon) = a + (A - a) \cos \tau + O(\varepsilon) + \ldots . \tag{4.50}$$

Thus, the solution corresponding to the linearized part ($\varepsilon = 0$) of eqn (4.41) follows from what may be called the *zeroth order perturbation problem.* Next, one proceeds to the first-order perturbation problem by differentiating eqn (4.41) with respect to ε and evaluating the derivatives at $\varepsilon = 0$. This operation yields

$$2\omega' x_{\tau\tau} + x'_{\tau\tau} + x' = x^2. \tag{4.51}$$

Substituting eqn (4.41) into eqn (4.51), and following the same procedure as before results in the derivatives

$$\begin{aligned} &p'_0 = a^2 + \tfrac{1}{2}(A - a)^2, \\ &p'_2 = -\tfrac{1}{6}(A - a)^2, \qquad r'_2 = 0, \\ &p'_m = r'_m = 0, \quad \text{for} \quad m \geq 3, \end{aligned} \tag{4.52}$$

and

assuming that $A \neq a$.

Again it is noted that the results for $m \geq 3$ do not require the solution of $2(M - 2)$ equations, but simply follow from two equations,

$$(1 - m^2) p'_m = 0 \quad \text{and} \quad (1 - m^2) r'_m = 0.$$

The derivatives (4.52) indicate that the coefficient (function) of ε in the first-order term of the solution, $x'(\tau; 0)$, is in the general form of

$$x'(\tau; 0) = p'_0 + p'_1 \cos \tau + r'_1 \sin \tau + p'_2 \cos 2\tau. \tag{4.53}$$

On the other hand, the initial conditions (4.43) yield $x'(0; 0) = 0$ and $x'_\tau(0; 0) = 0$, upon differentiating with respect to ε. Introducing these conditions into eqn (4.53) and using eqns (4.52) results in

$$p'_1 = -a^2 - \tfrac{1}{3}(A - a)^2 \quad \text{and} \quad r'_1 = 0. \tag{4.54}$$

With the aid of eqns (4.52) and (4.54), one then has

$$x'(\tau, 0) = a^2 + \tfrac{1}{2}(A - a)^2 - \{a^2 + \tfrac{1}{3}(A - a)\} \cos \tau - \tfrac{1}{6}(A - a)^2 \cos 2\tau. \tag{4.55}$$

This result is in complete agreement with that obtained by Mickens (1981) through a different approach.

The periodic solution of eqn (4.38) may now be expressed as

$$x(\tau, \varepsilon) = x(\tau, 0) + x'(\tau, 0)\varepsilon + O(\varepsilon^2) + \ldots \tag{4.56}$$

where $x'(\tau, 0)$ is given by eqn (4.55) and $x(\tau, 0)$ by eqn (4.50) with $\varepsilon = 0$.

One may proceed to the second-order perturbation problem by differentiating eqn (4.38) twice with respect to ε and evaluating at $\varepsilon = 0$ to obtain

$$2\{(\omega')^2 + \omega''\}x_{\tau\tau} + 4\omega' x'_{\tau\tau} + x''_{\tau\tau} + x'' = 4xx'. \tag{4.57}$$

Following the above procedure and using the initial conditions $x''(0; 0) = 0$ and $x''_\tau(0; 0) = 0$ as well as the derivatives obtained in the first perturbation problem yield the second derivatives of amplitudes which are then used to construct the terms of $O(\varepsilon^2)$. The procedure also yields (Atadan and Huseyin 1982*a*)

$$\omega'' = -3a^2(A - a) - \tfrac{5}{6}(A - a)^2, \tag{4.58}$$

which, together with $\omega' = -a$, leads to the $(\omega - \varepsilon)$ relationship

$$\omega(\varepsilon) = \omega_c + \Omega(\varepsilon) = 1 + \omega'\varepsilon + \tfrac{1}{2}\omega''\varepsilon^2 + \ldots \tag{4.59}$$

Problem 4.1. Determine the second derivatives of the amplitudes of all harmonics associated with the periodic solution of the differential equation (4.38) by using the second perturbation equation (4.57). Construct the terms of $O(\varepsilon^2)$.

Problem 4.2. Verify the expression (4.58) for the second derivative of $\omega(\varepsilon)$.

4.4 One-parameter systems

The intrinsic method of harmonic analysis described for scalar differential equations in the preceding section will now be generalized to n-dimensional vector equations to explore the dynamic bifurcation behaviour of autonomous systems—including the families of limit cycles, bifurcating paths and frequency-amplitude relationships. Attention will be focused directly on these non-linear characteristics of autonomous systems, and the reader is again referred to an earlier monograph (Huseyin 1978*b*) on linear systems which may be useful in conjunction with the developments here.

4.4.1 *Hopf bifurcation*

Consider the autonomous system (3.1),

$$\frac{\mathrm{d}z}{\mathrm{d}t} = \mathbf{Z}(z, \eta), \qquad z \in R^n, \qquad \eta \in R, \tag{4.60}$$

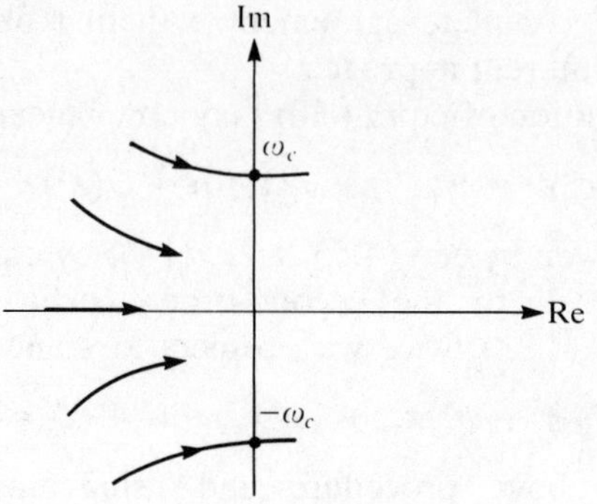

Fig. 4.1 Hopf bifurcation: when $\eta = \eta_c$, a pair of complex conjugate eigenvalues cross the imaginary axis with non-zero velocity. The remaining eigenvalues continue to have negative real parts.

where the vector function $\boldsymbol{Z}(\boldsymbol{z}, \eta)$ is assumed to be real and smooth—at least in a region G of interest.

Suppose eqn (4.60) has a single-valued equilibrium path, $\boldsymbol{z} = \boldsymbol{f}(\eta)$, in G which is initially stable so that the eigenvalues of the Jacobian evaluated on this path all have negative real parts. Suppose further that as η is increased gradually, at a critical value of the parameter ($\eta = \eta_c$) a pair of complex conjugate eigenvalues ($\lambda_{1,2}(\eta) = \alpha(\eta) \pm \mathrm{i}\omega(\eta)$) crosses the imaginary axis with non-zero velocity, in the sense that

$$\alpha' = \left.\frac{\mathrm{d}\alpha}{\mathrm{d}\eta}\right|_c \neq 0 \quad \text{(transversality condition)}, \tag{4.61}$$

while the remaining eigenvalues of the Jacobian continue to have negative real parts (Fig. 4.1). It follows that

$$\lambda_{1,2}(\eta_c) = \pm \mathrm{i}\omega(\eta_c) \equiv \pm \mathrm{i}\omega_c$$

and (4.62)

$$\alpha(\eta_c) = 0.$$

In order to facilitate the bifurcation analysis, introduce now the transformation

$$\boldsymbol{z} = \boldsymbol{f}(\eta) + \boldsymbol{Q}\boldsymbol{w}, \qquad |\boldsymbol{Q}| \neq 0, \tag{4.63}$$

as described in Chapter 3, such that the resulting system

$$\frac{\mathrm{d}\boldsymbol{w}}{\mathrm{d}t} = \boldsymbol{W}(\boldsymbol{w}, \eta) \tag{4.64}$$

has the following properties:

$$\boldsymbol{W}(\boldsymbol{0}, \eta) = \boldsymbol{W}'(\boldsymbol{0}, \eta) = \boldsymbol{W}''(\boldsymbol{0}, \eta) = \ldots = \boldsymbol{0} \tag{4.65}$$

and

$$[W_{ij}] = \left[\frac{\partial \boldsymbol{W}}{\partial \boldsymbol{w}}\right]_c = \operatorname{diag}\,[\boldsymbol{K}_1, \boldsymbol{K}_3, \boldsymbol{K}_5, \ldots, \alpha_s, \ldots], \tag{4.66}$$

where

$$\boldsymbol{K}_1 = \begin{bmatrix} 0 & \omega_c \\ -\omega_c & 0 \end{bmatrix}, \quad \boldsymbol{K}_q = \begin{bmatrix} \alpha_q & \omega_q \\ -\omega_q & \alpha_q \end{bmatrix}, \quad \begin{array}{l} (q = 3, 5, \ldots, Q), \\ (s = Q+1, \ldots, n), \end{array} \tag{4.67}$$

and $\alpha_t < 0$ $(t \neq 1)$.

This transformation is essentially similar to transformation (3.64), except that here there is no Jordan form associated with zero eigenvalues since the critical point is of a different nature. However, in addition to complex conjugate eigenvalues, real eigenvalues (α_s) are also assumed to be present in the Jacobian (4.66), and these eigenvalues appear on the main diagonal of the Jacobian. The transversality condition (4.61) plays a significant role in shaping up the bifurcation characteristics of the system in the vicinity of the critical point (4.62). For the perturbation analysis to be used here, it is therefore essential that the transversality condition be expressed in terms of the system coefficients.

In fact, it can be shown analytically that eqn (4.61) is equivalent to

$$W'_{11} + W'_{22} \neq 0, \tag{4.68}$$

where $W'_{ii} = \partial^2 W_i / \partial w^i\, \partial \eta|_c$ (Huseyin and Atadan 1983).

To this end, consider the expansion of the Jacobian around the critical point c on the equilibrium path $\boldsymbol{f}(\eta)$, which may be expressed as

$$[W_{ij}(\eta)] = [W_{ij} + W'_{ij}(\eta - \eta_c) + \tfrac{1}{2} W''_{ij}(\eta - \eta_c)^2 + \ldots]. \tag{4.69}$$

If $\lambda(\mu)$ is an eigenvalue of the Jacobian, crossing the imaginary axis at c, it should satisfy the characteristic equation

$$|W_{ij}(\mu) - \lambda(\mu) I| = 0 \tag{4.70}$$

identically, where $\mu = \eta - \eta_c$.

The characteristic equation (4.70) may be written in the representative form

$$\left|\begin{array}{cc|cc|c} W'_{11}\mu - \lambda & \omega_c + W'_{12}\mu & W'_{13}\mu & W'_{14}\mu & W'_{15}\mu \\ -\omega_c + W'_{21}\mu & W'_{22}\mu - \lambda & W'_{23}\mu & W'_{24}\mu & W'_{25}\mu \\ \hline W'_{31}\mu & W'_{32}\mu & \alpha_3 + W'_{33}\mu - \lambda & \omega_3 + W'_{34}\mu & W'_{35}\mu \\ W'_{41}\mu & W'_{42}\mu & -\omega_3 + W'_{43}\mu & \alpha_3 + W'_{44}\mu - \lambda & W'_{45}\mu \\ \hline W'_{51}\mu & W'_{52}\mu & W'_{53}\mu & W'_{54}\mu & \alpha_5 + W'_{55}\mu - \lambda \end{array}\right| + \ldots = 0 \tag{4.71}$$

where $\lambda = \lambda_{1,2} = \alpha(\mu) \pm \mathrm{i}\omega(\mu)$.

It, of course, follows that evaluating eqn (4.71) at the critical point $\mu = 0$ leads to eqn (4.66) with

$$
\begin{aligned}
&W_{12} = W_{21} = \omega_c, && \\
&W_{33} = W_{44} = \alpha_3, && \text{representing } \alpha_q, \\
&W_{34} = -W_{43} = \omega_3, && \text{representing } \omega_q, \text{ and} \\
&W_{55} = \alpha_5, && \text{representing } \alpha_s.
\end{aligned}
$$

Differentiating (4.71) with respect to μ and evaluating at the critical point c yields

$$
\left|\begin{array}{cc|cc|c}
W'_{11} - \lambda' & \omega_c & 0 & 0 & 0 \\
W'_{21} & -\mathrm{i}\omega_c & 0 & 0 & 0 \\
\hline
W'_{31} & 0 & \alpha_3 - \mathrm{i}\omega_c & \omega_3 & 0 \\
W'_{41} & 0 & -\omega_3 & \alpha_3 - \mathrm{i}\omega_c & 0 \\
\hline
W'_{51} & 0 & 0 & 0 & \alpha_5 - \mathrm{i}\omega_c
\end{array}\right| +
$$

$$
\left|\begin{array}{cc|cc|c}
-\mathrm{i}\omega_c & W'_{12} & 0 & 0 & 0 \\
-\omega_c & W'_{22} - \lambda' & 0 & 0 & 0 \\
\hline
0 & W'_{32} & \alpha_3 - \mathrm{i}\omega_c & \omega_3 & 0 \\
0 & W'_{42} & -\omega_3 & \alpha_3 - \mathrm{i}\omega_c & 0 \\
\hline
0 & W'_{52} & 0 & 0 & \alpha_5 - \mathrm{i}\omega_c
\end{array}\right| = 0 \qquad (4.72)
$$

where $\lambda' = \partial\lambda/\partial\mu|_c = \alpha' + \mathrm{i}\omega'$ and $\lambda(0) = \mathrm{i}\omega_c$. Note that, without loss of generality, eqn (4.72) is written only for $\lambda(0) = \mathrm{i}\omega_c$.

After some algebra, eqn (4.72) yields

$$
\alpha' = \tfrac{1}{2}(W'_{11} + W'_{22}) \quad \text{and} \quad \omega' = \tfrac{1}{2}(W'_{12} - W'_{21}), \qquad (4.73)
$$

demonstrating the transversality condition (4.68).

It then follows from Hopf's theorem (1944) that, under the condition (4.68), there exists a family of periodic solutions of system (4.64), bifurcating from the critical point c on the equilibrium path $f(\eta)$. Let this family be expressed in the parametric form

$$
\begin{aligned}
w^i &= w^i(t;\sigma), \\
\eta &= \eta(\sigma), \qquad (4.74)\\
\omega &= \omega(\sigma)
\end{aligned}
$$

such that, for a given σ,

$$w^i(t;\sigma) = w^i(t+T;\sigma)$$

where T is the period and σ a parameter.

If eqn (4.74) is a solution of eqn (4.64), then it satisfies the latter identically; that is,

$$\frac{\mathrm{d}}{\mathrm{d}t} w^i(t;\sigma) \equiv W_i\{w^j(t;\sigma), \eta(\sigma)\}, \tag{4.75}$$

which will form a basis for the perturbation procedure.

The periodic solution $w^i(t;\sigma)$ may be represented by the Fourier series

$$w^i(t;\sigma) = p_{i0}(\sigma) + \sum_{m=1}^{M} \{p_{im}(\sigma)\cos m\omega(\sigma)t + r_{im}(\sigma)\sin m\omega(\sigma)t\} \tag{4.76}$$

where $p_{i0}(\sigma)$, $p_{im}(\sigma)$, and $r_{im}(\sigma)$ are the ith components of the amplitude vectors corresponding to the mth harmonic which are assumed to be functions of σ.

Introducing now the time scaling

$$\tau = \omega(\sigma)t \tag{4.77}$$

into eqns (4.75) and (4.76), yields

$$\omega(\sigma)\frac{\mathrm{d}}{\mathrm{d}\tau} w^i(\tau;\sigma) \equiv W_i\{w^j(\tau;\sigma), \eta(\sigma)\} \tag{4.78}$$

and

$$w^i(\tau;\sigma) = p_{i0}(\sigma) + \sum_{m=1}^{M} \{p_{im}(\sigma)\cos m\tau + r_{im}(\sigma)\sin m\tau\}, \tag{4.79}$$

respectively, and the periodic solution becomes 2π-periodic with $\omega(0) = \omega_c$.

It is noted that at the critical point c, $w^i(\tau;0) = 0$, implying

$$p_{i0}(0) = p_{im}(0) = r_{im}(0) = 0 \tag{4.80}$$

for all i and m. Furthermore, since the system is autonomous, one may set

$$r_{11}(\sigma) \equiv 0, \tag{4.81}$$

without loss of generality.

Before proceeding with the perturbation procedure, it should be noted that instead of eqn (4.79), one may also consider the following alternative forms of the Fourier series for the periodic solution:

$$w^i(\tau;\sigma) = u_{i0}(\sigma) + \sum_{m=1}^{M} \{u_{im}(\sigma)\mathrm{e}^{\mathrm{i}m\tau} + \bar{u}_{im}(\sigma)\mathrm{e}^{-\mathrm{i}m\tau}\} \tag{4.82}$$

or

$$w^i(\tau;\sigma)=\rho(\sigma)+\sum_{m=1}^{M}\rho_{im}(\sigma)\cos m\{\tau+\varphi_m(\sigma)\} \tag{4.83}$$

where $\bar{u}_{im}$ is the complex conjugate of u_{im}.

Using eqn (4.82) results in complex algebra, while eqn (4.83) is essentially equivalent to eqn (4.79). In some specific problems one of these formulations may be more convenient than the others; the analysis here, however, will be based on eqn (4.79).

Following the familiar perturbation scheme, a series of perturbation equations may now be generated by differentiating eqn (4.78) with respect to σ and evaluating the derivatives at $\sigma=0$, the critical point. Thus, differentiating eqn (4.78) yields

$$\dot{\omega}w^i_\tau+\omega\dot{w}^i_\tau=W_{ij}\dot{w}^j+W'_i\dot{\eta}, \tag{4.84}$$

where a dot and subscript τ are used to denote differentiations with respect to σ and τ, respectively.

Evaluating eqn (4.84) at c, using eqn (4.65), and recognizing that $w^i_\tau=0$ at c, results in

$$\omega_c\dot{w}^i_\tau=W_{ij}\dot{w}^j. \tag{4.85}$$

Introducing now eqn (4.79) into eqn (4.85) leads to

$$\sum_{m=0}^{M}m(-\dot{p}_{im}\sin m\tau+\dot{r}_{im}\cos m\tau)=W_{ij}\sum_{m=0}^{M}\{\dot{p}_{im}\cos m\tau+\dot{r}_{jm}\sin m\tau\}, \tag{4.86}$$

where m ranges from zero to M, embracing $p_{i0}(\sigma)$.

Comparing the coefficients of $\cos m\tau$ and $\sin m\tau$ for all m results in the derivatives

$$\begin{aligned}\dot{p}_{11}&=-\dot{r}_{21},\\ \dot{p}_{21}&=\dot{r}_{11}=0,\\ \dot{p}_{i1}&=\dot{r}_{i1}=0 \qquad (i\geq 3)\end{aligned} \tag{4.87}$$

and

$$\dot{p}_{im}=\dot{r}_{im}=0 \qquad (m\neq 1,\ \forall i).$$

At this stage, it is recognized that p_{11} and r_{21} are suitable candidates for σ, and choosing p_{11} for this role, the solution of eqn (4.85) may be expressed as

$$\begin{bmatrix}\dot{w}^1\\ \dot{w}^2\\ \vdots\\ \dot{w}^q\\ \dot{w}^{q+1}\\ \vdots\\ \dot{w}^s\\ \vdots\end{bmatrix}=\begin{bmatrix}1\\0\\ \vdots\\0\\0\\ \vdots\\0\\ \vdots\end{bmatrix}\cos\tau+\begin{bmatrix}0\\-1\\ \vdots\\0\\0\\ \vdots\\0\\ \vdots\end{bmatrix}\sin\tau+\ldots, \tag{4.88}$$

where a dot now and henceforth denotes differentiation with respect to p_{11}.

The second perturbation equation is generated by differentiating eqn (4.78) twice with respect to $\sigma = p_{11}$,

$$\ddot{\omega} w^i_\tau + 2\dot{\omega}\dot{w}^i_\tau + \omega\ddot{w}^i_\tau = W_{ijk}\dot{w}^j\dot{w}^k + 2W'_{ij}\dot{w}^j\dot{\eta} + W_{ij}\ddot{w}^j + W''_i(\dot{\eta})^2 + W'_i\ddot{\eta}, \tag{4.89}$$

which, upon evaluation at c, results in

$$2\dot{\omega}\dot{w}^i_\tau + \omega_c\ddot{w}^i_\tau = W_{ijk}\dot{w}^j\dot{w}^k + 2W'_{ij}\dot{w}^j\dot{\eta} + W_{ij}\ddot{w}^j. \tag{4.90}$$

Substituting eqn (4.79) into eqn (4.90), comparing the coefficients of the first harmonics, and observing that $\dot{p}_{11} = 1$ while $\ddot{p}_{11} = \dddot{p}_{11} = \ldots = 0$ yields

$$\dot{\omega} = 0 \quad \text{and} \quad (W'_{11} + W'_{22})\dot{\eta} = 0, \tag{4.91}$$

which results in

$$\dot{\eta} = 0, \tag{4.92}$$

due to the transversality condition (4.68).

Also, comparing the coefficients of all harmonics and using eqn (4.92) yields the second derivatives

$$\begin{aligned}
&\ddot{p}_{i1} = \ddot{r}_{i1} = 0, \qquad \ddot{p}_{im} = \ddot{r}_{im} = 0 \qquad (m \geq 3;\ \forall i),\\
&\ddot{p}_{10} = p^{(2)}_{10} \triangleq \frac{1}{2\omega_c}(W_{211} + W_{222}),\\
&\ddot{p}_{20} = p^{(2)}_{20} \triangleq -\frac{1}{2\omega_c}(W_{111} + W_{122}),\\
&\ddot{p}_{q0} = p^{(2)}_{q0} \triangleq \frac{\alpha_q}{2(\alpha_q^2 + \omega_q^2)}\left\{W_{q11} + W_{q22} + \frac{\omega_q}{\alpha_q}(W_{(q+1)11} + W_{(q+1)22})\right\},\\
&\ddot{p}_{(q+1)0} = p^{(2)}_{(q+1)0} \triangleq \frac{\omega_q}{2(\alpha_q^2 + \omega_q^2)}\left\{W_{q11} + W_{q22} + \frac{\alpha_q}{\omega_q}(W_{(q+1)11} + W_{(q+1)22})\right\},\\
&\ddot{p}_{s0} = p^{(2)}_{s0} \triangleq -\frac{1}{2\alpha_s}(W_{s11} + W_{s22}),\\
&\ddot{p}_{12} = p^{(2)}_{12} \triangleq \frac{1}{6\omega_c}(W_{222} - W_{211} + 4W_{112}),\\
&\ddot{p}_{22} = p^{(2)}_{22} \triangleq \frac{1}{6\omega_c}(W_{111} - W_{122} + 4W_{212}),\\
&\ddot{p}_{q2} = p^{(2)}_{q2} \triangleq \frac{1}{2\omega_c}\left[W_{q12} - \frac{\omega_c}{c_4}\left\{\frac{1}{\omega_q}\left(\frac{\omega_c}{\omega_q\alpha_q}c_1c_2 - c_3\right)\right.\right.\\
&\qquad\left.\left. + \frac{1}{\alpha_q}\left(\frac{\omega_c}{\omega_q\alpha_q}c_1c_3 + c_2\right)\right\}\right],
\end{aligned} \tag{4.93}$$

$$\ddot{p}_{(q+1)2} = p^{(2)}_{(q+1)2} \triangleq \frac{1}{2\omega_c}\left[W_{(q+1)12} + \frac{\omega_c}{c_4}\left\{\frac{1}{\alpha_q}\left(\frac{\omega_c}{\omega_q\alpha_q}c_1c_2 - c_3\right)\right.\right.$$
$$\left.\left. - \frac{1}{\omega_q}\left(\frac{\omega_c}{\omega_q\alpha_q}c_1c_3 + c_2\right)\right\}\right],$$
$$\ddot{p}_{s2} = p^{(2)}_{s2} \triangleq \frac{1}{4\omega_c^2 + \alpha_s^2}\left\{2\omega_c W_{s12} + \frac{\alpha_s}{2}(W_{s22} - W_{s11})\right\},$$
$$\ddot{r}_{12} = r^{(2)}_{12} \triangleq \frac{1}{3\omega_c}(W_{111} - W_{122} + W_{212}),$$
$$\ddot{r}_{22} = r^{(2)}_{22} \triangleq \frac{1}{3\omega_c}(W_{211} - W_{222} - W_{112}),$$
$$\ddot{r}_{q2} = r^{(2)}_{q2} \triangleq \frac{1}{c_4}\frac{\omega_c}{\omega_q\alpha_q}\left(\frac{\omega_c}{\omega_q\alpha_q}c_1c_2 - c_3\right),$$
$$\ddot{r}_{(q+1)2} = r^{(2)}_{(q+1)2} \triangleq \frac{1}{c_4}\frac{\omega_c}{\omega_q\alpha_q}\left(\frac{\omega_c}{\omega_q\alpha_q}c_1c_3 + c_2\right),$$
$$\ddot{r}_{s2} = r^{(2)}_{s2} \triangleq \frac{1}{4\omega_c^2 + \alpha_s^2}\{\alpha_s W_{s12} + \omega_c(W_{s11} - W_{s22})\},$$

where

$$c_1 \triangleq (4\omega_c^2 + \alpha_q^2 - \omega_q^2)/2\omega_c,$$
$$c_2 \triangleq \frac{1}{2}\left(W_{q11} - W_{q22} + \frac{\alpha_q}{\omega_c}W_{q12} + \frac{\omega_q}{\omega_c}W_{(q+1)12}\right),$$
$$c_3 \triangleq \frac{1}{2}\left(W_{(q+1)11} - W_{(q+1)22} - \frac{\omega_q}{\omega_c}W_{q12} + \frac{\alpha_q}{\omega_c}W_{(q+1)12}\right),$$
$$c_4 \triangleq 1 + \left(\frac{\omega_c}{\omega_q\alpha_q}c_1\right)^2.$$

It is again noted that the derivatives $\ddot{p}_{im} = \ddot{r}_{im} = 0$ $(m \geq 3)$ obtained above do not require the solution of $n(M-3)$ equations, but actually follow from

$$\omega_c m\ddot{r}_{im} = W_{ij}\ddot{p}_{jm} \quad \text{and} \quad -\omega_c m\ddot{p}_{im} = W_{ij}\ddot{r}_{jm}$$

which are obtained by comparing the coefficients of $\cos m\tau$ and $\sin m\tau$ for $m \geq 3$.

The third-order perturbation equation evaluated at c takes the form

$$3\ddot{\omega}\dot{w}^i_\tau + 3\dot{\omega}\ddot{w}^i_\tau + \omega_c\dddot{w}^i_\tau = W_{ijkl}\dot{w}^j\dot{w}^k\dot{w}^l + 3W'_{ijk}\dot{w}^j\dot{w}^k\dot{\eta} + 3W''_{ij}\dot{w}^j(\dot{\eta})^2$$
$$+ 3W_{ijk}\ddot{w}^j\dot{w}^k + 3W'_{ij}(\dot{w}^j\ddot{\eta} + \ddot{w}^j\dot{\eta}) + W_{ij}\dddot{w}^j. \quad (4.94)$$

Substituting eqn (4.79) into eqn (4.94) and using eqns (4.91) and (4.92) yields

$$\ddot{\eta} = (\gamma_{22} - \gamma_{11})/6\alpha' \tag{4.95}$$

and

$$\ddot{\omega} = \{(\gamma_{22} - \gamma_{11})\omega' - (\gamma_{12} + \gamma_{21})\alpha'\}/6\alpha' \tag{4.96}$$

where α' and ω' are given by eqn (4.73) and the γ_{ij}s are

$$\begin{aligned} \gamma_{i1} &\triangleq \tfrac{3}{4}(W_{i111} + W_{i122}) + 3\sum_{j=1}^{n} \{W_{ij1}(\ddot{p}_{j0} + \tfrac{1}{2}\ddot{p}_{j2}) - \tfrac{1}{2}W_{ij2}\ddot{r}_{j2}\}, \\ \gamma_{i2} &\triangleq -\tfrac{3}{4}(W_{i112} + W_{i222}) + 3\sum_{j=1}^{n} \{\tfrac{1}{2}W_{ij1}\ddot{r}_{j2} - W_{ij2}(\ddot{p}_{j0} - \tfrac{1}{2}\ddot{p}_{j2})\}. \end{aligned} \tag{4.97}$$

The third perturbation also yields the third-order derivatives of the amplitudes (Huseyin and Atadan 1983) which will be given later since these derivatives are not required for a first-order approximation of the periodic solution.

Using the derivatives obtained so far, the Taylor's expansions

$$\eta(\sigma) \equiv \eta(p_{11}) = \eta_c + \dot{\eta}p_{11} + \tfrac{1}{2}\ddot{\eta}(p_{11})^2 + \ldots,$$

and

$$\omega(\sigma) \equiv \omega(p_{11}) = \omega_c + \dot{\omega}p_{11} + \tfrac{1}{2}\ddot{\omega}(p_{11})^2 + \ldots, \tag{4.98}$$

yield the first-order approximations

$$\mu = \frac{1}{12\alpha'}(\gamma_{22} - \gamma_{11})(p_{11})^2, \tag{4.99}$$

and

$$\Omega = \frac{1}{12\alpha'}\{(\gamma_{22} - \gamma_{11})\omega' - (\gamma_{12} + \gamma_{21})\alpha'\}(p_{11})^2 \tag{4.100}$$

for the *parameter-amplitude* and *frequency-amplitude* relationships, respectively, where $\mu = \eta - \eta_c$, $\Omega = \omega - \omega_c$, and it is assumed that

$$\gamma_{22} - \gamma_{11} \neq 0. \tag{4.101}$$

The parameter-amplitude relationship (4.99) describes a *post-critical* bifurcation path in the μ-p_{11} plane, branching off the critical point c on the fundamental equilibrium path, and representing a family of limit cycles (Fig. 4.2). This path always exists either for $\mu > 0$ (*super-critical*) or for $\mu < 0$ (*sub-critical*), depending on whether the curvature $\ddot{\eta}$ is positive or negative, respectively. It can be shown that the former (latter) situation corresponds to a stable (unstable) family of limit cycles branching off from the fundamental equilibrium path which is stable (unstable) for $\eta < \eta_c$ ($\eta > \eta_c$). As depicted in Fig. 4.3, *transient solutions* approach (leave) the stable (unstable) family of limit cycles as t tends to infinity, the equilibrium path acting as a *source* (*sink*).

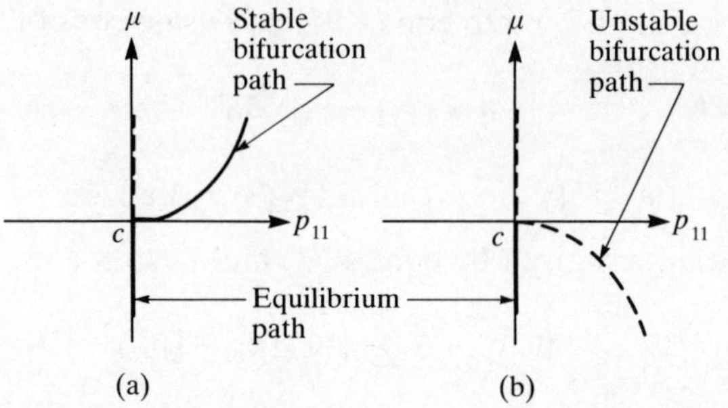

Fig. 4.2 Post-critical bifurcation path for a Hopf bifurcation.

Also note that, while the (Ω-p_{11}) relationship (eqn 4.100) is parabolic, like the bifurcation path (4.99), the first-order *parameter-frequency* relationship is linear and may be obtained by eliminating $(p_{11})^2$ between (4.99) and (4.100) to give

$$\Omega = [\{(\gamma_{22} - \gamma_{11})\omega' - (\gamma_{12} + \gamma_{21})\alpha'\}/(\gamma_{22} - \gamma_{11})]\mu. \qquad (4.102)$$

In order to construct ordered approximations for the periodic solution $w^i(\tau; p_{11})$, let the original Fourier series (4.79) be expressed as

$$w^i(\tau; p_{11}) = \dot{w}^i(\tau)p_{11} + \tfrac{1}{2}\ddot{w}^i(\tau)(p_{11})^2 + 0\{(p_{11})^3\} + \ldots \qquad (4.103)$$

where the functions $\dot{w}^i(\tau)$, $\ddot{w}^i(\tau)$, etc., follow from the Taylor's expansions of the amplitudes,

$$\begin{aligned} p_{im}(\sigma) &\equiv p_{im}(p_{11}) = p_{im}(0) + \dot{p}_{im}p_{11} + \tfrac{1}{2}\ddot{p}_{im}(p_{11})^2 + \ldots, \\ &\qquad\qquad (m = 0, \ldots, M) \\ r_{im}(\sigma) &\equiv r_{im}(p_{11}) = r_{im}(0) + \dot{r}_{im}p_{11} + \tfrac{1}{2}\ddot{r}_{im}(p_{11})^2 + \ldots \\ &\qquad\qquad (m = 1, \ldots, M). \end{aligned} \qquad (4.104)$$

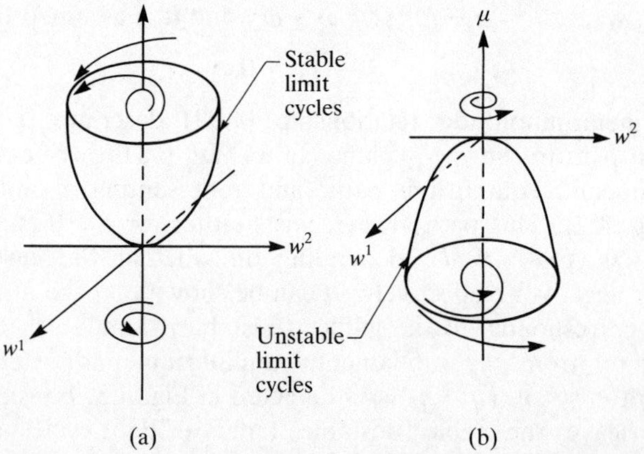

Fig. 4.3 Transient solutions (a) approaching the stable family of limit cycles and (b) leaving the unstable family of limit cycles, as t tends to infinity.

Thus,

$$\dot{\boldsymbol{w}}(\tau) = (\cos \tau, -\sin \tau, 0, \ldots, 0)^T, \tag{4.105}$$

and

$$\ddot{\boldsymbol{w}}(\tau) = \ddot{\boldsymbol{p}}_0 + \ddot{\boldsymbol{p}}_2 \cos 2\tau + \ddot{\boldsymbol{r}}_2 \sin 2\tau \tag{4.106}$$

in which

$$\ddot{\boldsymbol{p}}_0 = (\ddot{p}_{10}, \ddot{p}_{20}, \ldots, \ddot{p}_{n0})^T,$$
$$\ddot{\boldsymbol{p}}_2 = (\ddot{p}_{12}, \ddot{p}_{22}, \ldots, \ddot{p}_{n2})^T,$$
$$\ddot{\boldsymbol{r}}_2 = (\ddot{r}_{12}, \ddot{r}_{22}, \ldots, \ddot{r}_{n2})^T,$$

and T denotes the transpose of a vector.

In addition, one may proceed a step further to construct the third-order terms

$$\dddot{w}^i(\tau)(p_{11})^3$$

in eqn (4.103) without having to resort to further perturbations, since the third-order perturbation equation (4.94) yields the necessary derivatives of the amplitudes along with eqns (4.95) and (4.96), as remarked earlier. The derivatives which *do not vanish* are listed below:

$$\dddot{p}_{21} = p_{21}^{(3)} \triangleq -(\gamma_{22}W'_{11} + \gamma_{11}W'_{22})/2\omega_c\alpha',$$
$$\dddot{p}_{q1} = p_{q1}^{(3)} \triangleq (-g_1d_1 + 2\alpha_q\omega_qd_2)/(g_1^2 + 4\alpha_q^2\omega_q^2),$$
$$\dddot{p}_{(q+1)1} = p_{(q+1)1}^{(3)} \triangleq -(g_1d_2 + 2\alpha_q\omega_qd_1)/(g_1^2 + 4\alpha_q^2\omega_q^2),$$
$$\dddot{p}_{s1} = p_{s1}^{(3)} \triangleq -\{\omega_c\gamma_{s2} + \alpha_s\gamma_{s1} + (\omega_cW'_{s2} - \alpha_sW'_{s1})(\gamma_{11} - \gamma_{22})/2\alpha'\}/ (\omega_c^2 + \alpha_s^2),$$
$$\dddot{r}_{21} = r_{21}^{(3)} \triangleq \{(\gamma_{21} - \gamma_{12}) + (W'_{12} + W'_{21})(\gamma_{22} - \gamma_{11})/2\alpha'\}/2\omega_c,$$
$$\dddot{r}_{q1} = r_{q1}^{(3)} \triangleq (g_1d_3 - 2\alpha_q\omega_qd_4)/(g_1^2 + 4\alpha_q^2\omega_q^2),$$
$$\dddot{r}_{(q+1)1} = r_{(q+1)1}^{(3)} \triangleq (g_1d_4 + 2\alpha_q\omega_qd_3)/(g_1^2 + 4\alpha_q^2\omega_q^2),$$
$$\dddot{r}_{s1} = r_{s1}^{(3)} \triangleq -\{\alpha_s\gamma_{s2} - \omega_c\gamma_{s1} + (\alpha_sW'_{s2} + \omega_cW'_{s1})(\gamma_{11} - \gamma_{22})/2\alpha'\}/ (\omega_c^2 + \alpha_s^2),$$
$$\dddot{p}_{13} = p_{13}^{(3)} \triangleq -(3v_{12} + v_{21})/8\omega_c,$$
$$\dddot{p}_{23} = p_{23}^{(3)} \triangleq (v_{11} - 3v_{22})/8\omega_c,$$
$$\dddot{p}_{q3} = p_{q3}^{(3)} \triangleq (-g_2e_1 + 2\alpha_q\omega_qe_2)/(g_2^2 + 4\alpha_q^2\omega_q^2),$$
$$\dddot{p}_{(q+1)3} = p_{(q+1)3}^{(3)} \triangleq -(g_2e_2 + 2\alpha_q\omega_qe_1)/(g_2^2 + 4\alpha_q^2\omega_q^2),$$
$$\dddot{p}_{s3} = p_{s3}^{(3)} \triangleq -(\alpha_sv_{s1} + 3\omega_cv_{s2})/(9\omega_c^2 + \alpha_s^2),$$
$$\dddot{r}_{13} = r_{13}^{(3)} \triangleq (3v_{11} - v_{22})/8\omega_c,$$
$$\dddot{r}_{23} = r_{23}^{(3)} \triangleq (v_{12} + 3v_{21})/8\omega_c,$$
$$\dddot{r}_{q3} = r_{q3}^{(3)} \triangleq (g_2e_3 - 2\alpha_q\omega_qe_4)/(g^2 + 4\alpha_q^2\omega_q^2),$$
$$\dddot{r}_{(q+1)3} = r_{(q+1)3}^{(3)} \triangleq (g_2e_4 + 2\alpha_q\omega_qe_3)/(g^2 + 4\alpha_q^2\omega_q^2),$$
$$\dddot{r}_{s3} = r_{s3}^{(3)} \triangleq (3\omega_cv_{s1} - \alpha_sv_{s2})/(9\omega_c^2 + \alpha_s^2),$$

where

$$g_1 \triangleq \omega_c^2 + \alpha_q^2 - \omega_q^2, \qquad g_2 \triangleq 9\omega_c^2 + \alpha_q^2 - \omega_q^2,$$

$$d_1 \triangleq [\gamma_{q2} + (\alpha_q \gamma_{q1} + \omega_q \gamma_{(q+1)1})/\omega_c + \{W'_{q2} - (\alpha_q W'_{q1} + \omega_q W'_{(q+1)1})/\omega_c\}(\gamma_{11} - \gamma_{22})/2\alpha']\omega_c$$

$$d_2 \triangleq [\gamma_{(q+1)2} + (\alpha_q \gamma_{(q+1)1} - \omega_q \gamma_{q1})/\omega_c + \{W'_{(q+1)2} + (\omega_q W'_{q1} - \alpha_q W'_{(q+1)1})/\omega_c\}(\gamma_{11} - \gamma_{22})/2\alpha']\omega_c$$

$$d_3 \triangleq [\gamma_{q1} - (\alpha_q \gamma_{q2} + \omega_q \gamma_{(q+1)2})/\omega_c + \{W'_{q1} + (\alpha_q W'_{q2} + \omega_q W'_{(q+1)2})/\omega_c\}(\gamma_{22} - \gamma_{11})/2\alpha']\omega_c,$$

$$d_4 \triangleq [\gamma_{(q+1)1} + (\omega_q \gamma_{q2} - \alpha_q \gamma_{(q+2)2})/\omega_c + \{W'_{(q+1)1} + (-\omega_q W'_{q2} + \alpha_q W'_{(q+1)2})/\omega_c\}(\gamma_{22} - \gamma_{11})/2\alpha']\omega_c,$$

$$v_{i1} \triangleq \tfrac{1}{4}(W_{i111} - 3W_{i122}) + \tfrac{3}{2}\sum_{j=1}^{n} (W_{ij1}\ddot{p}_{j2} + W_{ij2}\ddot{r}_{j2}),$$

$$v_{i2} \triangleq -\tfrac{1}{4}(3W_{i112} - W_{i222}) + \tfrac{3}{2}\sum_{j=1}^{n} (W_{ij1}\ddot{r}_{j2} - W_{ij2}\ddot{p}_{j2}),$$

$$e_1 \triangleq 3\omega_c v_{q2} + \alpha_q v_{q1} + \omega_q v_{(q+1)1},$$

$$e_2 \triangleq 3\omega_c v_{(q+1)2} - \omega_q v_{q1} + \alpha_q v_{(q+1)1},$$

$$e_3 \triangleq 3\omega_c v_{q1} - \alpha_q v_{q2} - \omega_q v_{(q+1)2},$$

$$e_4 \triangleq 3\omega_c v_{(q+1)1} + \omega_q v_{q2} - \alpha_q v_{(q+1)2},$$

and it is noted that summation is not implied over repeated indices.

On the basis of these derivatives, one has

$$\dddot{\boldsymbol{w}}(\tau) = \dddot{\boldsymbol{p}}_1 \cos\tau + \dddot{\boldsymbol{r}}_1 \sin\tau + \dddot{\boldsymbol{p}}_3 \cos 3\tau + \dddot{\boldsymbol{r}}_3 \sin 3\tau \tag{4.107}$$

where

$$\dddot{\boldsymbol{p}}_1 = (0, \dddot{p}_{21}, \dddot{p}_{31}, \ldots, \dddot{p}_{n1})^T,$$

$$\dddot{\boldsymbol{r}}_1 = (0, \dddot{r}_{21}, \dddot{r}_{31}, \ldots, \dddot{r}_{n1})^T,$$

$$\dddot{\boldsymbol{p}}_3 = (\dddot{p}_{13}, \dddot{p}_{23}, \ldots, \dddot{p}_{n3})^T,$$

and

$$\dddot{\boldsymbol{r}}_3 = (\dddot{r}_{13}, \dddot{r}_{23}, \ldots, \dddot{r}_{n3})^T.$$

It is noted that, together with the transversality condition $\alpha' \neq 0$, the assumption (4.101) plays an important role in the post-critical oscillatory behaviour of the system. In fact, the bifurcation path (4.99) becomes invalid if any of these conditions is violated, and one has to proceed to higher order perturbations to obtain a new relationship. Such degenerate cases will be examined after illustrating the application of the formulas obtained in this section.

4.4.2 *An illustrative example*

Consider a mathematical model to demonstrate the applicability of the preceding theoretical results (Atadan 1983):

$$\frac{dz^1}{dt} = z^2 + \beta\mu z^1 - \beta\{2(z^1)^2 + z^1z^2 - 6z^2z^3\} - \beta^2\{-6(z^1)^2z^2 - 2(z^2)^3\},$$

$$\frac{dz^2}{dt} = -z^1 + \beta\mu z^2 + 5\mu z^3 + \beta\{10(z^1)^2 - 8z^1z^2 + z^1z^3 - 8z^2z^3 + 5(z^3)^2\}$$
$$- \beta^2\{60(z^1)^3 + 4(z^1)^2z^2 + 152z^1(z^2)^2 + (z^2)^3\},$$

$$\frac{dz^3}{dt} = -2z^3 + 3\beta\mu z^3 + \beta\{4(z^1)^2 + z^2z^3 + 3(z^3)^2\} - 5\beta^2z^1z^2z^3 \tag{4.108}$$

where β is a negative constant, and μ is a parameter.

It is first noted that $z^i = 0$ implies $dz^i/dt = 0$ for all μ, and the Jacobian along the equilibrium path $z^i = 0$ is in the form

$$\left[\frac{dz}{dt}\right]_{z=0} = \begin{bmatrix} \beta\mu & 1 & 0 \\ -1 & \beta\mu & 5\mu \\ 0 & 0 & -2+3\beta\mu \end{bmatrix}, \tag{4.109}$$

with the eigenvalues

$$\lambda_{1,2} = \beta\mu \pm i, \qquad \lambda_3 = -2 + 3\beta\mu. \tag{4.110}$$

It then follows that the transformation (4.63) is not required in this particular problem since the system is in the form of eqn (4.64) for $\mu = \mu_c = 0$. The complex conjugate eigenvalues, $\lambda_{1,2}$, cross the imaginary axis at $\mu = 0$ and all eigenvalues have negative (positive) real parts for $\mu > 0$ ($\mu < 0$), indicating that the equilibrium path is stable for $\mu > 0$ and becomes unstable for $\mu < 0$—a reversed stability property compared to what was assumed in the theory.

Introducing now the time scaling $\tau = \omega t$, eqn (4.108) is transformed into

$$\omega \begin{bmatrix} w^1_\tau \\ w^2_\tau \\ w^3_\tau \end{bmatrix} = \begin{bmatrix} \beta\mu & 1 & 0 \\ -1 & \beta\mu & 5\mu \\ 0 & 0 & -2+3\beta\mu \end{bmatrix} \begin{bmatrix} w^1 \\ w^2 \\ w^3 \end{bmatrix} + f(w^i), \tag{4.111}$$

which is in the notation used in the theory. The non-linear terms in eqn (4.108) are represented by $f(w^i)$.

The non-zero system coefficients evaluated at the critical equilibrium state—which is identified by $w^i = 0$ and $\mu = 0$—are readily determined as

$$\begin{aligned}
&W_{12} = -W_{21} = \omega_c = 1, \qquad W_{33} = \alpha_3 = -2,\\
&W'_{11} = W'_{22} = \tfrac{1}{3}W'_{33} = \beta, \qquad W'_{23} = 2,\\
&\tfrac{1}{4}W_{111} = W_{112} = -\tfrac{1}{6}W_{123} = -\beta,\\
&\tfrac{1}{20}W_{211} = -\tfrac{1}{8}W_{212} = W_{213} = -\tfrac{1}{8}W_{223} = \tfrac{1}{10}W_{233} = \beta,\\
&\tfrac{1}{8}W_{311} = -W_{323} = -\tfrac{1}{6}W_{333} = \beta,\\
&W_{1112} = W_{1222} = 12\beta^2,\\
&\tfrac{1}{360}W_{2111} = \tfrac{1}{8}W_{2112} = \tfrac{1}{304}W_{2122} = \tfrac{1}{6}W_{2222} = -\beta^2,\\
&W_{3123} = -5\beta^2.
\end{aligned} \tag{4.112}$$

Clearly, the transversality condition,

$$\alpha' = \tfrac{1}{2}(W'_{11} + W'_{22}) = \beta \neq 0, \tag{4.113}$$

is satisfied and there exists, therefore, a family of limit cycles bifurcating from the equilibrium path $w^i = 0$ at $\mu = 0$. According to the theory, $\dot{\mu} = \dot{\omega} = 0$, and one proceeds to determine the second derivatives (4.95) and (4.96). As a first step in this direction, one evaluates the derivatives in eqns (4.93) with the aid of eqns (4.112). Thus,

$$\begin{aligned}
&\ddot{p}_{10} = 10\beta, \qquad \ddot{p}_{20} = 2\beta,\\
&\ddot{p}_{30} = -\frac{1}{2\alpha_3}(W_{311} + W_{322}) = 2\beta;
\end{aligned} \tag{4.114}$$

$$\begin{aligned}
&\ddot{p}_{12} = -4\beta,\\
&\ddot{p}_{22} = -6\beta,\\
&\ddot{p}_{32} = \frac{1}{4\omega_c^2 + \alpha_3^2}(-\alpha_3 W_{311}/2) = \beta;
\end{aligned} \tag{4.115}$$

and

$$\begin{aligned}
&\ddot{r}_{12} = -4\beta,\\
&\ddot{r}_{22} = 7\beta,\\
&\ddot{r}_{32} = \frac{1}{4\omega_c^2 + \alpha_3^2}\{\alpha_3 W_{312} + \omega_c(W_{311} - W_{322})\} = \beta.
\end{aligned} \tag{4.116}$$

Introducing these derivatives into eqn (4.97) results in

$$\begin{aligned}
&\gamma_{11} = -108\beta^2, \qquad \gamma_{12} = \tfrac{9}{2}\beta^2,\\
&\gamma_{21} = -\tfrac{45}{2}\beta^2, \qquad \text{and} \quad \gamma_{22} = 132\beta^2.
\end{aligned} \tag{4.117}$$

It is now observed that $\gamma_{22} - \gamma_{11} = 240\beta^2 \neq 0$ and the condition (4.101) is satisfied.

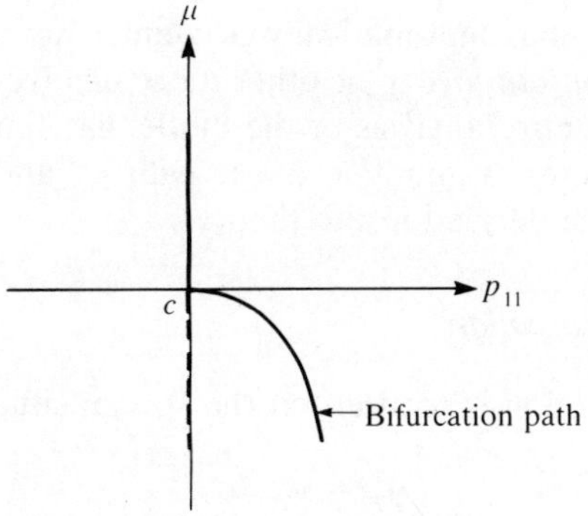

Fig. 4.4 Bifurcation path.

The relationships (4.99) and (4.100) follow readily as

$$\mu = 20\beta(p_{11})^2 \tag{4.118}$$

and

$$\Omega = \tfrac{3}{2}\beta^2(p_{11})^2, \qquad (\omega = \omega_c + \Omega), \tag{4.119}$$

respectively.

The bifurcation path (4.118) has a negative curvature since $\beta < 0$, and combining this result with the stability properties of the equilibrium path it can be demonstrated that the family of bifurcating limit cycles is stable (Fig. 4.4).

Substituting for the derivatives (4.114–4.116) into eqn (4.103) yields the limit cycles as

$$\begin{bmatrix} w^1 \\ w^2 \\ w^3 \end{bmatrix} = \begin{bmatrix} \cos\tau \\ -\sin\tau \\ 0 \end{bmatrix} p_{11} + \frac{1}{2}\left\{ \begin{bmatrix} 10 \\ 2 \\ 2 \end{bmatrix}\beta - \begin{bmatrix} 4 \\ 6 \\ -1 \end{bmatrix}\beta\cos 2\tau + \begin{bmatrix} -4 \\ 7 \\ 1 \end{bmatrix}\beta\sin 2\tau \right\}(p_{11})^2 + O\{(p_{11})^3\} + \ldots$$

where $\tau = \omega t$ and ω is given by eqn (4.119).

It is important to observe that the bifurcation path (4.118) as well as the frequency-amplitude relationship (4.119) are not independent of the non-critical state variable, w^3, despite the fact that the Jacobian (4.109) is in a block diagonal form—effectively uncoupling the critical pair of complex conjugate eigenvalues from the non-critical eigenvalue. In other words, by virtue of the formulae (4.97), the third state variable has a quantitative effect on the curvatures of the bifurcation path and frequency-parameter relationship. The intrinsic method of harmonic balancing, however, has automatically accounted for this effect without requiring *centre manifold theory* or other more involved techniques. As a matter of fact, the entire analysis of the model has simply been achieved mainly by determining a number of derivatives and substituting into appropriate formulae derived in the theory.

4.4.3 *Flat Hopf bifurcation*

In this section, attention is focused on the special situation arising under the condition

$$\gamma_{22} - \gamma_{11} = 0. \tag{4.120}$$

Thus, consider the transformed system (4.64) with all its basic properties described in Section 4.4.1, except for condition (4.101) which is replaced by condition (4.120) here. Under this condition, one has

$$\ddot{\mu} = 0 \quad \text{and} \quad \ddot{\omega} = -(\gamma_{12} + \gamma_{21})/6, \tag{4.121}$$

in addition to $\dot{\mu} = 0$ and $\dot{\omega} = 0$, indicating that for a first-order approximation of the bifurcation path, higher order perturbation problems will have to be solved. It is also noted that, by virtue of condition (4.120), the third-order derivatives $\dddot{p}_{21}$ and $\dddot{r}_{21}$ take the form

$$\dddot{p}_{21} = -\gamma_{11}/\omega_c \tag{4.122}$$

and

$$\dddot{r}_{21} = (\gamma_{21} - \gamma_{12})/2\omega_c, \tag{4.123}$$

respectively, and the derivatives which are not included in the list preceding eqn (4.107) are actually zero, that is

$$\dddot{p}_{i0} = \dddot{p}_{i2} = \dddot{r}_{i2} = \dddot{p}_{im} = \dddot{r}_{im} = 0 \qquad (m \geq 4,\ \forall i), \tag{4.124}$$

while $\dddot{p}_{11} \equiv 0$ and $\dddot{r}_{11} \equiv 0$.

Since the transversality condition (4.68) is valid here as well, the algebra related to the fourth perturbation of eqn (4.78) is essentially similar to that of the third-order perturbation carried out in Section 4.4.1—although considerably lengthier. If the fourth perturbation is carried out (Atadan and Huseyin 1984*b*) with the aid of the derivatives obtained so far, one has

$$\dddot{\eta} = 0, \qquad \dddot{\omega} = 0. \tag{4.125}$$

Similarly, a fifth perturbation of eqn (4.78) yields

$$\eta^{(4)} = (\delta_{22} - \delta_{11})/10\alpha' \tag{4.126}$$

where the superscript $^{(4)}$ denotes the fourth derivative of η with respect to p_{11} evaluated at c. This notation will be adopted in the following to indicate higher derivatives. The constants δ_{11} and δ_{22} are lengthy

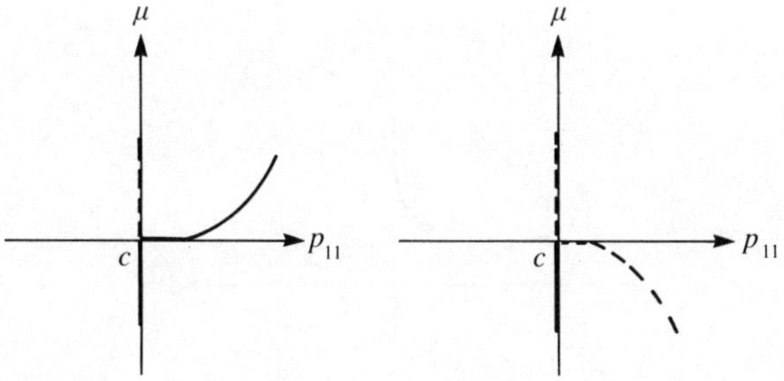

Fig. 4.5 Bifurcation path for a flat Hopf bifurcation.

expressions in terms of the system coefficients, and are not given here. However, a modified form of these constants will be obtained in Section 4.5 under a different formulation and they are listed in the Appendix.

The bifurcation path, representing the family of limit cycles branching off the point c, is now obtained as

$$\mu = \{(\delta_{22} - \delta_{11})/240\alpha'\}(p_{11})^4 \tag{4.127}$$

which describes a *flat* curve in the μ–p_{11} plane (Fig. 4.5). As in the case of the ordinary Hopf bifurcation, limit cycles exist either for $\mu > 0$ or $\mu < 0$, and if the equilibrium path is stable for $\mu < 0$ and unstable for $\mu > 0$, then the entire family of limit cycles in the vicinity of the critical point c is stable if $\eta^{(4)} > 0$ and unstable if $\eta^{(4)} < 0$.

The frequency-amplitude relationship follows from eqn (4.121) as

$$\Omega = -\{(\gamma_{12} - \gamma_{21})/12\}(p_{11})^2, \tag{4.128}$$

and the limit cycles are expressed as

$$w^i(\tau; p_{11}) = \dot{w}^i(\tau)p_{11} + \tfrac{1}{2}\ddot{w}^i(\tau)(p_{11})^2 + \frac{1}{3!}\dddot{w}^i(\tau)(p_{11})^3 + O\{(p_{11})^4\} + \dots \tag{4.129}$$

where $\dot{w}^i(\tau)$, $\ddot{w}^i(\tau)$, and $\dddot{w}^i(\tau)$ are as in eqns (4.105), (4.106), and (4.107) respectively, except that the coefficients associated with $\dddot{w}^i(\tau)$ are determined by taking into account condition (4.120) as indicated in eqns (4.121) and (4.122).

4.4.4 Symmetric bifurcation (tri-furcation)

So far, it has always been assumed that the transversality condition, $\alpha' \neq 0$, holds. If this condition is violated, however, several new situations may arise depending on the higher order derivatives of α as well as other coefficients.

Suppose now that a pair of complex conjugate eigenvalues, $\lambda(\eta) = \alpha(\eta) \pm i\omega(\eta)$, of the Jacobian of the fundamental equilibrium path

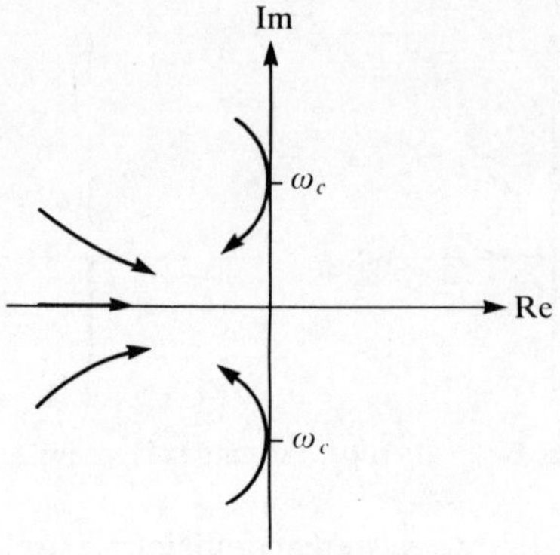

Fig. 4.6 Symmetric bifurcation: as $\eta = \eta_c$ the loci of a pair of complex conjugate eigenvalues in the λ plane becomes tangent to the imaginary axis.

associated with the autonomous system (4.64) varies with η such that at $\eta = \eta_c$ the loci of these eigenvalues in the complex λ-plane become tangent to the imaginary axis—instead of crossing it—while all the other eigenvalues remain in the left half-plane (Fig. 4.6). This occurs, for example, if

$$\alpha' = 0 \quad \text{and} \quad \alpha'' = \mathrm{d}^2\alpha/\mathrm{d}\eta^2|_c \neq 0, \tag{4.130}$$

which will be assumed to hold here. In fact, if the equilibrium path is stable for $\mu < 0$, then $\alpha'' < 0$ and the path remains stable for $\mu > 0$. All the other assumptions associated with the Hopf bifurcation, including assumption (4.101), are assumed to be valid here as well.

It can be shown by differentiating the determinantal identity (4.70) a second time with respect to η and evaluating at the critical point c with the aid of eqns (4.65), (4.66), (4.67), and $\alpha' = 0$, that

$$\alpha'' = \frac{1}{2}\left\{W''_{11} + W''_{22} + \sum_{i=3}^{n} (W'_{1i}p^{(2)}_{i1} + W'_{2i}r^{(3)}_{i1})\right\},$$

where $p^{(2)}_{i1}$ and $r^{(2)}_{i1}$ are given by eqns (4.93).

At this stage, it is not known whether p_{11} is an appropriate perturbation parameter to describe a possible family of limit cycles bifurcating from the equilibrium path at the critical point c; in fact, it is not even known whether such a family exists. Therefore, let the family—if it exists—be in the parametric form (4.76), and let the identity (4.78) serve again as the basis of the familiar perturbation procedure—with σ as the perturbation parameter.

Thus, differentiating eqn (4.78) with respect to σ and evaluating at c yields

$$\omega_c \dot{w}^i_\tau = W_{ij}\dot{w}^j \tag{4.131}$$

which is the same as eqn (4.85). As a matter of fact, introducing eqn (4.79) into eqn (4.131) again results in the first derivatives (4.87) which are repeated here for convenience:

$$\dot{p}_{11} = -\dot{r}_{21}, \qquad \dot{p}_{21} = \dot{r}_{11} = 0,$$
$$\dot{p}_{i1} = \dot{r}_{i1} = 0 \qquad (i \geq 3),$$

and
$$\dot{p}_{im} = \dot{r}_{im} = 0 \qquad (m \neq 1, \forall i). \tag{4.132}$$

There is strong indication once more—as in the case of the Hopf bifurcation—that p_{11} (or r_{21}) is a suitable candidate for σ. Nevertheless, if this observation is ignored, and eqn (4.79) is introduced into the second perturbation equation eqn (4.90), one obtains

$$\dot{p}_{11}\dot{\omega} = \omega'\dot{p}_{11}\dot{\eta}, \tag{4.133}$$
$$\alpha'\dot{p}_{11}\dot{\eta} = 0, \tag{4.134}$$
$$\ddot{p}_{21} = -\frac{1}{\omega_c}(W'_{11} - W'_{22})\dot{p}_{11}\dot{\eta}, \tag{4.135}$$
$$\ddot{r}_{21} = \frac{1}{\omega_c}(W'_{12} + W'_{21})\dot{p}_{11}\dot{\eta} - \ddot{p}_{11}, \tag{4.136}$$
$$\ddot{p}_{i1} = p^{(2)}_{i1}\dot{p}_{11}\dot{\eta}, \qquad \ddot{r}_{i1} = r^{(2)}_{i1}\dot{p}_{11}\dot{\eta} \qquad (i \geq 3), \tag{4.137}$$
$$\ddot{p}_{im} = p^{(2)}_{im}(\dot{p}_{11})^2, \qquad \ddot{r}_{i2} = r^{(2)}_{i2}(\dot{p}_{11})^2 \qquad (m = 0, 2; \forall i), \tag{4.138}$$

and
$$\ddot{p}_{im} = \ddot{r}_{im} = 0, \qquad (m \geq 3; \forall i), \tag{4.139}$$

where $p^{(2)}_{i1}$, $r^{(2)}_{i1}$, $r^{(2)}_{i2}$, $p^{(2)}_{im}$ $(m = 0, 2)$ are defined by (4.93). It is also noted that $\ddot{r}_{11} = 0$ due to eqn (4.81), and ω' is given by (4.73).

Now, it appears from these results that both p_{11} and η are suitable for the role of σ. Ignoring this indication once more, however, the third perturbation equation (4.94) yields

$$\{(\gamma_{11} - \gamma_{22})(\dot{p}_{11})^2 + 6\alpha''(\dot{\eta})^2\}\dot{p}_{11} = 0 \tag{4.140}$$

upon solving the algebraic equations corresponding to the first harmonics. This equation indicates that either $\dot{p}_{11} = 0$ or the expression in the brackets vanishes. The former situation corresponds to the equilibrium path while the latter yields the derivatives associated with the bifurcation path. If one sets $\sigma = p_{11}$, then

$$\dot{p}_{11} = 1 \quad \text{and} \quad \ddot{p}_{11} = \dddot{p}_{11} = \ldots = 0, \tag{4.141}$$

in which case the equilibrium path is automatically excluded and eqn (4.140) yields

$$\dot{\eta} = \pm\{(\gamma_{22} - \gamma_{11})/6\alpha''\}^{1/2} \tag{4.142}$$

where a dot now indicates differentiation with respect to p_{11}.

It is noted that, with $\sigma = p_{11}$, eqn (4.133) reduces to

$$\dot{\omega} = \omega' \dot{\eta}, \tag{4.143}$$

while eqn (4.134) is satisfied identically by virtue of assumption (4.130). Also, all the remaining derivatives, (4.135) to (4.138), can now be determined by introducing eqns (4.141) and (4.142) into these equations.

A first approximation for the bifurcation path follows from eqn (4.142) as

$$\mu = \pm\{(\gamma_{22} - \gamma_{11})/6\alpha''\}^{1/2} p_{11} \tag{4.144}$$

which indicates that a family of limit cycles branches off from the critical point c provided the *existence condition*

$$(\gamma_{22} - \gamma_{11})/\alpha'' > 0 \tag{4.145}$$

is satisfied. The amplitude-parameter relationship (4.144) describes a pair of intersecting straight lines in the (μ–p_{11}) space, and it can be shown that while the stable equilibrium path retains its stability through c, the pair of bifurcating paths is totally unstable in the vicinity of c. By extension, it is not difficult to reach the conclusion that if one starts with an unstable equilibrium path, it remains unstable through c while the bifurcation paths—if they exist—are totally stable (Fig. 4.7). This phenomenon has an unmistakable resemblance to the *tri-furcation* behaviour of equilibrium paths discussed in Section 3.1.2.

A first-order approximation for the frequency-amplitude relationship is also readily obtained from eqn (4.143) as

$$\Omega = \pm\omega'\{(\gamma_{22} - \gamma_{11})/6\alpha''\}^{1/2} p_{11} \tag{4.146}$$

which is real if condition (4.145) is satisfied.

On the other hand, the equations of the limit cycles are in the form

$$w^i(\tau; p_{11}) = \dot{w}^i(\tau) p_{11} + \tfrac{1}{2}\ddot{w}^i(\tau)(p_{11})^2 + \ldots, \tag{4.147}$$

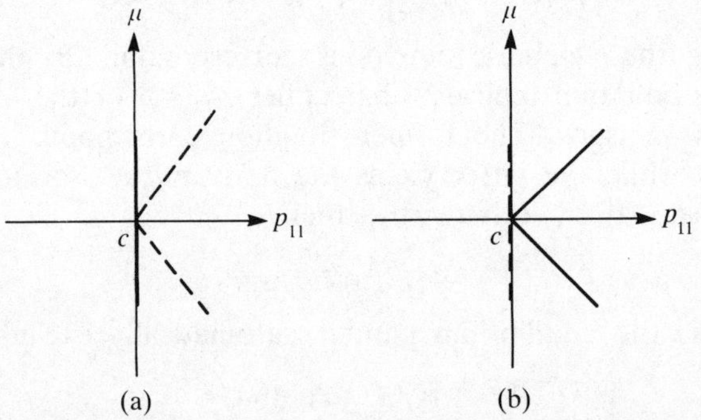

Fig. 4.7 Bifurcation paths for symmetric bifurcation (tri-furcation).

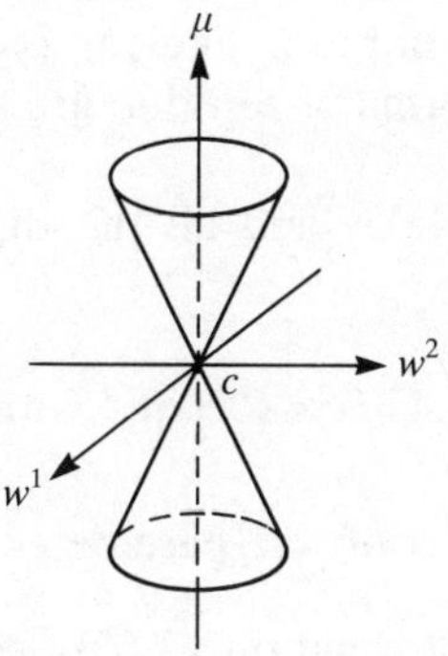

Fig. 4.8 The family of limit cycles in state space for symmetric bifurcation.

where $\dot{w}^i(\tau)$ is given by eqn (4.88) and $\ddot{w}^i(\tau)$ by

$$\ddot{w}^i(\tau) = p_{i0}^{(2)} + p_{i1}^{(2)}\dot{\eta}\cos\tau + r_{i1}^{(2)}\dot{\eta}\sin\tau + p_{i2}^{(2)}\cos 2\tau + r_{i2}^{(2)}\sin 2\tau. \tag{4.148}$$

In applications, $\dot{\eta}$ is substituted from eqn (4.142) while $p_{im}^{(2)}$ and $r_{im}^{(2)}$ are determined by means of eqn (4.93) as in the case of the Hopf bifurcation. The family of limit cycles in the state space is shown in Fig. 4.8.

Finally, it will be instructive to examine how the analysis would have developed if the parameter η were chosen as σ. To this end, it is first recognized that if $\sigma = \eta$, then

$$\dot{\eta} \equiv \eta' = 1 \quad \text{and} \quad \eta'' = \eta''' = \ldots = 0, \tag{4.149}$$

and the derivatives (4.133) to (4.139) are determined accordingly by substituting for eqn (4.149) and changing the 'dots' to 'primes', indicating derivatives with respect to η. Similarly, the relationship (4.140) now yields $p'_{11} = 0$ or

$$p'_{11} = \pm\{6\alpha''/(\gamma_{22} - \gamma_{11})\}^{1/2} \tag{4.150}$$

which again leads to eqns (4.144) and (4.146).

The limit cycles are in the form

$$w^i(\tau;\mu) = w^{i,\prime}(\tau)\mu + \tfrac{1}{2}w^{i,\prime\prime}(\tau)\mu^2 + \ldots, \tag{4.151}$$

where

$$w^{i,\prime}(\tau) = (\cos\tau, -\sin\tau, 0, \ldots, 0)^T p'_{11}, \tag{4.152}$$

$$\begin{aligned} w^{i,\prime\prime}(\tau) = {} & p_{i0}^{(2)}(p'_{11})^2 + p_{i1}^{(2)}p'_{11}\cos\tau + r''_{i1}\sin\tau \\ & + p_{i2}^{(2)}(p'_{11})^2\cos 2\tau + r_{i2}^{(2)}(p'_{11})^2\sin 2\tau, \end{aligned} \tag{4.153}$$

and r''_{i1} follows from eqns (4.136) and (4.137) as

$$\begin{aligned} r''_{21} &= \frac{1}{\omega_c}(W'_{12} + W'_{21})p'_{11} - p''_{11}, \\ r''_{i1} &= r_{i1}^{(2)}p'_{11} \qquad (i \geq 3). \end{aligned} \tag{4.154}$$

The solution (4.151) may not look like eqn (4.147), but it is not difficult to demonstrate that both forms of solution are equivalent.

Problem 4.3. Verify eqn (4.93) and all the third-order derivatives $p_{ij}^{(3)}$, $r_{ij}^{(3)}$.

Problem 4.4. Verify eqn (4.117) associated with the illustrative example.

4.5 Alternative formulation of one-parameter systems

It is recalled that the transformation (3.65) was introduced to facilitate the analyses of equilibrium path configurations and stability distributions in the preceding chapter. However, such a transformation from the W-system to the X-system is not always possible as noted before. Besides, it may not be desirable from an applications point of view, because the transformation itself requires additional work. The analyses of dynamic bifurcations, therefore, have so far been carried out on the basis of the W-system. Nevertheless, the structure of the Jacobian in all three cases analysed in Section 4.4 does not prevent the introduction of the X-system, and the latter is almost indispensable for the analyses of more degenerate cases. Furthermore, the stability analysis can best be performed on the basis of the X-system.

Thus, introduce the transformation

$$\boldsymbol{z} = \boldsymbol{f}(\eta) + \boldsymbol{Q}(\eta)\boldsymbol{x}, \qquad |\boldsymbol{Q}(\eta)| \neq 0, \tag{4.155}$$

into eqn (4.60) to obtain

$$\frac{\mathrm{d}\boldsymbol{x}}{\mathrm{d}t} = \boldsymbol{X}(\boldsymbol{x}, \eta), \tag{4.156}$$

with the properties

$$\boldsymbol{X}(\boldsymbol{0}, \eta) = \boldsymbol{X}'(\boldsymbol{0}, \eta) = \ldots = \boldsymbol{0} \tag{4.157}$$

and

$$\left[\frac{\partial \boldsymbol{X}}{\partial \boldsymbol{x}}\right]_{\boldsymbol{x}=\boldsymbol{0}} = \operatorname{diag}\,[\boldsymbol{K}_1, \boldsymbol{K}_2, \ldots, \boldsymbol{K}_q, \ldots, \alpha_s, \ldots], \tag{4.158}$$

such that

$$\boldsymbol{K}_q = \begin{bmatrix} X_{qq}(\eta) & X_{q(q+1)}(\eta) \\ X_{(q+1)q}(\eta) & X_{(q+1)(q+1)}(\eta) \end{bmatrix}, \tag{4.159}$$

where $(q = 1, 3, \ldots, Q)$, $(s = Q + 2, \ldots, n)$, and $\alpha_s < 0$.

As η varies in the vicinity of a critical point $\eta = \eta_c$, the block-diagonal form of the Jacobian (4.158) is preserved by adjusting $\boldsymbol{Q}(\eta)$ continuously and appropriately. This property is expressed by

$$X_{ij}(0, \eta) = X'_{ij}(0, \eta) = X''_{ij}(0, \eta) = \ldots = 0 \tag{4.160}$$

for $i \neq j$ provided $j \neq i+1$ when $i = 3, 5, \ldots, Q$ and $j \neq i-1$ when $i = 4, 6, \ldots, (Q+1)$.

Furthermore, if at $\eta = \eta_c$ the first pair of eigenvalues crosses the imaginary axis, $Q(\eta)$ is chosen such that

$$\begin{aligned} X_{11} = X_{22} &= \alpha = 0, \\ X_{12} = -X_{21} &= \omega_c, \\ X_{qq} = X_{(q+1)(q+1)} &= \alpha_q < 0, \\ X_{q(q+1)} = -X_{(q+1)q} &= \omega_q. \end{aligned}$$

One may insist that $X_{qq}(\eta) = X_{(q+1)(q+1)}(\eta) = \alpha_q(\eta)$ and $X_{q(q+1)}(\eta) = -X_{(q+1)q}(\eta) = \omega_q(\eta)$ for all η in the vicinity of c; however, this will not be enforced in order to render the theory more flexible in various applications.

The derivatives of α and ω with respect to η may be established by following the approach that led to eqn (4.73) and α'' earlier. Thus, considering the Jacobian $[X_{ij}(\eta)]$ associated with the fundamental equilibrium path $x = 0$ around c, and differentiating the characteristic equation

$$|X_{ij}(\mu) - \lambda(\mu)\boldsymbol{I}| = 0, \qquad (\eta = \eta_c + \mu)$$

successively yields, upon evaluation at c,

$$\begin{aligned} \alpha' &= \left.\frac{\mathrm{d}\alpha}{\mathrm{d}\eta}\right|_c = (X'_{11} + X'_{22})/2, \\ \alpha'' &= \left.\frac{\mathrm{d}^2\alpha}{\mathrm{d}\eta^2}\right|_c = (X''_{11} + X''_{22})/2, \end{aligned}$$

etc., (4.161)

$$\omega' = \left.\frac{\mathrm{d}\omega}{\mathrm{d}\eta}\right|_c = (X'_{12} - X'_{21})/2,$$

and

$$\begin{aligned} \omega'' = \left.\frac{\mathrm{d}^2\omega}{\mathrm{d}\eta^2}\right|_c &= -(\omega')^2/\omega_c + [4\{(X'_{11})^2 - X'_{12}X'_{21} \\ &\quad + 2\omega_c(X''_{12} - X''_{21}) - (\alpha')^2]/4\omega_c \end{aligned}$$

One of the simplifications achieved on the basis of the transformation (4.155) is clearly reflected in the expression for α''. It is also noted that if one insisted on having the real and imaginary parts of eigenvalues as the elements of each block in eqn (4.159), then α', α'', etc. would reduce to X'_{11}, X''_{11}, etc. respectively.

The family of limit cycles bifurcating from the critical point c—if it exists—is again assumed in the parametric form

$$x^i = x^i(t; \sigma),$$
$$\eta = \eta(\sigma), \tag{4.162}$$
$$\omega = \omega(\sigma),$$

in analogy with eqn (4.74). Furthermore, let

$$x^i(\tau; \sigma) = p_{i0}(\sigma) + \sum_{m=1}^{M} \{p_{im}(\sigma)\cos m\tau + r_{im}(\sigma)\sin m\tau\} \tag{4.163}$$

represent the periodic solution (4.162) which is rendered 2π-periodic by introducing the time scaling $\tau = \omega(\sigma)t$. Since $x^i(\tau; 0) = 0$, one has

$$p_{i0}(0) = p_{im}(0) = r_{im}(0) = 0, \qquad (\forall i, m), \tag{4.164}$$

and it is assumed, without loss of generality, that

$$r_{11}(\sigma) \equiv 0. \tag{4.165}$$

A sequence of perturbation equations can now be generated by successive differentiations of

$$\omega(\sigma)x^i_\tau(\sigma; \tau) \equiv X_i\{x^i(\tau; \sigma), \eta(\sigma)\}. \tag{4.166}$$

Both the Hopf bifurcation and the flat Hopf bifurcation may be analysed within the framework of this new formulation. The bifurcation path in the case of flat Hopf bifurcation, for instance, takes the form

$$\mu = \{(\bar{\delta}_{22} - \bar{\delta}_{11})/240\alpha'\}(p_{11})^4,$$

where $\bar{\delta}_{ii}$ are certain constants given in terms of the derivatives of X. Attention here, however, will be focused again on the symmetric bifurcation phenomenon.

Symmetric bifurcation (*tri-furcation*). Let

$$\alpha' = 0 \quad \text{and} \quad \alpha'' \neq 0 \tag{4.167}$$

where α' and α'' are now given by eqn (4.161). This implies that $X'_{11} = -X'_{22}$ (which would vanish if $X_{11}(\eta) = X_{22}(\eta) = \alpha(\eta)$ for all η).

The first perturbation equation evaluated at c,

$$\omega_c \dot{x}^i_\tau = X_{ij}\dot{x}^j, \tag{4.168}$$

yields precisely the same derivatives given by eqn (4.132).

The second perturbation equation evaluated at c is in the form

$$2\dot{\omega}\dot{x}^i_\tau + \omega_c \ddot{x}^i_\tau = X_{ijk}\dot{x}^j\dot{x}^k + 2X'_{ij}\dot{x}^j\dot{\eta} + X_{ij}\ddot{x}^j. \tag{4.169}$$

Introducing eqn (4.163) into eqn (4.169) and using eqn (4.160) leads to

$$\ddot{p}_{21} = p_{21}^{(2)}\dot{p}_{11}\dot{\eta}, \tag{4.170}$$

$$\ddot{r}_{21} = r_{21}^{(2)}\dot{p}_{11}\dot{\eta} - \ddot{p}_{11}, \tag{4.171}$$

$$\ddot{p}_{i1} = \ddot{r}_{i1} = 0 \qquad (i \geq 3), \tag{4.172}$$

$$\ddot{p}_{im} = p_{im}^{(2)}(\dot{p}_{11})^2 \qquad (m = 0, 2;\ \forall i), \tag{4.173}$$

$$\ddot{r}_{i2} = r_{i2}^{(2)}(\dot{p}_{11})^2 \qquad (\forall i), \tag{4.174}$$

$$\ddot{p}_{im} = \ddot{r}_{im} = 0 \qquad (m \geq 3,\ \forall i). \tag{4.175}$$

It is noted that while eqn (4.175) follows from $\ddot{r}_{11} = 0$ (because $r_{11}(\sigma) \equiv 0$), eqn (4.172) is a consequence of the new transformation (4.155). The coefficients in eqns (4.170) and (4.171) are given by

$$p_{21}^{(2)} = -2X'_{11}/\omega_c, \qquad r_{21}^{(2)} = (X'_{12} + X'_{21})/\omega_c, \tag{4.176}$$

and the coefficients $p_{i0}^{(2)}$, $p_{i2}^{(2)}$, $r_{i2}^{(2)}$ are listed in the Appendix in a general form which embraces many other coefficients required here as well as in the analyses of other dynamic bifurcation phenomena to be explored in the following.

In addition to the above results, the second perturbation also yields

$$\dot{\omega}\dot{p}_{11} = \omega'\dot{\eta}\dot{p}_{11} \tag{4.177}$$

and

$$\alpha'\dot{\eta}\dot{p}_{11} = 0. \tag{4.178}$$

The latter equation is identically satisfied and ω' is given by eqn (4.161).

The third perturbation finally yields

$$\{(\bar{\gamma}_{11} - \bar{\gamma}_{22})(\dot{p}_{11})^2 + 6\alpha''(\dot{\eta})^2\}\dot{p}_{11} = 0 \tag{4.179}$$

which results in the bifurcation paths

$$\mu = \pm\{(\bar{\gamma}_{22} - \bar{\gamma}_{11})/6\alpha''\}^{1/2}p_{11} \tag{4.180}$$

as before. Here, however, the coefficients $\bar{\gamma}_{ii}$ are in terms of X's and are given in the Appendix. Also the second derivative α'' is in the simple form (4.161).

The existence condition now takes the form

$$(\bar{\gamma}_{22} - \bar{\gamma}_{11})/\alpha'' > 0. \tag{4.181}$$

Similarly, the frequency-amplitude relationship (4.146) and the asymptotic equations of the limit cycles may be constructed in terms of the new coefficients.

Problem 4.5. Analyse the Hopf bifurcation phenomenon within the framework of the new formulation. Starting with the transformed system

(4.156), identify the basic Hopf conditions and show that the bifurcation path may be expressed as

$$\mu = \{(\bar{\gamma}_{22} - \bar{\gamma}_{11})/12\alpha'\}(p_{11})^2$$

Problem 4.6. Repeat Problem 4.5 for the case of the flat Hopf bifurcation ($\bar{\gamma}_{22} - \bar{\gamma}_{11} = 0$) and show that the bifurcation path is given by

$$\mu = \{(\bar{\delta}_{22} - \bar{\delta}_{11})/240\alpha'\}(p_{11})^4$$

where $\bar{\delta}_{11}$ and $\bar{\delta}_{22}$ are given in the Appendix.

4.5.1 *Double Hopf bifurcation*

It is observed that, within the framework of system (4.156), the flat Hopf bifurcation occurs when $(\bar{\gamma}_{22} - \bar{\gamma}_{11}) = 0$ while a symmetric bifurcation phenomenon takes place when the transversality condition is violated (i.e. $\alpha' = 0$). In this section, the case in which both of these key coefficients vanish simultaneously will be explored. Thus, let c be a critical point on the fundamental equilibrium path of system (4.156) where

$$\alpha' = 0, \qquad \alpha'' \neq 0, \tag{4.182}$$

and

$$\bar{\gamma}_{22} - \bar{\gamma}_{11} = 0. \tag{4.183}$$

It is recalled that α' and α'' are given by eqn (4.161), while $\bar{\gamma}_{11}$ and $\bar{\gamma}_{22}$ are listed in the Appendix.

It follows from eqn (4.182) that the loci of the eigenvalues associated with the symmetric bifurcation and this case are similar (Fig. 4.6).

In order to generate information concerning the post-critical oscillatory behaviour of the system under consideration, the identity (4.166) will be perturbed under the conditions described above. To this end, one notices that the first and second perturbation equations, (4.168) and (4.169), yield precisely the same results as before since the condition (4.183) does not come into the picture at this stage. Thus, one has the first derivatives, which look identical to those given in eqns (4.132), and the second derivatives eqns (4.170) to (4.178). Furthermore, eqn (4.178) is again identically satisfied since $\alpha' = 0$.

The third perturbation of eqn (4.166) yields

$$(\bar{\gamma}_{11} - \bar{\gamma}_{22})(\dot{p}_{11})^3 + 6\alpha''\dot{p}_{11}(\dot{\eta})^2 = 0 \tag{4.184}$$

which, in view of (4.183), results in

$$\dot{p}_{11}(\dot{\eta})^2 = 0, \tag{4.185}$$

implying that $\dot{p}_{11} = 0$ and/or $\dot{\eta} = 0$. The case of $\dot{p}_{11} = 0$ leads to $\ddot{p}_{11} = \dddot{p}_{11} = \ldots = 0$ with higher order perturbations—indicating that $\dot{p}_{11} = 0$ is associated with the equilibrium path. In order to obtain the limit cycles bifurcating off the equilibrium path (that is, if they exist), one assumes at this stage that

$$\dot{p}_{11} \neq 0 \quad \text{and} \quad \dot{\eta} = 0. \tag{4.186}$$

This observation enables one to assign the role of σ to p_{11} in which case

$$\dot{p}_{11} = 1, \qquad \ddot{p}_{11} = \dddot{p}_{11} = \ldots = 0, \tag{4.187}$$

where a dot now, and henceforth, denotes differentiation with respect to p_{11}.

In view of eqns (4.186) and (4.187), the results given by eqns (4.170), (4.171), (4.173), (4.174), and (4.177) take the form

$$\ddot{p}_{21} = 0, \qquad \ddot{r}_{21} = 0, \tag{4.188}$$

$$\ddot{p}_{im} = p_{im}^{(2)}, \qquad \ddot{r}_{i2} = r_{i2}^{(2)} \qquad (m = 0, 2; \forall i) \tag{4.189}$$

and

$$\dot{\omega} = 0. \tag{4.190}$$

Returning back to the third-order perturbation process, and taking into account eqns (4.186) and (4.190), the full third-order perturbation equation—which is similar to eqn (4.9) in format—may be reduced to the simpler form

$$3\ddot{\omega}\dot{x}_\tau^i + \omega_c \dddot{x}_\tau^i = X_{ijkl}\dot{x}^j\dot{x}^k\dot{x}^l + 3X_{ijk}\ddot{x}^j\dot{x}^k + 3X'_{ij}\dot{x}^j\ddot{\eta} + X_{ij}\dddot{x}^j. \tag{4.191}$$

Introducing Fourier's expansion (with $\sigma \equiv p_{11}$) into eqn (4.191) results in

$$\left.\begin{aligned}
&\dddot{p}_{21} = -(\bar{\gamma}_{11} + 3X'_{11}\ddot{\eta})/\omega_c, \\
&\dddot{r}_{21} = \{(\bar{\gamma}_{21} - \bar{\gamma}_{12}) + 3(X'_{12} + X'_{21})\ddot{\eta}\}/2\omega_c, \\
&\dddot{p}_{i1} = p_{i1}^{(3)}, \qquad \dddot{r}_{i1} = r_{i1}^{(3)} \qquad (i \geq 3), \\
&\dddot{p}_{i3} = p_{i3}^{(3)}, \qquad \dddot{r}_{i3} = r_{i3}^{(3)} \qquad (\forall i). \\
&\dddot{p}_{im} = \dddot{r}_{im} = 0, \qquad (m \geq 4; \forall i), \\
&\dddot{p}_{im} = 0, \qquad (m = 0, 2; \forall i), \qquad \dddot{r}_{i2} = 0,
\end{aligned}\right\} \tag{4.192}$$

where $p_{i1}^{(3)}$, $r_{i1}^{(3)}$, $p_{i3}^{(3)}$, and $r_{i3}^{(3)}$ are given in the Appendix.

Furthermore, $\dddot{p}_{11} = \dddot{r}_{11} = 0$ as before, and one obtains

$$\ddot{\omega} = \{6\omega'\ddot{\eta} - (\bar{\gamma}_{12} + \bar{\gamma}_{21})\}/6. \tag{4.193}$$

It is noted in passing that keeping σ unidentified until the situation becomes clear and subsequent back-tracking (as has just been done) is essential, particularly in the analysis of highly degenerate cases.

At this stage, one observes from the above results that although many derivatives have already been determined for the problem under consideration, two crucial derivatives, namely $\ddot{\omega}$ and $\ddot{\eta}$, have not yet emerged. These derivatives are obviously required for a first-order approximation of the bifurcating path and the frequency-amplitude relationship, and it appears that one has to proceed to higher order perturbations to obtain this information.

By proceeding to the fourth-order perturbation problem and using the derivatives obtained so far, one obtains certain additional derivatives—but not $\ddot{\eta}$ or $\ddot{\omega}$. The derivatives obtained at this stage are

$$p_{im}^{(4)} = r_{im}^{(4)} = 0, \qquad (m \geq 5), \tag{4.194}$$

and $p_{im}^{(4)}$, $r_{im}^{(4)}$ for $m = 0, 2$ which are listed in the Appendix. It is recalled that a superscript ^{k} indicates a derivative of order k replacing dots which become impractical.

Finally, a fifth perturbation yields

$$90\alpha''(\ddot{\eta})^2 + 3(\zeta_{11} - \zeta_{22})\ddot{\eta} + 5(\bar{\delta}_{11} - \bar{\delta}_{22}) = 0 \tag{4.195}$$

which follows from the equations associated with the first harmonics (Huseyin and Atadan 1984*a*). The constants $\bar{\delta}_{ii}$ have already appeared in the analysis of the flat Hopf bifurcation, and the new constants ζ_{ii} are given in the Appendix along with $\bar{\delta}_{ii}$.

Solving eqn (4.195) for $\ddot{\eta}$ yields

$$\ddot{\eta} = -[(\zeta_{11} - \zeta_{22}) \pm \{(\zeta_{11} - \zeta_{22})^2 - 200\alpha''(\bar{\delta}_{11} - \bar{\delta}_{22})\}^{1/2}]/60\alpha''. \tag{4.196}$$

It follows that the bifurcating path

$$\mu = \tfrac{1}{2}\ddot{\eta}(p_{11})^2, \qquad (\eta = \eta_c + \mu), \tag{4.197}$$

generally consists of two branches which share the same slope $\dot{\eta} = 0$ but have different curvatures (eqn 4.196). The bifurcating paths (4.197) are, of course, real only if the *existence condition*

$$(\zeta_{11} - \zeta_{22})^2 - 200\alpha''(\bar{\delta}_{11} - \bar{\delta}_{22}) \geq 0 \tag{4.198}$$

holds; otherwise the family of limit cycles represented by eqn (4.197) does not exist.

A number of topologically distinct phenomena may be identified when the existence condition (4.198) is valid:

(a) Let $(\bar{\delta}_{11} - \bar{\delta}_{22})\alpha'' > 0$, then
  (i) both the curvatures $\ddot{\eta} > 0$ if $(\zeta_{11} - \zeta_{22})\alpha'' < 0$; and
  (ii) both the curvatures $\ddot{\eta} < 0$ if $(\zeta_{11} - \zeta_{22})\alpha'' > 0$.

(b) Let $(\bar{\delta}_{11} - \bar{\delta}_{22})\alpha'' < 0$, then one of the curvatures is positive $(\ddot{\eta}_1 > 0)$ and the other is negative $(\ddot{\eta}_2 < 0)$.

(c) Let $(\zeta_{11} - \zeta_{22})^2 - 200\alpha''(\bar{\delta}_{11} - \bar{\delta}_{22}) = 0$, then
  (i) $\ddot{\eta}_1 = \ddot{\eta}_2 < 0$ if $(\zeta_{11} - \zeta_{22})\alpha'' > 0$; and
  (ii) $\ddot{\eta}_1 = \ddot{\eta}_2 > 0$ if $(\zeta_{11} - \zeta_{22})\alpha'' < 0$.

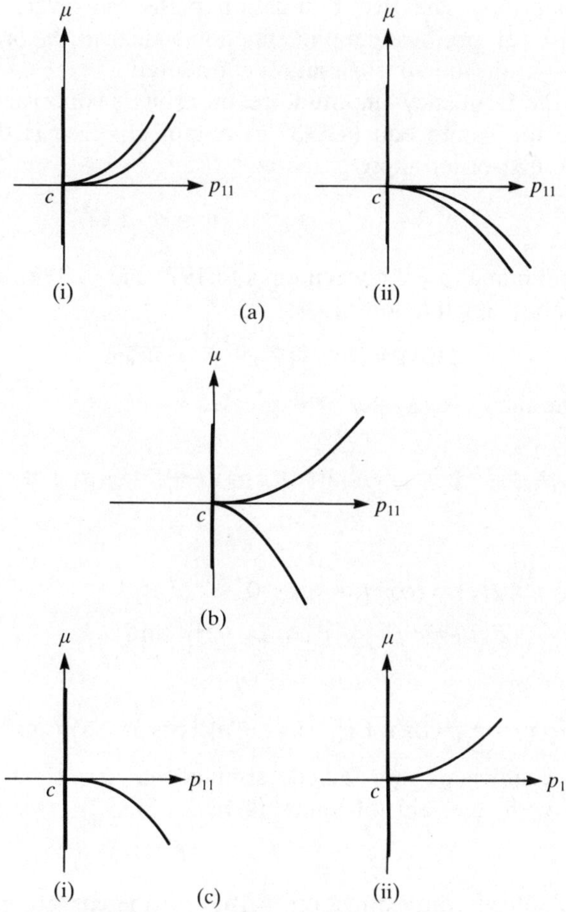

Fig. 4.9 Double Hopf bifurcations when existence condition (eqn 4.198) is valid but constants δ_{ii} and ζ_{ii} are varied. (a) $(\bar{\delta}_{11}-\bar{\delta}_{22})\alpha''>0$, (b) $(\bar{\delta}_{11}-\bar{\delta}_{22})\alpha''<0$, and (c) $(\zeta_{11}-\zeta_{22})^2-200\alpha''(\bar{\delta}_{11}-\bar{\delta}_{22})=0$.

Each of these cases is illustrated in Figs. 4.9*a, b,* and *c,* respectively. Consider, for example, Fig. 4.9*b*; in this case two branches of the bifurcation path emerge from the critical point *c* initially in the direction of p_{11}, but follow different curvatures with opposite signs later on. Both of the branches are tangent to p_{11} as well as to each other, and it follows that limit cycles exist for both $\mu>0$ and $\mu<0$. A stability analysis is not performed here, but it can be shown that if the fundamental equilibrium path remains stable (unstable) through *c,* then the entire family of limit cycles branching off the equilibrium path at *c* is totally unstable (stable) in the vicinity of *c.*

On the other hand, limit cycles can only exist for $\mu>0$ or $\mu<0$ in Cases (a) and (c). Furthermore, in Case (a), the limit cycles are concentric for a given value of $\mu\neq 0$ in the region *G,* with different stability properties. The two bifurcation paths, however, collapse together in Case (c), producing a phenomenon similar to the ordinary Hopf bifurcation—within the approximation considered.

As far as the frequency-amplitude relationship is concerned, it suffices to substitute for $\ddot{\eta}$ into eqn (4.193) to obtain $\ddot{\omega}$ which is then used to construct the first-order approximation

$$\Omega(p_{11}) = \tfrac{1}{2}\ddot{\omega}(p_{11})^2, \qquad (\omega = \omega_c + \Omega). \tag{4.199}$$

One may eliminate p_{11} between eqns (4.197) and (4.199) to obtain the frequency-parameter relationship

$$\Omega(\mu) = \{\omega' - (\bar{\gamma}_{12} + \bar{\gamma}_{21})/6\ddot{\eta}\}\mu. \tag{4.200}$$

Finally, the limit cycles may be expressed as

$$x^i(\tau; p_{11}) = \dot{x}^i(\tau)p_{11} + \frac{1}{2!}\ddot{x}^i(\tau)(p_{11})^2 + \frac{1}{3!}\dddot{x}^i(\tau)(p_{11})^3 + \ldots \tag{4.201}$$

where

$$\dot{\boldsymbol{x}}(\tau) = (\cos\tau, -\sin\tau, 0, \ldots, 0)^T,$$
$$\ddot{x}^i(\tau) = p_{i0}^{(2)} + p_{i2}^{(2)}\cos 2\tau + r_{i2}^{(2)}\sin 2\tau,$$

and

$$\dddot{x}^i(\tau) = p_{i1}^{(3)}\cos\tau + r_{i1}^{(3)}\sin\tau + p_{i3}^{(3)}\cos 3\tau + r_{i3}^{(3)}\sin 3\tau$$

where the coefficients are all determined in terms of the system coefficients with the aid of eqns (4.189), (4.192), (4.196) and the Appendix.

Problem 4.7. Verify the expression (4.196) for the curvature $\ddot{\eta}$.

4.5.2 Cusp bifurcation

Another degenerate bifurcation phenomenon is exhibited by system (4.156) if the critical point c on the fundamental equilibrium path $\boldsymbol{x} = 0$ is characterized by

$$\alpha' = \alpha'' = 0 \tag{4.202}$$

and

$$\alpha''' = (X_{11}''' + X_{22}''')/2 \neq 0 \tag{4.203}$$

As a consequence of these conditions, the eigenvalues $\lambda_{1,2}(\eta) = \alpha(\eta) \pm i\omega(\eta)$ cross the imaginary axis such that their trajectory in the complex λ-plane has a point of inflexion at $\alpha(\eta_c) = 0$ on the imaginary axis (Fig. 4.10).

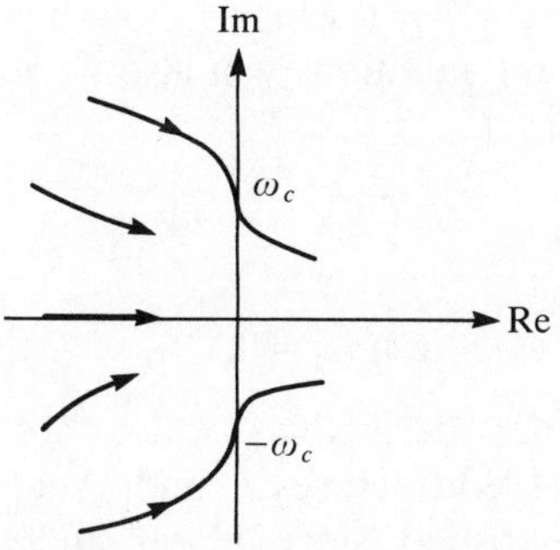

Fig. 4.10 Cusp bifurcation: the eigenvalues $(\lambda_{1,2}(\eta))$ cross the imaginary axis such that their trajectory in the complex λ-plane has a point of inflexion at $\alpha(\eta_c)=0$ on the imaginary axis.

It is understood that (4.203) is readily obtained by perturbing the characteristic equation as in the case of α' and α'' given by eqn (4.161). In fact, the canonical form of the Jacobian is such that a derivative of arbitrary order (say k) can be easily established as

$$\alpha^{(k)} = (X_{11}^{(k)} + X_{22}^{(k)})/2. \tag{4.204}$$

It is also understood that all the other coefficients, including $(\bar{\gamma}_{22}-\bar{\gamma}_{11})$, are assumed to be non-zero.

Considering again the initial perturbation equations generated from eqn (4.166), it is observed that the derivatives obtained from the first three perturbation equations remain essentially valid here as well; i.e. one has the derivatives (4.132) and (4.170) to (4.179), while (4.178) is identically satisfied. However, (4.179) now yields

$$(\bar{\gamma}_{11}-\bar{\gamma}_{22})(\dot{p}_{11})^2 = 0, \tag{4.205}$$

since $\alpha''=0$, and one concludes immediately that

$$\dot{p}_{11} = 0, \tag{4.206}$$

indicating clearly that p_{11} is no longer suitable for the role of σ. Furthermore, by virtue of eqn (4.206), the derivatives (4.132) may now be expressed simply as

$$\dot{p}_{im} = \dot{r}_{im} = 0, \qquad (\forall i, \forall m), \tag{4.207}$$

and the derivatives (4.170) to (4.175) take the form

$$\begin{aligned} &\ddot{p}_{21} = 0, && \ddot{r}_{21} = -\ddot{p}_{11}, \\ &\ddot{p}_{i1} = \ddot{r}_{i1} = 0, && (i \geq 3), \\ &\ddot{p}_{im} = \ddot{r}_{im} = 0, && (\forall i,\ m \neq 1). \end{aligned} \tag{4.208}$$

The third-order perturbation equation reduces to a remarkably simple form,

$$3\dot{\omega}\ddot{x}_\tau^i + \omega_c \dddot{x}_\tau^i = 3X_{ij}'\ddot{x}^j\dot{\eta} + X_{ij}\dddot{x}^j, \tag{4.209}$$

upon recognizing that $\dot{x}^i_\tau = \dot{x}^i = 0$ which follows from eqn (4.207). All the higher order perturbation equations will also be simplified on this basis.

This equation leads to

$$\dot{\omega}\ddot{p}_{11} = \omega'\dot{\eta}\ddot{p}_{11} \tag{4.210}$$

and

$$\alpha'\dot{\eta}\ddot{p}_{11} = 0, \tag{4.211}$$

among other third-order derivatives which are not given here (see Atadan and Huseyin 1983*a*; Huseyin and Atadan 1984*b*). The latter equation is identically satisfied, since $\alpha' = 0$, and eqn (4.210) yields

$$\dot{\omega} = \omega'\dot{\eta}, \tag{4.212}$$

upon recognizing and excluding $\ddot{p}_{11} = 0$ as the equilibrium path.

Proceeding to the fourth and fifth perturbation problems leads to

$$\dot{\eta} = \dot{\omega} = 0$$

and (4.213)

$$\ddot{\omega} = \omega'\ddot{\eta},$$

after some back-tracking as in the case of the double Hopf bifurcation.

It is observed that η too is not an appropriate perturbation parameter, and the analysis has to be carried on with σ unidentified.

The sixth perturbation equation leads to

$$(\bar{\gamma}_{11} - \bar{\gamma}_{22})(\ddot{p}_{11})^2 = 0 \tag{4.214}$$

which results in

$$\ddot{p}_{11} = 0. \tag{4.215}$$

This is rather interesting since $\ddot{p}_{11} = 0$ was earlier discarded as the equilibrium path. The analysis now asserts that not only the equilibrium path but also the family of periodic solutions being sought is associated with the derivative $\ddot{p}_{11} = 0$. In view of this, one may choose to carry on with further perturbations under condition (4.215) or (4.212), or both. It can be demonstrated (Atadan 1983), however, that under both conditions, one eventually arrives at identical conclusions.

It turns out that the final results are extremely elusive in this problem, and a number of further perturbations are required. In fact, only after a total of nine perturbations and a considerable number of delicate manoeuvers can one finally see the light at the end of the tunnel (see Huseyin and Atadan 1984*b*). Thus, the ninth perturbation yields

$$\{(\bar{\gamma}_{11} - \bar{\gamma}_{22})(\dddot{p}_{11})^2 + 9\alpha'''(\ddot{\eta})^3\}\dddot{p}_{11} = 0. \tag{4.216}$$

It is now clear that, in addition to

$$\dddot{p}_{11} = 0, \tag{4.217}$$

which is associated with the fundamental equilibrium path, one has

$$(\ddot{\eta})^3 = \{(\bar{\gamma}_{22} - \bar{\gamma}_{11})/9\alpha'''\}(\ddot{p}_{11})^2 \tag{4.218}$$

where the constants $(\bar{\gamma}_{22} - \bar{\gamma}_{11})$ and α''' are as defined before.

The first-order approximations for $p_{11}(\sigma)$ and $\eta(\sigma)$ are, then, constructed as

$$p_{11} = \frac{1}{3!}\dddot{p}_{11}\sigma^3 \tag{4.219}$$

and

$$\mu = \frac{1}{2!}\ddot{\eta}\sigma^2 \qquad (\eta = \eta_c + \mu), \tag{4.220}$$

respectively.

Eliminating σ between eqns (4.219) and (4.220) yields the bifurcation path

$$\mu = \{(\bar{\gamma}_{22} - \bar{\gamma}_{11})/2\alpha'''\}^{1/3}(p_{11})^{2/3} \tag{4.221}$$

which describes a cusp shaped curve in the $(\mu\text{–}p_{11})$ space as depicted in Fig. 4.11.

Using eqn (4.213), the frequency $\Omega(\sigma)$ may be expressed as

$$\Omega = \frac{1}{2!}\ddot{\omega}\sigma^2 = \frac{1}{2!}\omega'\ddot{\eta}\sigma^2, \tag{4.222}$$

and eliminating σ between eqns (4.222) and (4.220) yields

$$\Omega = \omega'\mu. \tag{4.223}$$

Similarly, eliminating σ between eqns (4.222) and (4.219) gives the frequency-amplitude relationship

$$\Omega = \omega'\{(\bar{\gamma}_{22} - \bar{\gamma}_{11})/2\alpha'''\}^{1/3}(p_{11})^{2/3} \tag{4.224}$$

which is also in the shape of a cusp in the $(\Omega\text{–}p_{11})$ space.

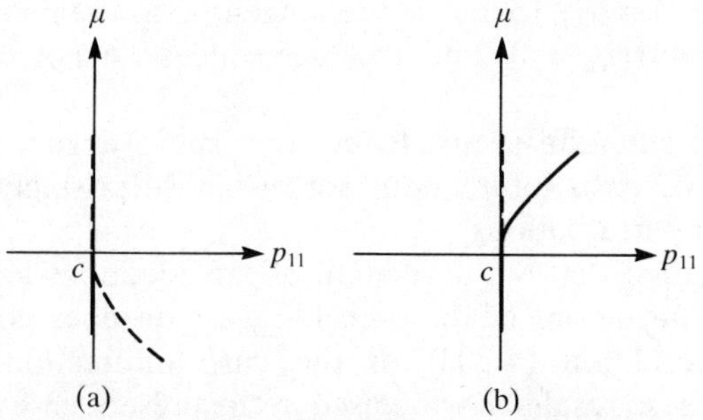

Fig. 4.11 Bifurcation path for cusp bifurcation.

Finally, it can be shown (Atadan 1983) that the limit cycles may be expressed as

$$x^i(\tau) = x_1^i(\tau)p_{11} + \tfrac{1}{2}x_2^i(\tau)p_{11}\mu + \tfrac{1}{2}x_3^i(\tau)(p_{11})^2 + \ldots, \tag{4.225}$$

where

$$\boldsymbol{x}_1(\tau) = (\cos\tau, -\sin\tau, 0, \ldots, 0)^T,$$
$$\boldsymbol{x}_2(\tau) = (0, (p_{21}^{(2)}\cos\tau + r_{21}^{(2)}\sin\tau), 0, \ldots, 0)^T,$$
$$x_3^i(\tau) = p_{i0}^{(2)} + p_{i2}^{(2)}\cos 2\tau + r_{i2}^{(2)}\sin 2\tau,$$

and $p_{21}^{(2)}$, $r_{21}^{(2)}$ are given by eqn (4.176). The remaining coefficients, namely $p_{i0}^{(2)}$, $p_{i2}^{(2)}$, and $r_{i2}^{(2)}$, are listed in the Appendix.

It is deduced from eqn (4.221) that the family of limit cycles can only exist for either $\mu > 0$ or $\mu < 0$ depending on the sign of $\{(\bar{\gamma}_{22} - \bar{\gamma}_{11})/\alpha'''\}$ as depicted in Fig. 4.11. A stable equilibrium path becomes unstable upon passing through the critical point and the family of limit cycles can be shown to be stable or unstable according to whether the family exists for $\mu > 0$ or $\mu < 0$, respectively.

Problem 4.8. Verify the equations (4.216) and (4.225). (Caution: the derivation of these results is a rather lengthy process!)

4.5.3 *Tangential bifurcation*

A tangential dynamic bifurcation phenomenon may take place, in analogy with the corresponding static bifurcation (Section 3.1.2), if the critical point on the equilibrium path $\boldsymbol{x} = 0$ of system (4.156) is characterized by

$$\alpha' = \alpha'' = \alpha''' = 0$$

and

$$\alpha^{(4)} = (X_{11}^{(4)} + X_{22}^{(4)})/2 \neq 0, \tag{4.226}$$

where $\alpha^{(4)} \triangleq \alpha''''$.

It is deduced on the basis of eqns (4.226) that the loci of the eigenvalues $\lambda_{1,2}(\eta) = \alpha(\eta) \pm \mathrm{i}\omega(\eta)$ are tangential to the imaginary axis at the critical point $\alpha(\eta_c) = 0$, and the eigenvalues do not cross the axis (Fig. 4.12).

Although this situation seems to be even more degenerate than the preceding case of cusp bifurcation, somewhat surprisingly it does not require as many perturbations.

It is first observed that the derivation of the equations associated with the first five perturbations of the preceding section does not involve the distinguishing condition (4.203) of the cusp bifurcation. It follows, therefore, that these results may be used in the present case as well—with appropriate modifications if necessary.

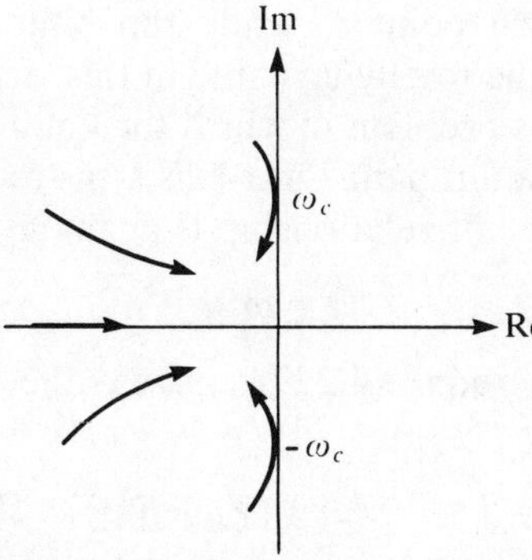

Fig. 4.12 Tangential bifurcation: the loci of the eigenvalues $\lambda_{1,2}(\eta)$ are tangential to the imaginary axis at the critical point $\alpha(\eta_c)$ and do not cross the axis.

It becomes clear in the sixth perturbation that η is a suitable parameter for the role of σ, and setting $\sigma = \eta$ leads to

$$\{(\bar{\gamma}_{11} - \bar{\gamma}_{22})(p''_{11})^2 + 2\alpha^{(4)}\}p''_{11} = 0, \tag{4.227}$$

where primes replace dots according to the convention.

Assuming again that $(\bar{\gamma}_{11} - \bar{\gamma}_{22}) \neq 0$ results in $p''_{11} = 0$ (which is associated with the equilibrium path) or

$$p''_{11} = \pm\{2\alpha^{(4)}/(\bar{\gamma}_{22} - \bar{\gamma}_{11})\}^{1/2}, \tag{4.228}$$

correspond to the periodic solutions.

One can now construct the bifurcation path as

$$p_{11} = \{\alpha^{(4)}/2(\bar{\gamma}_{22} - \bar{\gamma}_{11})\}^{1/2}\mu^2. \tag{4.229}$$

Clearly, real solutions can only exist if

$$\alpha^{(4)}/(\bar{\gamma}_{22} - \bar{\gamma}_{11}) > 0 \tag{4.230}$$

which may be treated as an *existence condition*. The bifurcation path (4.229) is a parabolic curve in the (p_{11}–μ) space which is tangent to the equilibrium path $p_{11} = 0$ (Fig. 4.13). The latter retains its stability as it

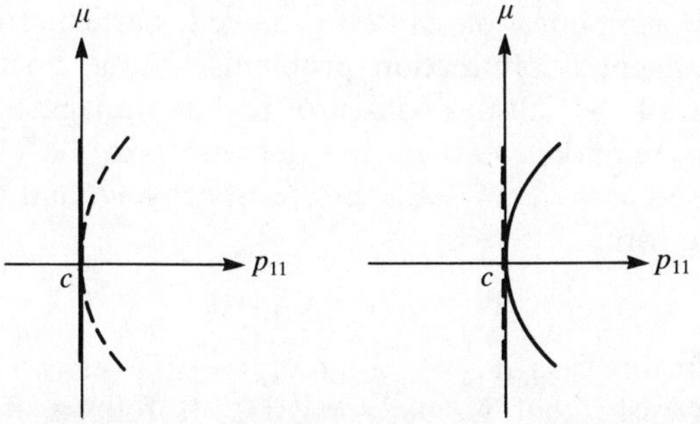

Fig. 4.13 Bifurcation path for tangential bifurcation.

goes through the critical point c, while the bifurcating family of limit cycles can be shown to be totally unstable in this vicinity. It is not difficult to imagine an analogous problem in which the equilibrium path is entirely unstable but the bifurcation path—if it exists—is totally stable.

The frequency-parameter relationship is again given by

$$\Omega = \omega'\mu \tag{4.231}$$

which, together with eqn (4.229), gives the frequency-amplitude relationship

$$p_{11} = \{\alpha^{(4)}/2(\omega')^4(\bar{\gamma}_{22} - \bar{\gamma}_{11})\}^{1/2}\Omega^2. \tag{4.232}$$

It can be demonstrated (Atadan 1983) that the limit cycles may be expressed as

$$x^i(\tau) = x_1^i(\tau)p_{11} + \tfrac{1}{2}x_2^i(\tau)p_{11}\mu + \frac{1}{4!}\{x_4^i(\tau) + \underline{x}_4^i(\tau)\}\mu^4 \tag{4.233}$$

where

$$\begin{aligned}
\boldsymbol{x}_1(\tau) &= (\cos\tau, -\sin\tau, 0, \ldots, 0)^T,\\
\boldsymbol{x}_2(\tau) &= (0, p_{21}^{(2)}\cos\tau + r_{21}^{(2)}\sin\tau, 0, \ldots, 0)^T,\\
\boldsymbol{x}_4(\tau) &= (0, (\bar{p}_{21}^{(2)}\cos\tau + \bar{r}_{21}^{(2)}\sin\tau), 0, \ldots, 0)^T p_{11}'',\\
\underline{x}_4^i(\tau) &= 3(p_{i0}^{(2)} + p_{i2}^{(2)}\cos 2\tau + r_{i2}^{(2)}\sin 2\tau).
\end{aligned}$$

In these equations $p_{21}^{(2)}$ and $r_{21}^{(2)}$ are given by eqn (4.176); p_{11}'' by eqn (4.228); $p_{i0}^{(2)}$, $p_{i2}^{(2)}$, and $r_{i2}^{(2)}$ are listed in the Appendix;

$$\bar{p}_{21}^{(2)} = 6(2X_{11}'X_{12}' - \omega_c X_{11}'')/\omega_c^2;$$

and

$$\bar{r}_{21}^{(2)} = -3\{3(X_{12}')^2 - (X_{21}')^2 + 2X_{12}'X_{21}' - 4(X_{11}')^2 - 2\omega_c(X_{12}'' + X_{21}'')\}/2\omega_c^2.$$

For the details of the above analysis, the reader is referred to the publications mentioned earlier in the case of the cusp bifurcation (e.g. Huseyin and Atadan 1984*b*).

A noteworthy simplification can be achieved, particularly in the cases of cusp and tangential bifurcation problems, if one assumes that the elements of eqn (4.159) always consist of real and imaginary parts of the eigenvalues. As remarked earlier, the derivatives α', α'', α''', etc. may then be expressed as X_{11}', X_{11}'', X_{11}''', etc., respectively, and the conditions (4.226) take the form

$$X_{11}' = X_{11}'' = X_{11}''' = 0.$$

Furthermore, in this case $X_{12}' = -X_{21}'$, $X_{12}'' = -X_{21}''$, etc., and the derivatives (4.176) vanish; that is, $p_{21}^{(2)} = r_{21}^{(2)} = 0$. It follows that the terms involving the cross product $p_{11}\mu$ in the families of limit cycles (4.225) and

(4.233)—associated with the cusp and tangential bifurcations, respectively—vanish since $x_2(\tau)=0$ in both cases. In addition, it is observed that $\boldsymbol{x}_4(\tau)=0$ in the latter case since both $\bar{p}_{21}^{(2)}$ and $\bar{r}_{21}^{(2)}$ also vanish under these conditions. Clearly, all the derivatives involving α', α'', etc., and ω', ω'', etc., in the text as well as in the Appendix will have to be adjusted according to the new conditions.

Finally, it is noted that many more degenerate bifurcation problems may be generated by continuing to specify the combinations of key coefficients which vanish simultaneously at a critical point. A summary of the phenomena explored hitherto is given in Table 4.1 which illustrates typical bifurcation paths.

Problem 4.9. Verify eqns (4.227) and (4.233).

Problem 4.10. Discuss the analogies (and differences) between the static and dynamic bifurcation phenomena explored in Chapter 3 and in this chapter respectively.

4.6 Stability of limit cycles

The stability of equilibrium paths and the stability distribution on the fundamental and post-critical paths in the vicinity of a critical point have been examined analytically in Section 3.1.3. In the dynamic branching case, one has a fundamental equilibrium path and a family of limit cycles bifurcating from a critical point. The stability of the fundamental path can readily be established since it depends on $\alpha(\mu)$, and attention here is focused on the stability of limit cycles.

Consider a particular limit cycle $\boldsymbol{z}=\boldsymbol{p}(t)$ associated with the non-linear autonomous system

$$\boldsymbol{z}_t=\boldsymbol{Z}(\boldsymbol{z}), \qquad (\boldsymbol{z}_t \triangle \mathrm{d}\boldsymbol{z}/\mathrm{d}t) \tag{4.234}$$

corresponding to a fixed value of the parameter η which need not be specified explicitly here. The equation of the first variation about the periodic motion $\boldsymbol{p}(t)$ is

$$\boldsymbol{z}_t=\boldsymbol{A}(t)\boldsymbol{z} \tag{4.235}$$

where

$$\boldsymbol{A}(t)=\left[\frac{\partial \boldsymbol{Z}}{\partial \boldsymbol{z}}\right]_{\boldsymbol{z}=\boldsymbol{p}(t)}.$$

It is noted that the Jacobian matrix $\boldsymbol{A}(t)$ is not a constant matrix as discussed in Section (1.4). In fact, $\boldsymbol{A}(t)$ is periodic with the same period T as the limit cycle $\boldsymbol{p}(t)$.

Table 4.1. Summary of dynamic bifurcations

Type of bifurcation	Main conditions	Bifurcation path	
Hopf bifurcation	$\alpha' \neq 0$ $\gamma_{22} - \gamma_{11} \neq 0$	$\mu = \{(\gamma_{22} - \gamma_{11})/12\alpha'\}(p_{11})^2$	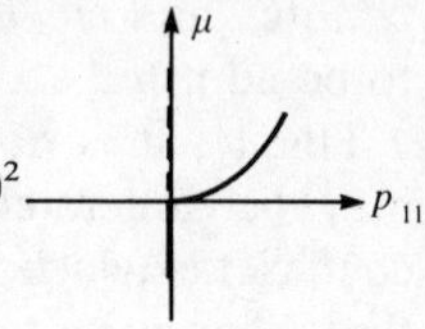
Flat Hopf bifurcation	$\alpha' \neq 0$ $\gamma_{22} - \gamma_{11} = 0$ $\delta_{22} - \delta_{11} \neq 0$	$\mu = \{(\delta_{22} - \delta_{11})/240\alpha'\}(p_{11})^4$	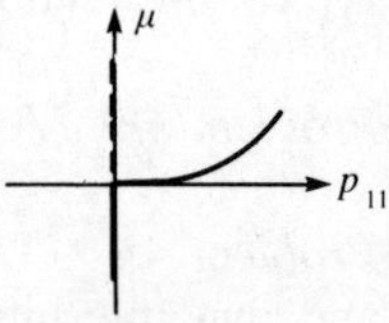
Symmetric bifurcation (tri-furcation)	$\alpha' = 0$ $\alpha'' \neq 0$ $\gamma_{22} - \gamma_{11} \neq 0$ or $\bar{\gamma}_{22} - \bar{\gamma}_{11} \neq 0$	$\mu = \{(\gamma_{22} - \gamma_{11})/6\alpha''\}^{1/2} p_{11}$ or $\mu = \{(\bar{\gamma}_{22} - \bar{\gamma}_{11})/6\alpha''\}^{1/2} p_{11}$	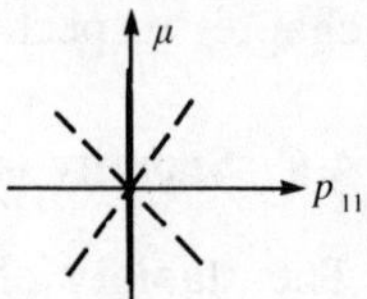
Double Hopf bifurcation	$\alpha' = 0$ $\alpha'' \neq 0$ $\bar{\gamma}_{22} - \bar{\gamma}_{11} = 0$ $\ddot{\eta} \neq 0$	$\mu = \frac{1}{2}\ddot{\eta}(p_{11})^2$, $\ddot{\eta}$ given by eqn (4.196)	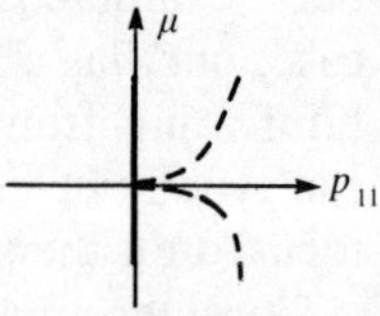
Cusp bifurcation	$\alpha' = \alpha'' = 0$ $\alpha''' \neq 0$ $\bar{\gamma}_{22} - \bar{\gamma}_{11} \neq 0$	$\mu = \{(\bar{\gamma}_{22} - \bar{\gamma}_{11})/2\alpha'''\}^{1/3}(p_{11})^{2/3}$	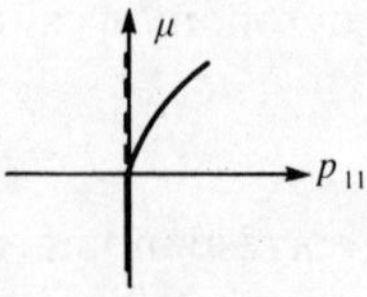
Tangential bifurcation	$\alpha' = \alpha'' = \alpha''' = 0$ $\alpha^{(4)} \neq 0$ $\bar{\gamma}_{22} - \bar{\gamma}_{11} \neq 0$	$p_{11} = \{\alpha^{(4)}/2(\bar{\gamma}_{22} - \bar{\gamma}_{11})\}^{1/2}\mu^2$	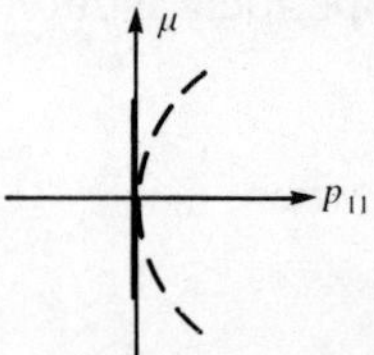

The solutions of the linearized equation (4.235) are often expressed in terms of a *fundamental matrix* $\boldsymbol{F}(t)$ which satisfies

$$\boldsymbol{F}_t = \boldsymbol{A}(t)\boldsymbol{F} \tag{4.236}$$

and $\boldsymbol{F}(0) = \boldsymbol{I}$ (the unit matrix).

A particular solution determined by the initial condition $\boldsymbol{x}(0) = \boldsymbol{x}_0$ is then given by $\boldsymbol{x}(t) = \boldsymbol{F}(t)\boldsymbol{x}_0$. According to standard Floquet theory (Hale 1969, Hartman 1973), there exists a non-singular matrix $\boldsymbol{P}(t)$ such that

$$\boldsymbol{F}(t) = \boldsymbol{P}(t)\mathrm{e}^{\boldsymbol{B}t}, \tag{4.237}$$

and the matrix $\boldsymbol{P}(t)$ is actually T-periodic as $\boldsymbol{A}(t)$. It follows that

$$\boldsymbol{P}(0) = \boldsymbol{P}(0 + T) = \boldsymbol{I}. \tag{4.238}$$

Hence, the behaviour of the solutions of (4.235) depends on the eigenvalues of the constant matrix $\boldsymbol{B}$, which are called the *characteristic exponents* of eqn (4.235). The trivial solution of eqn (4.235) is asymptotically stable if the eigenvalues of $\boldsymbol{B}$ have negative real parts. The matrices $\boldsymbol{P}$ and $\boldsymbol{B}$, however, are not uniquely defined, and one may consider the eigenvalues of $\mathrm{e}^{\boldsymbol{B}T}$, the so-called *characteristic multipliers,* instead of the characteristic exponents. In this case, asymptotic stability is ensured if all the multipliers have an absolute value less than one. If a multiplier (an exponent) with an absolute value greater than one (with a positive real part) exists, then the trivial solution of eqn (4.235) is unstable. Similarly, if there are no multipliers (exponents) with an absolute value greater than one (with positive real parts) but at least one multiplier (exponent) has an absolute value of one (a zero real part), then the trivial solution of eqn (4.235) may be stable or unstable.

In general, the determination of the characteristic multipliers or the characteristic exponents is surprisingly hard, and one often resorts to numerical calculations on the basis of

$$\boldsymbol{F}(T) = \boldsymbol{P}(T)\mathrm{e}^{\boldsymbol{B}T} = \mathrm{e}^{\boldsymbol{B}T}. \tag{4.239}$$

On the other hand, it can be shown that a limit cycle, $\boldsymbol{p}(t)$, can never be asymptotically stable, and one of the multipliers in this case always lies on the unit circle. To this end, note that $\boldsymbol{p}(t)$ is a solution of eqn (4.234), and

$$\boldsymbol{p}_t(t) = \boldsymbol{Z}\{\boldsymbol{p}(t)\} \tag{4.240}$$

which, upon differentiation with respect to t, yields

$$\boldsymbol{p}_{tt}(t) = \boldsymbol{A}(t)\boldsymbol{p}_t(t), \tag{4.241}$$

indicating that $\boldsymbol{p}_t(t)$ is a solution of eqn (4.235). But $\boldsymbol{p}_t(t)$ is periodic, and the trivial solution of eqn (4.235) cannot, therefore, be asymptotically stable (e.g. Willems 1970).

A more relevant stability property for periodic orbits is the *orbital stability* of closed trajectories discussed in Section 1.4. It appears from the above discussion that the limit cycle $\boldsymbol{p}(t)$ is *asymptotically orbitally stable* if $(n-1)$ characteristic multipliers (exponents) associated with eqn (4.235) have an absolute value smaller than one (negative real parts). The nth multiplier is of absolute value one. The proof of this theorem may be found in a number of books (e.g. Coddington and Levinson 1955).

In the special case of a two-dimensional system $(n=2)$, a criterion for the asymptotic orbital stability of $\boldsymbol{p}(t)$ is given by

$$\int_0^T \operatorname{Tr} \boldsymbol{A}(t)\,\mathrm{d}t < 0, \tag{4.242}$$

since the product of the multipliers is then less than one (Willems 1970).

In the analysis of the Hopf bifurcation as well as various degenerate dynamic bifurcations, it was assumed that all the eigenvalues of the corresponding Jacobian, except a pair, remained on the left half-plane of the complex λ-space in the vicinity of a critical point. In view of this assumption, the above criterion gains significance in stability analyses of bifurcation paths.

4.6.1 Hopf bifurcation

The stability of the family of limit cycles given by eqns (4.99) and (4.103) may be examined by determining the characteristic exponents (or multipliers) of the family. As remarked earlier, this is not an easy task but it can be accomplished since the limit cycles are defined explicitly. Nevertheless, here a two-dimensional system will be examined via the criterion (4.242), assuming that system (4.64) is two-dimensional and the Jacobian (4.66) consists of $\boldsymbol{K}_1$.

Note that system (4.4) has a parameter and the family of limit cycles is, of course, in terms of a parameter. Then, the equation of the first variation about the limit cycles can also be expressed in terms of this parameter as

$$\omega(\sigma)\boldsymbol{w}_t = \boldsymbol{A}(\tau;\sigma)\boldsymbol{w}, \qquad \boldsymbol{A}(\tau;\sigma) = \left[\frac{\partial \boldsymbol{W}}{\partial \boldsymbol{w}}\right]_{\text{L.C.}}, \tag{4.243}$$

where the time scaling $\tau = \omega(\sigma)t$ is introduced for convenience, and the solutions of eqn (4.243) are in the form $\boldsymbol{w} = \boldsymbol{w}(\tau;\sigma)$.

The corresponding criterion (4.242) may be expressed as

$$\frac{1}{\omega(\sigma)} \sum_{i=1}^{2} \int_0^{2\pi} A_{ii}(\tau;\sigma)\,\mathrm{d}\tau < 0, \tag{4.244}$$

where

$$A_{ii}(\tau;\sigma) = W_{iij}w^j + W'_{ii}\mu + \tfrac{1}{2}(W_{iijk}w^jw^k + 2W'_{iij}w^j\mu + W''_{ii}\mu^2) + \ldots,$$

and

$$w^j = w^j(\tau;\sigma) \quad \text{together with} \quad \mu = \mu(\sigma) \tag{4.245}$$

describes the family of limit cycles.

It is understood that, since this family was obtained by assigning the role of σ to amplitude p_{11} of the first harmonic, substituting for μ and w^j from eqns (4.99) and (4.103), respectively, into eqn (4.244) results in a criterion in terms of p_{11}. Indeed, if this operation is performed, (4.244) yields

$$\frac{1}{\omega(p_{11})}\sum_{i=1}^{2}\int_0^{2\pi} A_{ii}(\tau;p_{11})\,\mathrm{d}\tau = \frac{\pi}{3\omega(p_{11})}(\gamma_{11}-\gamma_{22})(p_{11})^2 + O\{(p_{11})^3\} + \ldots < 0, \tag{4.246}$$

where $\omega(p_{11}) = \{2\pi/T(p_{11})\} > 0$.

It follows that the family of limit cycles described by eqns (4.99) and (4.103) is *asymptotically orbitally stable* or *unstable* according to whether

$$\gamma_{11} - \gamma_{22} < 0 \tag{4.247}$$

or

$$\gamma_{11} - \gamma_{22} > 0, \tag{4.248}$$

respectively.

The stability distributions on the fundamental path $w = 0$ and the bifurcation path

$$\mu = \{(\gamma_{22} - \gamma_{11})/12\alpha'\}(p_{11})^2$$

are now readily established, confirming the results depicted in Fig. 4.2. Indeed, if the path $\boldsymbol{w} = 0$ is stable for $\mu < 0$ and unstable for $\mu > 0$, then $\alpha' > 0$, and the bifurcating path is stable (unstable) if its curvature is positive (negative).

4.6.2 *Symmetric bifurcation (tri-furcation)*

In this case, the $A_{ii}(\tau;p_{11})$ in eqn (4.244) are obtained by substituting for μ and w^j from eqns (4.144) and (4.147), respectively, into eqn (4.244). Performing the integration and summation again yields (Atadan and Huseyin 1983*b*)

$$\sum_{i=1}^{2}\int_0^{2\pi} A_{ii}(\tau;p_{11})\,\mathrm{d}\tau = \frac{\pi}{3}(\gamma_{11}-\gamma_{22})(p_{11})^2 + \ldots, \tag{4.249}$$

leading to precisely the same criteria as given by eqns (4.247) and (4.248). Thus, the bifurcation paths described by eqn (4.144) are *asymptotically orbitally stable* (*unstable*) if $\gamma_{11}-\gamma_{22}<0$ $(\gamma_{11}-\gamma_{22}>0)$. On the other hand, if the fundamental equilibrium path is stable for $u<0$, it remains stable for $\mu>0$ since $\alpha'=0$ and

$$\alpha(\mu)=\tfrac{1}{2}\alpha''\mu^2.$$

In other words, the equilibrium path is stable (unstable) if $\alpha''<0$ $(\alpha''>0)$.

Combining these results with the existence condition

$$(\gamma_{22}-\gamma_{11})/\alpha''>0$$

of the bifurcation paths

$$\mu=\pm\{(\gamma_{22}-\gamma_{11})/6\alpha''\}^{1/2}p_{11},$$

one concludes that:

i) If both α'' and $(\gamma_{11}-\gamma_{22})$ are negative (positive), then the equilibrium path is stable (unstable) and the bifurcating limit cycles do *not* exist.

ii) On the other hand, if α'' and $(\gamma_{11}-\gamma_{22})$ have opposite signs, then the bifurcating family exists and it is totally unstable (stable) if the equilibrium path is stable (unstable), (Fig. 4.7).

It is recalled that the symmetric bifurcation phenomenon was also explored through an alternative formulation in Section 4.5. It is not difficult to show that, in this case, eqn (4.249) takes the form

$$\sum_{i=1}^{2}\int_0^{2\pi}A_{ii}(\tau;p_{11})\,d\tau=\frac{\pi}{3}(\bar{\gamma}_{11}-\bar{\gamma}_{22})(p_{11})^2+\ldots, \tag{4.250}$$

and the $\bar{\gamma}_{ii}$ replace γ_{ii} in the stability criterion for the family of bifurcating limit cycles.

Finally, if a full analysis of the characteristic multipliers (or exponents) of an n-dimensional system is undertaken (Atadan 1983), it can be shown that the criteria for the stability of limit cycles remain identical to those obtained here for a two-dimensional system.

Problem 4.11. Following the procedure described in this section examine the stability of the limit cycles associated with the phenomena of the double Hopf bifurcation and tangential bifurcation.

4.7 Multiple-parameter systems

Consider the autonomous system (3.203),

$$\frac{dz}{dt}=\mathbf{Z}(z,\boldsymbol{\eta}),\qquad z\in R^n,\qquad \boldsymbol{\eta}\in R^m, \tag{4.251}$$

where the vector function $\boldsymbol{Z}(\boldsymbol{z}, \boldsymbol{\eta})$ is assumed to be analytic in the z^i and η^k—at least in a region G of interest. In this section, the analysis will be restricted to $n=2$ and $m=2$ for simplicity. Generalizations to higher dimensions, however, do not pose fundamental difficulties. Indeed, generalizations and extensions become rather straight-forward if the patterns developed so far are followed.

Suppose now that there exists a fundamental (initial) equilibrium surface in the region of interest obtained from

$$Z_i(z^j, \eta^\alpha) = 0, \qquad (i, j = 1, 2;\ \alpha, \beta = 1, 2), \tag{4.252}$$

as $z^i = f_i(\eta^\alpha)$, where the functions $f_i(\eta^\alpha)$ are single-valued.

By virtue of this assumption, and in analogy with eqn (3.305), one may introduce the transformation

$$\boldsymbol{z} = \boldsymbol{f}(\boldsymbol{\eta}) + \boldsymbol{Q}\boldsymbol{w}, \qquad |\boldsymbol{Q}| \neq 0, \tag{4.253}$$

into (4.251) such that the resulting system

$$\frac{\mathrm{d}\boldsymbol{w}}{\mathrm{d}t} = \boldsymbol{W}(\boldsymbol{w}, \boldsymbol{\eta}) \tag{4.254}$$

has the properties

$$\boldsymbol{W}(\boldsymbol{0}, \boldsymbol{\eta}) = \boldsymbol{W}_\alpha(\boldsymbol{0}, \boldsymbol{\eta}) = \boldsymbol{W}_{\alpha\beta}(\boldsymbol{0}, \boldsymbol{\eta}) = \ldots = \boldsymbol{0}, \tag{4.255}$$

and

$$[W_{ij}] = \left[\frac{\partial \boldsymbol{W}}{\partial \boldsymbol{w}}\right]_c = \begin{bmatrix} 0 & \omega_c \\ -\omega_c & 0 \end{bmatrix}, \tag{4.256}$$

where the point c of interest is again a critical point, $\eta^\alpha = \eta^\alpha_c$, on the fundamental equilibrium surface at which the real part of a complex conjugate pair of eigenvalues vanishes. Since the system now has two independent parameters, the critical point c is located on a *critical zone* which is a curve on the equilibrium surface. As in the case of one-parameter systems, the derivatives of the real part of the eigenvalues $\lambda_{1,2}(\eta^\alpha) = \alpha(\eta^\alpha) \pm \mathrm{i}\omega(\eta^\alpha)$ with respect to η^α (at c) are expected to play a major role in shaping the behaviour characteristics of the system under consideration, and various distinct phenomena may arise.

The derivatives of $\alpha(\eta^\alpha)$ and $\omega(\eta^\alpha)$ may be expressed in terms of the system coefficients by following the procedure that has led to eqn (4.73). Thus, if $\lambda(\eta^\alpha)$ is an eigenvalue of the Jacobian $[W_{ij}(\eta^\alpha)]$, it should satisfy the characteristic equation

$$|W_{ij}(\eta^\alpha) - \lambda(\eta^\alpha)\boldsymbol{I}| = 0 \tag{4.257}$$

identically, which yields

$$\alpha(\eta^\alpha) = \{W_{11}(\eta^\alpha) + W_{22}(\eta^\alpha)\}/2, \tag{4.258}$$

and

$$\omega(\eta^\alpha) = [4\{W_{11}(\eta^\alpha)W_{22}(\eta^\alpha) - W_{12}(\eta^\alpha)W_{21}(\eta^\alpha)\} \\ - \{W_{11}(\eta^\alpha) + W_{22}(\eta^\alpha)\}^2]^{1/2}/2. \tag{4.259}$$

Differentiating $\alpha(\eta^\alpha)$ and $\omega(\eta^\alpha)$ with respect to η^α $(\alpha = 1, 2)$ and evaluating at c, where $\alpha(\eta_c^\alpha) = 0$ and $\lambda_{1,2}(\eta_c^\alpha) = \pm i\omega(\eta_c^\alpha) \equiv \pm i\omega_c$, results in

$$\partial\alpha/\partial\eta^\alpha|_c = (W_{11}^\alpha + W_{22}^\alpha)/2 \tag{4.260}$$

and

$$\partial\omega/\partial\eta^\alpha|_c = (W_{12}^\alpha - W_{21}^\alpha)/2, \tag{4.261}$$

where the real part $\alpha(\eta^\alpha)$ and the superscript on η are not to be confused (the notation is not changed for consistency with the rest of the book). The above derivatives will simply be indicated by a corresponding superscript as α^1, α^2, ω^1, and ω^2.

A second differentiation of eqn (4.258) with respect to η^β $(\beta = 1, 2)$ yields

$$\partial^2\alpha/\partial\eta^\alpha\partial\eta^\beta|_c = (W_{11}^{\alpha\beta} + W_{22}^{\alpha\beta})/2, \tag{4.262}$$

upon evaluation at c.

Let now a prospective two-parameter family of limit cycles bifurcating off the fundamental equilibrium surface (if it exists) be expressed in the parametric form

$$\begin{aligned} w^i &= w^i(\tau; \sigma^a), \\ \eta^\alpha &= \eta^\alpha(\sigma^a), \\ \omega &= \omega(\sigma^a), \end{aligned} \tag{4.263}$$

where $a = 1, 2$; $\omega(\sigma^a)$ is the frequency of the limit cycles with $\omega(0) = \omega_c$, and the time scaling $\tau = \omega(\sigma^a)t$ is introduced.

Substituting this assumed solution back into eqn (4.254) yields the identity

$$\omega(\sigma^a)w_\tau^i(\tau; \sigma^a) \equiv W_i\{w^j(\tau; \sigma^a), \eta^\alpha(\sigma^a)\}. \tag{4.264}$$

A sequence of perturbation equations is then generated by differentiating eqn (4.264) with respect to σ^a $(a = 1, 2)$ successively. Thus,

$$\omega^{,a}w_\tau^i + \omega w_\tau^{i,a} = W_{ij}w^{j,a} + W_{i\alpha}\eta^{\alpha,a}, \tag{4.265}$$

$$\begin{aligned} &\omega^{,ab}w_\tau^i + \omega^{,a}w_\tau^{i,b} + \omega^{,b}w_\tau^{i,a} + \omega w_\tau^{i,ab} \\ &= (W_{ijk}w^{k,b} + W_{ij\alpha}\eta^{\alpha,b})w^{j,a} + W_{ij}w^{j,ab} \\ &\quad + (W_{ij\alpha}w^{j,b} + W_{i\alpha\beta}\eta^{\beta,b})\eta^{\alpha,a} + W_{i\alpha}\eta^{\alpha,ab}, \end{aligned} \tag{4.266}$$

etc.,

which are written in the familiar notation with the aid of the summation convention. In particular, note that the derivatives of ω associated with the limit cycles are indicated with the aid of a comma as distinct from the derivatives of ω related to the equilibrium surface.

The family of limit cycles (eqns 4.263) is further assumed to be in a more explicit form expressed by an essentially multiple-parameter Fourier series

$$w^i(\tau;\sigma^a)=\sum_{m=0}^{M}\{p_{im}(\sigma^a)\cos m\tau+r_{im}(\sigma^a)\sin m\tau\},\qquad(4.267)$$

where

$$p_{im}(0)=r_{im}(0)=0,\qquad(\forall i,m),\qquad(4.268)$$

and $w^i(\tau;0)=0$ at c (Atadan and Huseyin 1983c).

It will further be assumed without loss of generality that

$$r_{11}(\sigma^a)\equiv 0.\qquad(4.269)$$

A number of interesting cases will next be identified and explored within the framework of this formulation.

4.7.1 *Generalized Hopf bifurcation No. 1*

Let the derivatives (4.260) satisfy the conditions

$$\partial\alpha/\partial\eta^\alpha|_c\neq 0,\qquad(\alpha=1,2).\qquad(4.270)$$

Evaluating now the first perturbation equations (4.265) at c yields

$$\omega_c w^{i,a}_{\tau}=W_{ij}w^{j,a},\qquad(4.271)$$

where W_{ij} is in the canonical form (4.256). Introducing eqn (4.267) into (4.271) results in

$$\omega_c\sum_{m=0}^{M}m(-p_{im}^{,a}\sin m\tau+r_{im}^{,a}\cos m\tau)=W_{ij}\sum_{m=0}^{M}(p_{jm}^{,a}\cos m\tau+r_{jm}^{,a}\sin m\tau),\qquad(4.272)$$

where the superscripts a on p's and r's, together with commas, indicate differentiation with respect to the σ^a. Comparing the coefficients of $\cos m\tau$ and $\sin m\tau$ for each m leads to the derivatives

$$p_{11}^{,a}=-r_{21}^{,a},\qquad p_{21}^{,a}=r_{11}^{,a}=0,$$

and (4.273)

$$p_{im}^{,a}=r_{im}^{,a}=0\qquad(m\neq 1,\ \forall i,a).$$

It appears now, and it can be shown, that either pair (p_{11},η^1) or (p_{11},η^2) may be selected as σ^a. Selecting the first pair, for example, leads to

$$\dot{p}_{11} = 1, \qquad p_{11}^{,1} = 0, \tag{4.274}$$

$$\dot{\eta}^1 = 0 \quad \text{and} \quad \eta^{1,1} = 1, \tag{4.275}$$

since p_{11} and η^1 are independent parameters. It is understood that here a 'dot' and a superscript '1' are used to denote differentiation with respect to p_{11} and η^1, respectively.

Evaluating the second perturbation equation (4.266) at c results in a simplified form,

$$\omega^{,a} w_{\tau}^{i,b} + \omega^{,b} w_{\tau}^{i,a} + \omega w_{\tau}^{i,ab} = W_{ijk} w^{j,a} w^{k,b} + W_{ij\alpha} \eta^{\alpha,b} w^{j,a} + W_{ij} w^{j,ab} + W_{ij\alpha} w^{j,b} \eta^{\alpha,a}, \tag{4.276}$$

which represents six equations ($a, b = 1, 2$). Introducing eqn (4.267) into eqn (4.276) and using the results of the first perturbation yields (Atadan and Huseyin 1985*a*)

$$\begin{aligned}
\dot{p}_{21}^{,1} &= (W_{11}^2 W_{22}^1 - W_{11}^1 W_{22}^2)/2\omega_c \alpha^2 \\
\dot{r}_{21}^{,1} &= \{(W_{12}^1 + W_{21}^1)\alpha^2 - (W_{12}^2 + W_{21}^2)\alpha^1\}/2\omega_c \alpha^2, \\
\ddot{p}_{21} &= p_{21}^{,11} = \ddot{r}_{21} = r_{21}^{,11} = 0, \\
\ddot{p}_{im} &= p_{im}^{(2)}, \qquad (m = 0, 2), \\
\ddot{r}_{i2} &= r_{i2}^{(2)}, \\
\dot{p}_{im}^{,1} &= p_{im}^{,11} = 0, \qquad (m = 0, 2) \\
\dot{r}_{i2}^{,1} &= r_{i2}^{,11} = 0, \\
p_{im}^{,ab} &= r_{im}^{,ab} = 0, \qquad (m \geq 3, \ \forall i, \ a),
\end{aligned} \tag{4.277}$$

The last result follows from

$$p_{11}^{,ab} = r_{11}^{,ab} = 0$$

which is due to eqn (4.269) and the fact that $p_{11} = \sigma^1$.

Here, $p_{im}^{(2)}$ and $r_{i2}^{(2)}$ are precisely as defined by eqn (4.93) for the Hopf bifurcation, while α^1 and α^2 are given by eqn (4.260).

It is noted that the derivatives for $m \geq 3$ are obtained for all i, a, b, and $m \geq 3$ solving only four equations, given by

$$m\omega_c r_{im}^{,ab} = W_{ij} p_{im}^{,ab} \quad \text{and} \quad -m\omega_c p_{im}^{,ab} = W_{ij} r_{im}^{,ab}$$

where the order of differentiation is of course immaterial.

In addition to eqns (4.277), the equations associated with the first harmonics also yield

$$\begin{aligned}
\dot{\eta}^2 &= 0, \\
\eta^{2,1} &= -\alpha^1/\alpha^2, \\
\dot{\omega} &= 0,
\end{aligned} \tag{4.278}$$

and

$$\omega_{,1} = \{(W_{12}^1 - W_{21}^1)\alpha^2 - (W_{12}^2 - W_{21}^2)\alpha^1\}/2\alpha^2.$$

It is clear from these derivatives that in order to obtain a meaningful first-order characterization of the post-critical behaviour, one has to proceed to a third perturbation. Thus, differentiating the identity (4.264) with respect to the independent variables σ^c (that is, p_{11} and η^1) for a third time and evaluating at c yields

$$\ddot{\eta}^2 = (\gamma_{22} - \gamma_{11})/6\alpha^2$$

and (4.279)

$$\ddot{\omega} = \{(\gamma_{22} - \gamma_{11})(W_{12}^2 - W_{21}^2) - 2(\gamma_{12} - \gamma_{21})\alpha^2\}/12\alpha^2.$$

The γ_{ij} are determined from eqn (4.97) for $i, j = 1, 2$ as in the case of the Hopf bifurcation, and it is assumed that

$$\gamma_{22} - \gamma_{11} \neq 0. \tag{4.280}$$

One can now construct the Taylor's expansions of the relationships $\eta^2 = \eta^2(p_{11}, \eta^1)$ and $\omega = \omega(p_{11}, \eta^1)$ on the basis of the above derivatives. Thus, one has

$$\mu^2 = \eta^2{}_{,1}\mu^1 + \tfrac{1}{2}\ddot{\eta}^2(p_{11})^2 \tag{4.281}$$

and

$$\Omega = \omega_{,1}\mu^1 + \tfrac{1}{2}\ddot{\omega}(p_{11})^2, \tag{4.282}$$

where $\mu^\alpha = \eta^\alpha - \eta_c^\alpha$ and $\Omega = \omega - \omega_c$.

Eliminating p_{11} between eqns (4.281) and (4.282) also yields

$$\mu^2 = \{(\eta^2{}_{,1}\ddot{\omega} - \ddot{\eta}^2\omega_{,1})/\ddot{\omega}\}\mu^1 + (\ddot{\eta}^2/\ddot{\omega})\Omega. \tag{4.283}$$

The surface described by eqn (4.281) is a parabolic surface (Fig. 4.14), and has an unmistakable resemblance to the fold catastrophe associated

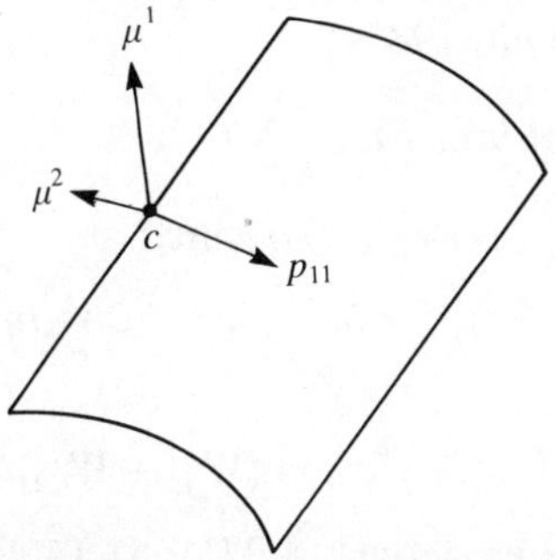

Fig. 4.14 The behaviour surface described by eqn (4.281).

with equilibrium surfaces (Chapters 2 and 3). It will be called the *behaviour surface,* treating the amplitude p_{11} as a behaviour variable. It is noted that the behaviour surface (4.281) would have looked topologically identical to the equilibrium surfaces associated with the fold catastrophe if the φ-transformation of the parameters were not introduced in the latter case. Indeed, the fold catastrophe (given by eqn (20) in Huseyin (1977)) becomes identical to eqn (4.281), with p_{11} replacing the generalized coordinate u_1. The essential difference between the fold catastrophe and the phenomenon described by eqn (4.281) is, of course, physical and not mathematical. While the fold catastrophe explored in the preceding chapters are associated with equilibrium states, each point on the behaviour surface (4.281) represents a limit cycle. Furthermore, the *stability boundary* in the vicinity of c is now obtained as the intersection of the *fundamental equilibrium surface* and the *behaviour surface.* In other words, setting $p_{11} = 0$ in eqn (4.281) yields the stability boundary which is a straight line to the approximation considered here.

The periodic solutions (limit cycles) associated with the behaviour surface are also constructed from the derivatives obtained above, and may be expressed as

$$w^i(\tau; p_{11}, \mu^1) = \dot{w}^i p_{11} + \tfrac{1}{2}\{\ddot{w}^i (p_{11})^2 + 2\dot{w}^{i,1} p_{11}\mu^1\} + \ldots, \tag{4.284}$$

where

$$\dot{\boldsymbol{w}} = (\cos\tau,\ -\sin\tau)^T,$$
$$\dot{\boldsymbol{w}}^{,1} = (0,\ (\dot{p}^{,1}_{21}\cos\tau + \dot{r}^{,1}_{21}\sin\tau))^T,$$
$$\ddot{w}^i = \ddot{p}_{i0} + \ddot{p}_{i2}\cos 2\tau + \ddot{r}_{i2}\sin 2\tau,$$

and $\dot{p}^{,1}_{21}$, $\dot{r}^{,1}_{21}$, $\ddot{p}_{i0}$, $\ddot{p}_{i2}$, $\ddot{r}_{i2}$ are all given by eqns (4.277).

It is observed that setting $\mu^1 = 0$ in eqns (4.281) to (4.284) yields the corresponding equations of the Hopf bifurcation studied in Section 4.4.1. A final note on the perturbation procedure: one may wonder what would have happened if p_{11} and η^2 were selected as perturbation parameters. As a matter of fact, such a choice would have resulted in a direct analogue of the present situation leading precisely to the same conclusions (Atadan and Huseyin 1985*a*).

4.7.2 *Generalized Hopf bifurcation No. 2*

Let the derivatives (4.260) now be specified as

$$\alpha^1 = \partial\alpha/\partial\eta^1|_c = 0, \qquad \alpha^2 = \partial\alpha/\partial\eta^2|_c \neq 0,$$

and (4.285)

$$\alpha^{11} = \partial^2\alpha/\partial(\eta^1)^2|_c = (W^{11}_{11} + W^{11}_{22})/2 \neq 0.$$

Under these conditions, eqns (4.271) to (4.278) remain essentially valid, requiring only minor adjustments in view of $\alpha^1 = 0$. Thus, since this condition implies $W^1_{11} = -W^1_{22}$, one obtains from eqns (4.277) the derivatives

$$\dot{p}^{\,1}_{21} = -W^1_{11}/\omega_c,$$
$$\dot{r}^{\,1}_{21} = (W^1_{12} + W^1_{21})/2\omega_c, \tag{4.286}$$

and from eqns (4.278) the derivatives

$$\eta^{2,1} = 0 \quad \text{and} \quad \omega^{,1} = (W^1_{12} - W^1_{21})/2. \tag{4.287}$$

It is noted that $\omega^{,1}$ is now equal to ω^1 which is associated with the equilibrium surface, as indicated by eqn (4.261). It is also observed that the only difference from the previous case arises due to a vanishing derivative, namely $\eta^{2,1} = 0$, necessitating a higher derivative. Thus, proceeding to a third-order perturbation under the new conditions leads to

$$\ddot{\eta}^2 = (\gamma_{22} - \gamma_{11})/6\alpha^2,$$
$$\dot{\eta}^{2,1} = 0, \qquad \eta^{2,11} = -\alpha^{11}/\alpha^2,$$
$$\ddot{\omega} = \{(\gamma_{22} - \gamma_{11})\omega^2 - (\gamma_{12} + \gamma_{21})\alpha^2\}/6\alpha^2, \tag{4.288}$$

and

$$\dot{\omega}^{,1} = 0,$$

where $\omega^2 = (w^2_{12} - w^2_{21})/2$, and $\ddot{\eta}^2$ and $\ddot{\omega}$ are obviously identical to those given by eqns (4.279).

Under the assumption (4.280), the behaviour surface may now be constructed as

$$\mu^2 = \tfrac{1}{2}\{\ddot{\eta}^2(p_{11})^2 + \eta^{2,11}(\mu^1)^2\}, \tag{4.289}$$

which is either an *elliptic paraboloid* ($\ddot{\eta}^2\eta^{2,11} > 0$) or a *saddle surface* ($\ddot{\eta}^2\eta^{2,11} < 0$), depicted in Figs. 4.15 and 4.16 respectively. Both of these surfaces are akin to the fold catastrophe shown in Figs 2.1 and 3.19. The behaviour surfaces here are, of course, meaningful only for positive amplitude p_{11}, and these surfaces actually branch off from the fundamental equilibrium surface defined by $p_{11} = 0$. The intersection of the behaviour surface with the parameter plane gives the stability boundary which is defined by eqn (4.289) with $p_{11} = 0$. Clearly, the boundary is a parabolic curve in the vicinity of c. The elliptic paraboloid surface depicted in Fig. 4.15 is analogous to the elliptic fold catastrophe (Figs. 2.1*a* and 3.19*a*), and the periodic solutions exist only for $\mu^2 > 0$ in this case, since it has been assumed that both $\ddot{\eta}^2 > 0$ and $\eta^{2,11} > 0$.

If one considers a ray in the parameter space defined by $\mu^1 = l^1\rho$ and $\mu^2 = l^2\rho$, where l^1 and l^2 are the direction cosines, then on a plot of ρ versus p_{11}, one observes a stable Hopf bifurcation provided $l^2 \neq 0$ (Fig. 4.15*c*). For $l^2 = 0$, eqn (4.289) yields an isolated critical point on a totally

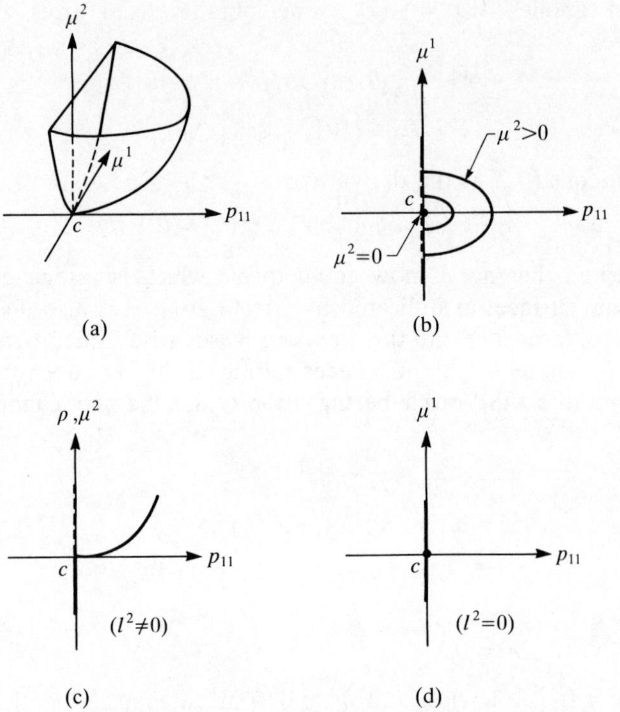

Fig. 4.15 Elliptic paraboloid behaviour surface.

stable equilibrium path (Fig. 4.15*d*). In fact, the entire behaviour surface in the vicinity of *c* is stable according to the criteria developed in Section 4.6. Clearly, a reversed situation arises if both $\ddot{\eta}^2 < 0$ and $\eta^{2,11} < 0$, in which case the entire behaviour surface is unstable, and unstable Hopf bifurcations are exhibited with regard to all rays (with $l^2 \neq 0$) in the parameter space. As far as the fundamental equilibrium surface is concerned, it is evident that it loses its stability on the stability boundary in both cases. It is also noted that, for constant values of μ^2, one has a concentric family of ellipses (Fig. 4.15*b*), and on a plot of μ^1 versus p_{11} the corresponding equilibrium path first loses and then regains its stability at critical points of Hopf bifurcation.

Consider next the saddle surface (Fig. 4.16) which is shown for $\ddot{\eta}^2 > 0$ and $\eta^{2,11} < 0$. This behaviour surface is analogous to the *anticlastic* fold catastrophe depicted in Figs. 2.1*b* and 3.19*b*. Periodic solutions now exist for both $\mu^2 > 0$ and $\mu^2 < 0$. If one again considers the parameter ray ρ (with $l^2 \neq 0$), it is observed that (Fig. 4.16*c*) a corresponding equilibrium path along ρ loses its stability at a critical point where a stable family of limit cycles branches off from the equilibrium path (stable Hopf bifurcation). On the other hand, if one specifies $l^2 = 0$, a *symmetric*

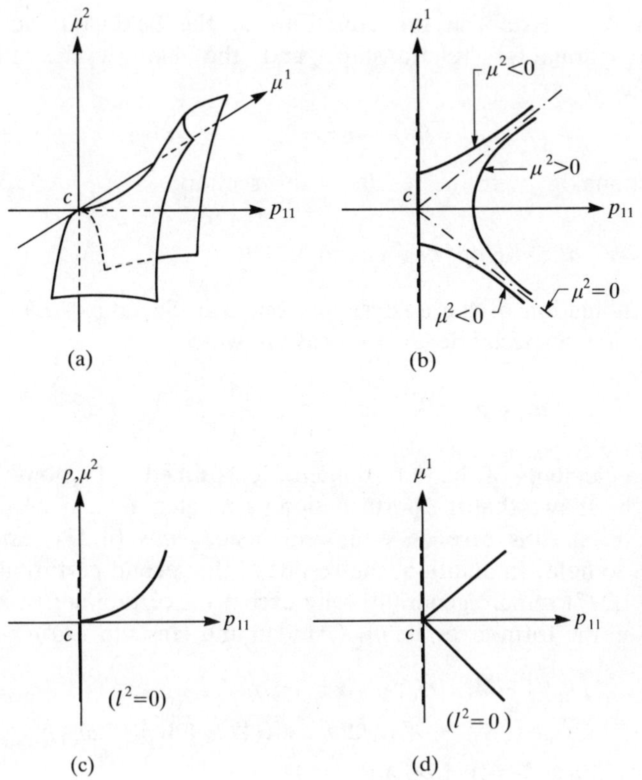

Fig. 4.16 Saddle surface.

bifurcation (*tri-furcation*) occurs (Fig. 4.16*d*). In other words, the critical point c appears as a point of Hopf bifurcation on all rays except for $\rho \equiv \mu^1$, in which case c appears as a point of *tri-furcation* from where a stable family of limit cycles branches off (Fig. 4.16*d*). At constant negative values of μ^2, an unstable equilibrium path becomes stable with increasing μ^1, and then unstable again at critical points of Hopf bifurcation (Fig. 4.16*b*). A direct analogue of the saddle surface depicted in Fig. 4.16 emerges when the signs of the principal curvatures are reversed such that $\ddot{\eta}^2 < 0$ and $\eta^{2,11} > 0$. In this case the behaviour surface is totally unstable.

The asymptotic equations of the family of limit cycles associated with the behaviour surface (4.289) are again given by eqn (4.284), except that in this case $\dot{p}_{21}^{,1}$ and $\dot{r}_{21}^{,1}$ are to be determined from eqn (4.286). Also, the frequency-parameter relationship is in the form

$$\Omega = \omega^{,1}\mu^1 + \tfrac{1}{2}\ddot{\omega}(p_{11})^2, \tag{4.290}$$

where $\omega^{,1}$ and $\ddot{\omega}$ follow from eqns (4.287) and (4.288), respectively.

Problem 4.12. Establish the equations of the behaviour surface, the frequency-parameter relationship, and the limit cycles under the conditions

$$\alpha^1 \neq 0, \qquad \alpha^2 = 0, \qquad \alpha^{22} \neq 0,$$

by direct analogy with the results of this section.

4.7.3 *Generalized Hopf bifurcation No. 3*

As a continuation of the analysis presented in Sections 4.7.1 and 4.7.2, consider now a special degenerate case in which

$$\alpha^1 = \alpha^2 = 0, \quad \text{and} \quad \begin{vmatrix} \alpha^{11} & \alpha^{12} \\ \alpha^{21} & \alpha^{22} \end{vmatrix} = 0 \quad \text{at } c, \tag{4.291}$$

while the elements of the determinant are assumed to be non-zero.

It can be shown that the perturbation parameters σ^a may be selected as p_{11} and η^1 in this case as well, and hence eqns (4.274) and (4.275) continue to hold. In addition, the results of the second perturbation given by eqn (4.277) remain generally valid except for certain derivatives which are now in the form given below (Atadan and Huseyin 1985*a*):

$$\begin{aligned}
\dot{p}_{21}^{,1} &= -(W_{11}^1/\omega_c) - (W_{11}^2/\omega_c)\eta^{2,1},\\
\dot{r}_{21}^{,1} &= \{(W_{12}^1 + W_{21}^1)/2\omega_c\} + \{(W_{12}^2 + W_{21}^2)/2\omega_c\}\eta^{2,1},\\
\ddot{p}_{21} &= -(2W_{11}^2/\omega_c)\dot{\eta}^2,\\
\ddot{r}_{21} &= \{(W_{12}^2 + W_{21}^2)/\omega_c\}\dot{\eta}^2,\\
p_{21}^{,11} &= r_{21}^{,11} = 0,\\
\dot{\omega} &= \omega^2\dot{\eta}^2,\\
\omega^{,1} &= \omega^1 + \omega^2\eta^{2,1}.
\end{aligned} \tag{4.292}$$

The third-order perturbation problem yields

$$\dot{\eta}^2 = \pm\{(\gamma_{22} - \gamma_{11})/6\alpha^{22}\}^{1/2},$$

and

$$\eta^{2,1} = -\alpha^{12}/\alpha^{22}, \tag{4.293}$$

which lead to a first-order approximation for the behaviour surface as

$$\mu^2 = \eta^{2,1}\mu^1 + \dot{\eta}^2 p_{11}. \tag{4.294}$$

For the existence of this solution, however, the condition

$$(\gamma_{22} - \gamma_{11})\alpha^{22} > 0 \tag{4.295}$$

has to be satisfied.

It is recognized from eqns (4.293) that the behaviour surface is an improper one, consisting of two intersecting planes in the parameter-amplitude space.

The frequency-amplitude relationship is constructed as

$$\Omega = \omega'^{1}\mu^{1} + \dot{\omega} p_{11}, \tag{4.296}$$

where ω'^{1} and $\dot{\omega}$ are determined by the expressions given in eqns (4.292).

The corresponding limit cycles are expressed as

$$\begin{bmatrix} w^1 \\ w^2 \end{bmatrix} = \begin{bmatrix} \cos\tau \\ -\sin\tau \end{bmatrix} p_{11} + \frac{1}{2}\left\{ \begin{bmatrix} \ddot{p}_{10} \\ \ddot{p}_{20} \end{bmatrix} + \begin{bmatrix} 0 \\ \ddot{p}_{21} \end{bmatrix} \cos\tau + \begin{bmatrix} 0 \\ \ddot{r}_{21} \end{bmatrix} \sin\tau \right.$$
$$\left. + \begin{bmatrix} \ddot{p}_{12} \\ \ddot{p}_{22} \end{bmatrix} \cos 2\tau + \begin{bmatrix} \ddot{r}_{12} \\ \ddot{r}_{22} \end{bmatrix} \sin 2\tau \right\} (p_{11})^2$$
$$+ \left\{ \begin{bmatrix} 0 \\ \dot{p}^{1}_{21} \end{bmatrix} \cos\tau + \begin{bmatrix} 0 \\ \dot{r}^{1}_{21} \end{bmatrix} \sin\tau \right\} p_{11}\mu^{1} + \ldots \tag{4.297}$$

where the derivatives $\ddot{p}_{i0}$, $\ddot{p}_{i2}$, and $\ddot{r}_{i2}$ are given by eqns (4.277), and the rest by eqns (4.292).

Finally, it is important to note that although the transformation matrix in eqn (4.253) is a constant matrix, system (4.254) is, in fact, equivalent to the X-system because the problems considered in Section 4.7 are all two-dimensional. The formulae derived in terms of the X coefficients (Appendix) are, therefore, equally applicable. Consider the following problem.

Problem 4.13. Suppose Q-transformation is chosen as $\boldsymbol{Q}(\eta^1, \eta^2)$ such that the Jacobian of the transformed system $\boldsymbol{x}_t = \boldsymbol{X}(\boldsymbol{x}, \boldsymbol{\eta})$ remains in the form

$$[X_{ij}(\eta^{\alpha})] = \begin{bmatrix} \alpha(\eta^1, \eta^2) & \omega(\eta^1, \eta^2) \\ -\omega(\eta^1, \eta^2) & \alpha(\eta^1, \eta^2) \end{bmatrix}_{x=0}, \qquad (\alpha = 1, 2), \tag{4.298}$$

as the η^{α} vary. It follows that α^1, α^2, α^{11}, α^{12}, α^{22}, etc., may be expressed simply as X^{1}_{11}, X^{2}_{11}, X^{11}_{11}, X^{12}_{11}, X^{22}_{11}, etc., respectively. Also, $X^{\alpha}_{12} = -X^{\alpha}_{21}$ and $X^{\alpha\beta}_{12} = -X^{\alpha\beta}_{21}$ (for $\alpha, \beta = 1, 2$). On the basis of this information, cast all the results of Section 4.7 in terms of the X-system. Discuss the simplifications achieved by this formulation.

4.7.4 *More generalized bifurcations*

In order to utilize the formulae given in the Appendix directly, the X-system will be used in this section, in its general form—without enforcing the property (4.298).

As remarked earlier, all that it takes to shift from the W-system to the X-system is to replace W's by X's, since the analysis is restricted to two-dimensional systems.

Thus, consider now the case in which

$$\alpha^1 = \alpha^2 \neq 0 \quad \text{and} \quad \bar{\gamma}_{22} - \bar{\gamma}_{11} = 0 \tag{4.299}$$

where the $\bar{\gamma}_{ii}$ are defined in the Appendix. This situation may be viewed as a generalization of the flat Hopf bifurcation. The analysis requires a total of five perturbations which, when completed, yield the necessary derivatives to construct the behaviour surface as

$$\mu^2 = \eta^{2,1}\mu^1 + \frac{1}{4!}\eta^{2,(4)}(p_{11})^4 \tag{4.300}$$

where

$$\eta^{2,1} = -\alpha^1/\alpha^2,$$
$$\eta^{2,(4)} = \partial^4(\eta^2)/\partial(p_{11})^4|_c = (\bar{\delta}_{22} - \bar{\delta}_{11})/10\alpha^2,$$

and $\bar{\delta}_{ii}$ are given in the Appendix.

Similarly, one has

$$\Omega = \omega'^1\mu^1 + \tfrac{1}{2}\ddot{\omega}(p_{11})^4 \tag{4.301}$$

where

$$\omega'^1 = \{(X^1_{12} - X^1_{21})\alpha^2 - (X^2_{12} - X^2_{21})\alpha^1\}/2\alpha^2$$

and

$$\ddot{\omega} = -(\bar{\gamma}_{12} + \bar{\gamma}_{21})/6.$$

Details of this analysis and the equations of the limit cycles are not given here (see Yu 1984). It is noted, however, that setting $\mu^1 = 0$ in eqn (4.300) leads to the *flat Hopf bifurcation* described in Problem 4.6.

Next, consider the case in which

$$\begin{aligned} \alpha^1 = \alpha^{11} = 0, \qquad & \bar{\gamma}_{22} - \bar{\gamma}_{11} = 0, \\ \alpha^2 \neq 0, \qquad \alpha^{111} \neq 0, \qquad & \text{and} \quad \bar{\delta}_{22} - \bar{\delta}_{11} \neq 0 \end{aligned} \tag{4.302}$$

These conditions imply

$$X^1_{11} = -X^1_{22} \quad \text{and} \quad X^{11}_{11} = -X^{11}_{22}.$$

Again, after five perturbations and using p_{11} and η^1 as perturbation parameters, one obtains the behaviour surface

$$\mu^2 = \tfrac{1}{2}\ddot{\eta}^{2,1}\mu^1(p_{11})^2 + \tfrac{1}{2}\dot{\eta}^{2,11}p_{11}(\mu^1)^2 + \frac{1}{4!}\eta^{2,(4)}(p_{11})^4, \tag{4.303}$$

where $\ddot{\eta}^{2,1}$ and $\dot{\eta}^{2,11}$ are certain expressions in terms of the derivatives of X (Yu 1984), and $\eta^{2,(4)}$ is exactly as given by eqn (4.300).

A smooth transformation of the form

$$\begin{aligned} \bar{\mu}^2 &= \mu^2 \\ \bar{p}_{11} &= p_{11} \\ \bar{\mu}^1 &= \mu^1 + \frac{\dot{\eta}^{2,11}(\mu^1)^2}{\ddot{\eta}^{2,1}p_{11}} \end{aligned} \tag{4.304}$$

reduces eqn (4.303) to the simpler form

$$\bar{\mu}^2 = \tfrac{1}{2}\ddot{\eta}^{2,1}\bar{\mu}^1(p_{11})^2 + \frac{1}{4!}\,\eta^{2,(4)}(\bar{p}_{11})^4$$

which may further be reduced to

$$\varphi^2 = \varphi^1 u + (u)^2 \tag{4.305}$$

by specifying

$$u = (\bar{p}_{11})^2, \qquad \varphi^1 = \frac{12\ddot{\eta}^{2,1}}{\eta^{2,(4)}}\,\bar{\mu}^1, \qquad \varphi^2 = \frac{4!}{\eta^{2,(4)}}\,\bar{\mu}^2. \tag{4.306}$$

Clearly, by treating $u = (p_{11})^2$ as a *behaviour variable,* and φ^1 and φ^2 as the new parameters, one observes that the behaviour surface described by (4.305) is akin to an *anticlastic fold catastrophe* (Fig. 4.17).

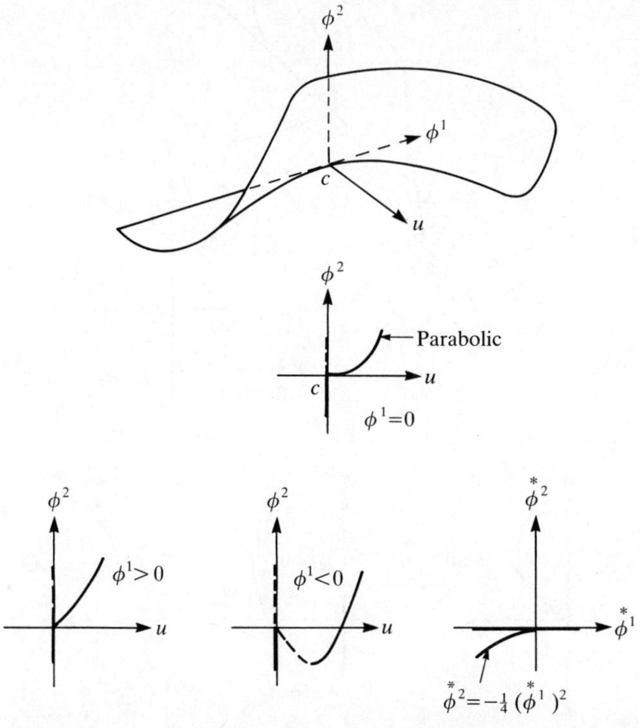

Fig. 4.17 The behaviour surface described by eqn (4.305), which is similar to an anticlastic fold catastrophe.

Finally, consider the situation arising under the conditions

$$\alpha^1 = \alpha^{11} = 0, \qquad \bar{\gamma}_{22} - \bar{\gamma}_{11} = 0, \qquad \bar{\delta}_{22} - \bar{\delta}_{11} = 0, \\ \alpha^2 \neq 0, \qquad \alpha^{111} \neq 0. \tag{4.307}$$

After a lengthy process of seven perturbations (Yu 1984), the behaviour surface emerges as

$$\mu^2 = \tfrac{1}{2}\ddot{\eta}^{2,1}\mu^1(p_{11})^2 + \tfrac{1}{2}\dot{\eta}^{2,11}p_{11}(\mu^1)^2 + \frac{1}{6!}\,\eta^{2,(6)}(p_{11})^6 \tag{4.308}$$

where $\ddot{\eta}^{2,1}$ and $\dot{\eta}^{2,11}$ are the same as before, and $\eta^{2,(6)}$ is a complicated expression in terms of the derivatives of X, which has been obtained by Yu (1984) explicitly.

Using the smooth transformation (4.304) yields

$$\bar{\mu}^2 = \tfrac{1}{2}\ddot{\eta}^{2,1}\bar{\mu}^1(\bar{p}_{11})^2 + \frac{1}{6!}\,\eta^{2,(6)}(\bar{p}_{11})^6 \tag{4.309}$$

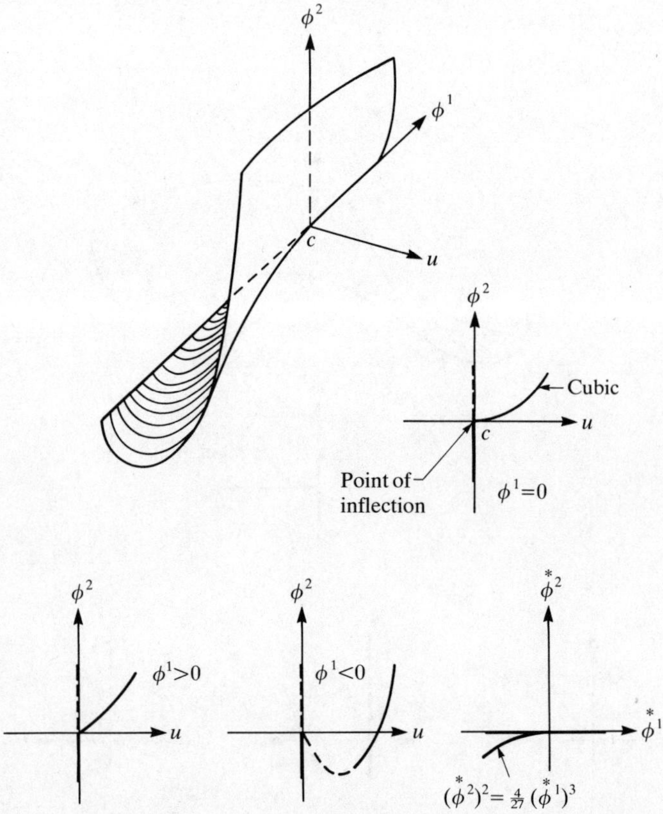

Fig. 4.18 The behaviour surface described by eqn (4.310), which is an analogue of the cusp catastrophe.

which reduces to

$$\varphi^2 = \varphi^1 u + (u)^3 \tag{4.310}$$

by a suitable scalar transformation similar to (4.306).

This behaviour surface is then recognized as an analogue of the *cusp catastrophe* associated with equilibrium states (Fig. 4.18).

It is also recognized, however, that in the case of both eqn (4.305) and eqn (4.310), $u > 0$, and the behaviour surfaces bifurcate off from the fundamental equilibrium surface. In addition to bifurcation points on the fundamental equilibrium surface ($u = 0$), however, there now exist *secondary bifurcation points* on both the behaviour surfaces as illustrated in Figs. 4.17 and 4.18.

Problem 4.14. Identify the secondary critical points on the behaviour surfaces (4.305) and (4.310) by analogy with the corresponding equilibrium surfaces (that is, *general* critical points of order 2 and 3, respectively), and then confirm your results analytically in a more formal way.

4.8 Concluding remarks

In this chapter, the phenomenon of Hopf bifurcation and several other dynamic bifurcation phenomena associated with autonomous systems have been explored via an intrinsic perturbation technique which parallels the methodology of the preceding chapters on static instabilities. There are, of course, many other dynamic bifurcation phenomena that can be studied through this approach, such as bifurcations associated with *non-simple* eigenvalues (Atadan and Huseyin 1985*b*) and bifurcations into invariant tori (Yu and Huseyin 1985), which will not be presented here. The literature on the subject is rapidly expanding, and the references noted below are by no means complete.

As emphasized in the introduction, the phenomenon of Hopf bifurcation is related to flutter instability in mechanics. However, while linear theories concerning flutter—both of a general as well as a specific nature—are substantial, in the non-linear post-flutter range the amount of work is rather limited. In this context, two investigations on panel flutter problems (Morino 1969; Kuo, Morino, and Dugundji 1972) should be noted. Also, Burgess and Levinson (1972) studied some aspects of the post-flutter behaviour of undamped systems. The activities in this area increased (e.g. Smith and Morino 1976; Sethna and Schapiro 1977; Bajaj, Sethna and Lundgren 1980; Bajaj and Sethna 1982; Scheidl, Troger and Zeman 1984) with the publication of a book totally devoted to the study of Hopf bifurcation and its applications (Marsden and McCracken 1976). Holmes (1977), and Holmes and Marsden (1978) studied the bifurcations to divergence and flutter in flow-induced oscillations.

Bio-chemical systems have been analysed by Kopell and Howard (1974); Auchmuty and Nicolis (1975); Nicolis and Prigogine (1977); Herschkowitz-Kaufman (1975); Merrill (1978); Cohen, Coutsias, and Neu (1979), and many others. Several interesting papers and references concerning non-linear oscillations in biology may be found in a book edited by Hoppensteadt (1979).

Examples of applications in electrical circuits and systems include the studies by Hirsch and Smale (1974), Mees and Chua (1979), Mees (1981), and Allwright (1977).

Basic developments are wide-ranging and include Hopf-Friederich's bifurcation theory by Poore (1976); degenerate bifurcations by Kielhöfer (1980), Flockerzi (1981), and Golubitsky and Langford (1981); global bifurcations by Rosenblat (1977), Alexander and Yorke (1978), and Chow, Paret and Yorke (1978); generalized Hopf bifurcations, multiple-parameter and/or multiple eigenvalues by Takens (1973), Jost and Zehnder (1972), Chafee (1968), Kielhöfer (1979), and Guckenheimer (1984); periodic and stationary-state mode interactions by Langford (1979), and Sprig (1983).

The list of references can be very lengthy, which is not attempted here; however, the reader is referred to several recent books which, along with numerous additional references, also provide excellent expositions of a variety of methods, phenomena, and applications (Iooss and Joseph 1980; Chow and Hale 1982; Hassard, Kazaninoff, and Wan 1981; Guckenheimer and Holmes 1983; Haken 1983; Abraham and Shaw 1982; Gurel and Rössler 1979). Also, a substantial amount of background information is available in a number of other books (e.g. Abraham and Marsden 1978; Arnold 1983).

5

APPLICATIONS

In the preceding chapters a general non-linear theory concerning the static and dynamic instabilities associated with autonomous systems has been developed from a unified point of view. In this chapter, several examples in a variety of fields will be analysed to illustrate the main features of the theory as well as the analytical value and advantages of the general results in applications.

5.1 Potential systems in mechanics

A large number of problems concerning the instability behaviour of conservative systems in mechanics have been studied with particular reference to simple critical points (e.g. Huseyin 1975), and attention in this section will, therefore, be focused on the imperfection sensitivities of coincident critical points.

5.1.1 A structural model

Consider the structural model shown in Fig. 5.1. This two-degree-of-freedom, rigid-link model was introduced by Chilver (1967), and it was also adapted (Huseyin 1975) to illustrate the behaviour of a conservative system in the vicinity of a coincident critical point. Here, certain initial imperfections, described by ε^1 and $\bar{\varepsilon}^2$ (Fig. 5.1), are introduced into the model in order to explore the effect of imperfections. The model consists of three rigid links hinged at the joints and restrained by four elastic non-linear springs as shown. An axial force P and a small couple $\bar{\varepsilon}^2/2$ act on the system.

The potential energy function of the system may be expressed (Huseyin 1975, *pp.* 101–106; Huseyin and Mandadi 1977*b*) as

$$\begin{aligned} S(u^i, \eta, \varepsilon^\alpha) = {} & \tfrac{1}{2}(2A + A_1)(u^1)^2 + \tfrac{1}{2}(2A + A_2)(u^2)^2 \\ & + \frac{1}{3!}\{(B + B_1 + B_2)(u^1)^3 + 3(B_1 - B_2)(u^1)^2 u^2 \\ & + 3(B_1 + B_2)u^1(u^2)^2 + (B_1 - B_2 + D)(u^2)^3\} \\ & + \frac{1}{4!}\{(2C + C_1)(u^1)^4 + 12C(u^1)^2(u^2)^2 + (2C + C_2)(u^2)^4\} \\ & - \eta(2A + A_1)[(u^1)^2 + 3(u^2)^2 + \tfrac{1}{4}\{(u^1)^4 + 6(u^1)^2(u^2)^2 \\ & + 9(u^2)^4\}]/2 - \varepsilon^1(2A + A_1)u^1 - \varepsilon^2(2A + A_1)u^2, \end{aligned} \tag{5.1}$$

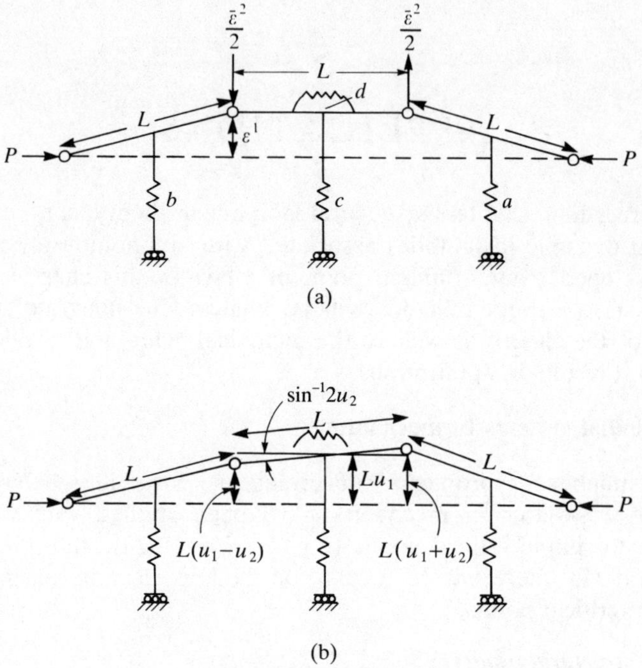

Fig. 5.1 Structural model.

where As, Bs and Cs are certain constants, and

$$C_2 = 16A_2, \qquad \eta = 2PL/(2A + A_1), \qquad \varepsilon^2 = \bar{\varepsilon}^2 L/(2A + A_1). \quad (5.2)$$

Note that the potential energy function $S(u^i, \eta, \varepsilon^\alpha)$ has been used instead of $V(q^i, \eta, \varepsilon^\alpha)$, since the transformation (2.122) is not required here. This follows of course from the fact that $u^1 = u^2 = 0$, satisfying $S_i(0, \eta, 0) = 0$, describes the fundamental equilibrium path of the perfect system ($\varepsilon^\alpha = 0$), and the quadratic form of the energy function is diagonalized on this path for all values of η.

By adjusting the structural constants of the system appropriately, many aspects of the theory in Chapter 2 can now be illustrated. Thus, let

$$A_2 - 3A_1 = 4A. \quad (5.3)$$

It is then readily verified that for $\eta = \eta_c = 1$, the stability coefficients S_{11} and S_{22} on the fundamental equilibrium path ($u^i = 0$) vanish simultaneously. In other words, the point c is a two-fold critical point under the condition (5.3). Indeed,

$$S_{11}(0, \eta, 0) = 2A + A_1 - \eta(2A + A_1),$$
$$S_{22}(0, \eta, 0) = 2A + A_2 - 3\eta(2A + A_1),$$

and one has

$$\begin{aligned} S_{11}(0, \eta_c, 0) &= 0, \\ S_{22}(0, \eta_c, 0) &= 0. \end{aligned} \tag{5.4}$$

Now let

$$B = -4, \qquad B_1 = B_2 = \tfrac{3}{2}, \qquad D = 0, \qquad 2A + A_1 = \tfrac{1}{2}. \tag{5.5}$$

It then follows that

$$\begin{aligned} S_{211} &= B_1 - B_2 = 0, \\ S_{222} &= B_1 - B_2 + D = 0, \\ S'_{11} &= -(2A + A_1) = -\tfrac{1}{2}, \\ S'_{22} &= -3(2A + A_1) = -\tfrac{3}{2}, \\ S_{122} &= B_1 + B_2 = 3, \\ S_{111} &= B + B_1 + B_2 = -1, \end{aligned} \tag{5.6}$$

and the problem under consideration falls within the scope of the theory presented in Section 2.6, Case 1. In other words, since $S_{211} = S_{222} = 0$, the system has symmetry with regard to u^2. The asymptotic equations of the equilibrium surface in the vicinity of c are given by eqns (2.130) and (2.131) which can be determined readily as

$$\varepsilon^1 = -(u^1)^2 + 3(u^2)^2 - u^1\mu$$

and

$$\varepsilon^2 = 6u^1u^2 - 3u^2\mu \tag{5.7}$$

respectively, upon observing that eqn (2.126) is given by

$$[S_{\alpha\alpha}] = -\frac{1}{2}\begin{bmatrix} 1 & 0 \\ 0 & 1 \end{bmatrix}. \tag{5.8}$$

The second fundamental tensors associated with eqn (5.7) are represented by

$$\begin{bmatrix} -2 & 0 & -1 \\ 0 & 6 & 0 \\ -1 & 0 & 0 \end{bmatrix} \quad \text{and} \quad \begin{bmatrix} 0 & 6 & 0 \\ 6 & 0 & -3 \\ 0 & -3 & 0 \end{bmatrix},$$

with the sets of eigenvalues $(6, -1 \pm \sqrt{2})$ and $(0, \pm\sqrt{45})$. By comparing these eigenvalues with those of eqn (2.112), one may deduce that the equilibrium surface (5.7) is associated with an *elliptic umbilic*. The critical zone and the stability boundary may be obtained by applying the procedure of Section 2.5.2. However, attention will now be focused on a special situation in which $\varepsilon^2 = 0$. In this case, eqn (5.7) yields

$$u^2 = 0 \quad \text{or} \quad \mu = 2u^1 \tag{5.9}$$

which results in two equilibrium surfaces

$$\left.\begin{aligned} u^2 &= 0 \\ \varepsilon^1 &= -(u^1)^2 - u^1\mu \end{aligned}\right\} \tag{5.10}$$

and

$$\left.\begin{aligned} \mu &= 2u^1 \\ \varepsilon^1 &= -(u^1)^2 + 3(u^2)^2 - u^1\mu \end{aligned}\right\} \tag{5.11}$$

as described in Section 2.6, Case 1.

It is now established that eqn (2.137) takes the form

$$\alpha = \frac{S'_{11}S_{122}}{S'_{22}S_{111}} = -1, \tag{5.12}$$

indicating that the behaviour of the surfaces (5.10) and (5.11) is as described under the category $\alpha < \frac{1}{2}$ (Fig. 2.9). The intersecting line is given by $\mu = 2u^1$, which results in the bifurcation boundary (eqn 2.140)

$$\overset{*}{\mu} = \pm(-\tfrac{4}{3}\varepsilon^1)^{1/2}, \tag{5.13}$$

upon taking into account eqn (5.8).

On the other hand, the locus of the limit points on the surface (5.10) constitutes another critical line in the $(\mu\text{–}\varepsilon^1)$ plane and is given by eqn (2.139) which takes the form

$$\overset{*}{\mu} = \pm(4\varepsilon^1)^{1/2}, \tag{5.14}$$

in view of eqn (5.8).

It is observed that eqns (5.13) and (5.14) are real only for $\varepsilon^1 < 0$ and $\varepsilon^1 > 0$ respectively, and together they constitute the *stability boundary* of the system (Fig. 5.2). As emphasized in the theory, the imperfection sensitivity curve would consist of eqn (5.14) if c were a *simple* critical

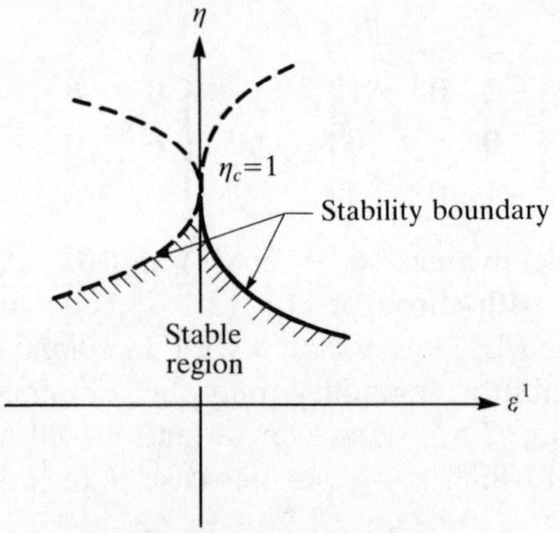

Fig. 5.2 Stability boundary.

point. The effect of coincidence of eigenvalues is reflected in the presence of the *bifurcation boundary* (5.13) in a region which would otherwise be totally stable.

Problem 5.1. If the structural parameters of the system are modified, the behaviour of the system may of course change. Show that the model will exhibit the phenomena discussed under the categories (ii), (iii), and (iv), in Section 2.6, if

ii) $B_1 = B_2 = -1, \quad B = 1, \quad 2A + A_1 = \frac{1}{2}, \quad D = 0;$
iii) $B_1 = B_2 = -3, \quad B = 4, \quad 2A + A_1 = \frac{1}{2}, \quad D = 0;$ and
iv) $B_1 = B_2 = -3, \quad B = 5, \quad 2A + A_1 = \frac{1}{2}, \quad D = 0,$

hold, respectively.

Problem 5.2. Adapt the model described by eqn (5.1) to illustrate the phenomena discussed in Section 2.6, Case 2, by adjusting the structural parameters appropriately.

5.1.2 *Columns on elastic foundations*

Consider an axially compressed elastic column of length l supported over its entire length by an elastic foundation as shown in Fig. 5.3. It is assumed that the column is inextensible and the reaction of the foundation at any point is proportional to the deflection. Furthermore, the column is assumed to be geometrically imperfect, with a small deviation $w_0(x)$ from the straight line configuration.

Let k_1 and k_2 denote the bending stiffness of the column and the modulus of the elastic foundation, respectively. In the light of the assumption that the imperfection is small, only the smallest order terms in the imperfection parameter will be considered. The total potential energy of this continuum problem is, therefore, expressed as follows:

$$V = \tfrac{1}{2}k_1 \int_0^l \{(w'' - w_0'')^2 + (w'')^2(w')^2\}\, dx + \tfrac{1}{2}k_2 \int_0^l (w - w_0)^2\, dx - P \int_0^l \{\tfrac{1}{2}(w')^2 + \tfrac{1}{8}(w')^4\}\, dx. \qquad (5.15)$$

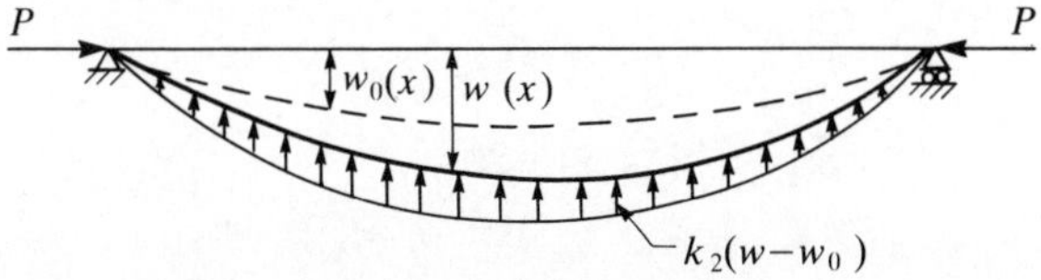

Fig. 5.3 Elastic column on elastic foundations.

The continuum represented by the functional (5.15) may be discretized by assuming that the displacements and the initial imperfections are in the form of

$$w = q^1 \sin\frac{\pi x}{l} + q^2 \sin\frac{2\pi x}{l}$$

and

$$w_0 = \bar{\varepsilon}^1 \sin\frac{\pi x}{l} + \bar{\varepsilon}^2 \sin\frac{2\pi x}{l}, \tag{5.16}$$

respectively.

Substituting from eqns (5.16) into eqn (5.15) and evaluating the integrals yields the discretized approximation for the potential energy

$$\begin{aligned} S(u^i, \eta, \varepsilon^\alpha) = {} & (1+k)(u^1)^2 + (16+k)(u^2)^2 \\ & + \tfrac{1}{4}\pi^2\{(u^1)^4 + 40(u^1)^2(u^2)^2 + 64(u^2)^4\} - \eta\{(u^1)^2 + 4(u^2)^2\} \\ & - \tfrac{3}{16}\pi^2\eta\{(u^1)^4 + 16(u^1)^2(u^2)^2 + 16(u^2)^4\} + \varepsilon^1 u^1 + \varepsilon^2 u^2, \end{aligned} \tag{5.17}$$

where

$$S = \frac{4lV}{\pi^4 k_1}, \qquad q^1 = lu^1, \qquad q^2 = lu^2, \qquad k = \frac{k_2 l^4}{k_1 \pi^4},$$

$$\eta = \frac{pl^2}{\pi^2 k_1}, \qquad \varepsilon^1 = -\frac{2(1+k)}{l}\bar{\varepsilon}^1, \qquad \varepsilon^2 = -\frac{2(16+k)}{l}\bar{\varepsilon}^2.$$

Here, function V has been replaced by function S, because it possesses the properties of S described in Section 2.6. Indeed, the equilibrium equations $S_i(u^j, \eta, 0) = 0$ indicate clearly that $u^i(\eta) = 0$ is a fundamental equilibrium path along which the quadratic form of the energy function maintains its diagonal form—thus satisfying eqns (2.124) and (2.125). The Hessian matrix

$$[S_{ij}(0, \eta, 0)] = \begin{bmatrix} 2(1+k) - 2\eta & 0 \\ 0 & 2(16+k) - 8\eta \end{bmatrix} \tag{5.18}$$

becomes singular at the critical point $\eta_c = 5$ if one lets $k = 4$. It is also noticed that the critical point c is a *two-fold* critical point at which both the stability coefficients vanish.

Following the theory of Section 2.6, Case 2, and setting $\eta = 5 + \mu$ yields the derivatives

$$\begin{aligned} \bar{S}_{1111} &= S_{1111} = -\tfrac{33}{2}\pi^2, \\ \bar{S}_{1122} &= S_{1122} = -20\pi^2, \\ S'_{11} &= -2, \qquad S'_{22} = -8, \\ \bar{S}_{2222} &= S_{2222} = 24\pi^2, \end{aligned} \tag{5.19}$$

and

$$S_{111} = S_{112} = S_{122} = S_{222} = 0.$$

Using eqns (2.154) and (2.155) yields the equilibrium surface

$$\begin{aligned} \varepsilon^1 &= 2.75\pi^2(u^1)^3 + 10\pi^2 u^1(u^2)^2 + 2u^1\mu, \\ \varepsilon^2 &= -4\pi^2(u^2)^3 + 10\pi^2 u^2(u^1)^2 + 8u^2\mu, \end{aligned} \tag{5.20}$$

in the vicinity of the critical point c. It is clear that the surface has the full complement of normals, and c is a *general* critical point.

Consider now the special situation arising when $\varepsilon^2 = 0$. In this case, eqn (5.20) degenerates into two surfaces

$$\left.\begin{aligned} u^2 &= 0 \\ \varepsilon^1 &= 2.75\pi^2(u^1)^3 + 2u^1\mu \end{aligned}\right\} \tag{5.21}$$

and

$$\left.\begin{aligned} (u^2)^2 &= 2.5(u^1)^2 + (2/\pi^2)\mu \\ \varepsilon^1 &= 27.75\pi^2(u^1)^3 + 22u^1\mu \end{aligned}\right\} \tag{5.22}$$

These surfaces intersect along the parabola

$$\mu = -\tfrac{5}{4}\pi^2(u^1)^2, \tag{5.23}$$

and the asymptotes associated with the equilibrium surface (5.21), that is the equilibrium paths of the perfect system ($\varepsilon^1 = 0$) on this surface, are given by

$$u^1 = 0 \quad \text{and} \quad 2.75\pi^2(u^1)^2 + 2\mu = 0. \tag{5.24}$$

The key coefficients 'a' and 'b', defined in the theory, are now in the form of

$$a = -2\frac{S'_{11}}{S_{1111}} = -\frac{8}{33\pi^2} \tag{5.25}$$

and

$$b = -2\frac{S'_{22}}{S_{1122}} = -\frac{4}{5\pi^2}. \tag{5.26}$$

It follows that the system under consideration falls within the scope of Section 2.6, Case 2, Category (iv). Furthermore, $|b| > 3\,|a|$.

The locus of the limit points on the surface (5.21) is described by

$$\mu = -4.125\pi^2(u^1)^2, \tag{5.27}$$

and the corresponding stability boundary takes the form

$$\overset{*}{\mu}{}^3 = -2.32\pi^2(\varepsilon^1)^2, \tag{5.28}$$

which is the familiar cusp, describing the imperfection sensitivity in the (μ–ε^1) plane. This phenomenon was illustrated in Fig. 2.17.

Suppose now that $\varepsilon^1 = 0$; in this case surface (5.20) degenerates into the surfaces

$$\left.\begin{aligned} u^1 &= 0 \\ \varepsilon^2 &= -4\pi^2(u^2)^3 + 8u^2\mu \end{aligned}\right\} \tag{5.29}$$

and

$$\left.\begin{aligned} (u^1)^2 &= -3.64(u^2)^2 - (0.73/\pi^2)\mu \\ \varepsilon^2 &= -40.4\pi^2(u^2)^3 + 0.7u^2\mu. \end{aligned}\right\} \tag{5.30}$$

These surfaces intersect along the parabola

$$\mu = -4.99\pi^2(u^2)^2, \tag{5.31}$$

and the asymptotes associated with eqns (5.29) are

$$u^2 = 0 \quad \text{and} \quad -4\pi^2(u^2)^2 + 8\mu = 0,$$

which indicate that the complementary paths will be intersected by the parabola (5.31), resulting in an initial loss of stability by bifurcations. The *bifurcation boundary* may be obtained by substituting for u^2 from eqn (5.31) into (5.29):

$$\overset{*}{\mu}{}^3 = 0.065\pi^2(\varepsilon^2)^2. \tag{5.32}$$

It can readily be demonstrated, by analogy with the case of $\varepsilon^2 = 0$, that in this case $a > 0$ and $b < 0$, thus placing the system under the category (ii). This phenomenon was illustrated in Fig. 2.15.

It is noted that the behaviour of the system is not significantly different from that associated with a *simple* critical point if the imperfection is in the u^1 mode. On the other hand, the presence of an imperfection in the u^2 mode induces a loss of stability at a parameter value less than the critical η_c associated with the perfect system. This indicates that the coincidence of eigenvalues may result in a higher imperfection sensitivity.

5.1.3 *Experimental results*

A number of experimental results concerning *general* critical points have been reported previously (Huseyin 1975, *pp.* 155–156). These experiments involved conservative structural systems under the influence of more than one independent parameter, and exhibited *tilted cusps*. Here, an additional experimental result which was not reported before is presented. It is concerned with a shallow, pre-stressed arch which was produced by providing simple support conditions for an initially straight steel strip of section 1 in. by $\frac{1}{16}$ in. such that the distance between the supports was shorter than the length of the strip, resulting in a rise of

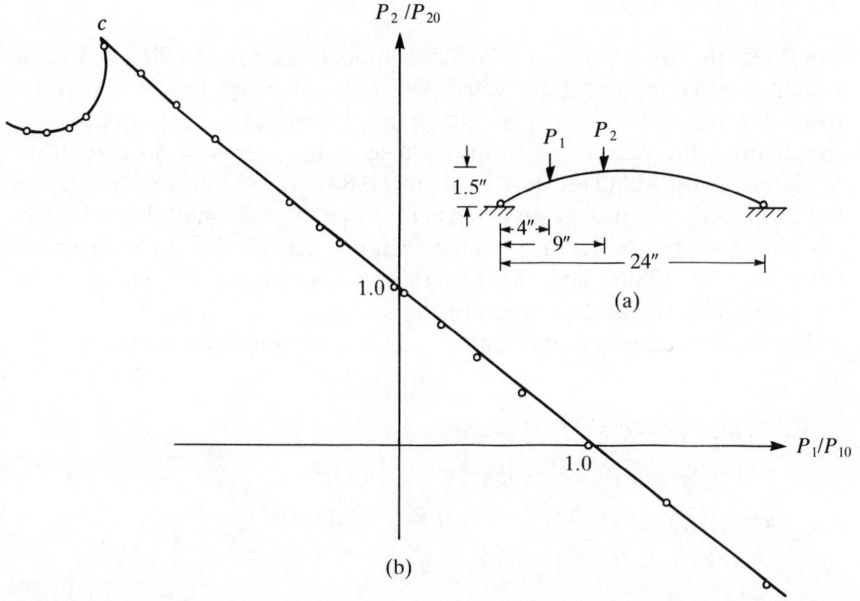

Fig. 5.4 Experimental results for a shallow pre-stressed arch (a). The stability boundary of the system is shown in (b).

1.5 in. at the crown (Fig. 5.4*a*). Vertical negative as well as positive loads were applied by means of pullies located above the arch. One of the loads was applied first and kept constant while the other was increased gradually until *snap-through* occurred after large deflections. This procedure was repeated for different but constant values of the first load. Thus, the critical combinations of P_1 and P_2 were obtained and plotted as shown in Fig. 5.4*b*, demonstrating the *stability boundary* of the system. P_{10} and P_{20} in this figure are the critical values of P_1 and P_2, respectively, when they were applied individually. All points on this boundary except *c* are associated with *general* critical points of order 2. In other words, the local equilibrium surface in the vicinity of these points (except for *c*) is described by the *fold* catastrophe. On the other hand, the critical point *c* is associated with a *singular general* point, and the figure shows the corresponding *tilted cusp* in the parameter space.

5.2 Autonomous mechanical and electrical systems

Many mechanical systems have interesting analogues in electrical networks, and this analogy is demonstrated in some of the problems analysed in this section. Both static and dynamic instabilities are illustrated.

5.2.1 *A mechanical system*

Consider the two-degree-of-freedom model shown in Fig. 5.5. The system consists of two rigid weightless links of equal length l, carrying two concentrated masses $2m$ and m as shown. The constants a, b, c characterize the non-linear springs whose strain energy (4th order in the θ_i) is given by Mandadi and Huseyin (1980). In addition to a follower force P_1 and a conservative vertical force P_2, a small lateral force $\bar{\varepsilon}$—which will be treated as an imperfection—acts on the model (Huseyin and Mandadi 1980), and the generalized coordinates θ_1 and θ_2 completely specify the configuration of the system.

The first-order equations of motion in non-dimensional form are

$$\begin{aligned}
z_t^1 &= z^2,\\
z_t^2 &= \tfrac{1}{2}(\eta^1+\eta^2-f_1-2f_2)z^1-\tfrac{1}{2}(\eta^1-2f_2)z^3\\
&\quad+\tfrac{1}{4}\{(2h_2-h_1)(z^1)^2-4h_2z^1z^3+2h_2(z^3)^2\}\\
&\quad+\tfrac{1}{12}\{(3f_3-k_1-2k_2)(z^1)^3+(6k_2+3f_3)(z^1)^2z^3\\
&\quad-(6k_2+3f_3)z^1(z^3)^2+(2k_2-3f_3)(z^3)^3\}+\tfrac{1}{2}\varepsilon,\\
z_t^3 &= z^4,\\
z_t^4 &= \tfrac{1}{2}(f_1+4f_2-2f_3-\eta^1-\eta^2)z^1+\tfrac{1}{2}(\eta^1-4f_2-2f_3)z^3\\
&\quad-\tfrac{1}{4}\{(2h_3+4h_2-h_1)(z^1)^2+(4h_3-8h_2)z^1z^3+(4h_2+2h_3)(z^3)^2\}\\
&\quad-\tfrac{1}{12}\{(k_1+4k_2)-2k_3-f_3)(z^1)^3-3(4k_2+2k_3+f_3)(z^1)^2z^3\\
&\quad+3(4k_2-2k_3+3f_3)z^1(z^3)^2+(11f_3-4k_2-2k_3)(z^3)^3\}-\tfrac{1}{2}\varepsilon
\end{aligned} \tag{5.33}$$

where

$$z^1=\theta_1,\qquad z^2=\dot{\theta}_1,\qquad z^3=\theta_2,\qquad z^4=\dot{\theta}_2,$$

η^1, η^2, and ε are non-dimensionalized parameters; and f_i, h_i, and k_i are certain constants.

Note that this example is concerned with a *non-dissipative system which is non-potential due to the presence of a follower force.* The eigenvalues of the Jacobian along an equilibrium path cannot be expected, therefore, to have negative real parts even when the system is stable—and asymptotic stability is ruled out. This model is capable of demonstrating a variety of phenomena, including interactions between static and dynamic instability modes. In fact, the model has been adapted by Scheidl, Troger, and Zeman (1984) for this purpose. Here, the model will be used to demonstrate certain static instabilities, and it is noted that the theory of Chapter 3 is applicable to this case even though the non-critical eigenvalues may be purely imaginary. It is, of course, assumed that such imaginary eigenvalues do not change their character as the parameters vary in the vicinity of a divergence point, and do not represent the onset of a flutter instability.

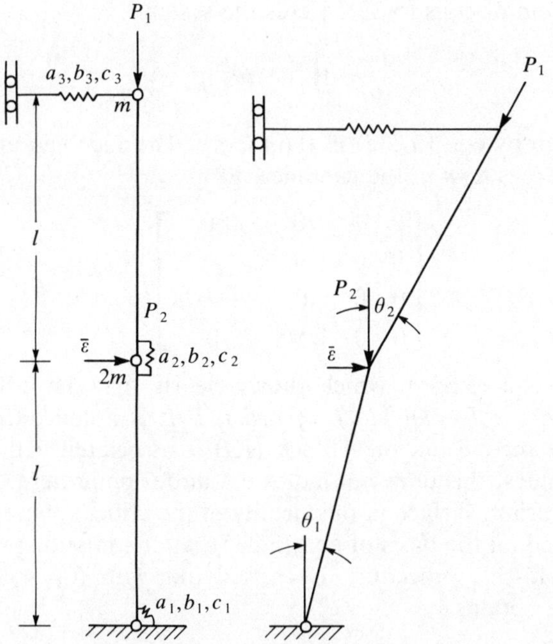

Fig. 5.5 A two-degree-of-freedom model.

It is first observed that the perfect system ($\varepsilon = 0$) possesses a fundamental equilibrium surface defined by $z^i = 0$ which satisfies $z^i_t = 0$. Let $f_1 = 2$, $f_2 = 0.125$, $f_3 = 0.25$, and consider a parameter ray $\eta^i = l^i\rho$ with $l^1 = l^2 = 1/\sqrt{2}$. It is easy to verify that $\rho_c = \sqrt{2}$ is a critical value of ρ at which point the Jacobian matrix, associated with the surface $z^i = 0$, takes the form

$$[Z_{ij}] = \begin{bmatrix} 0 & 1 & 0 & 0 \\ -\frac{1}{8} & 0 & -\frac{3}{8} & 0 \\ 0 & 0 & 0 & 1 \\ 0 & 0 & 0 & 0 \end{bmatrix}, \tag{5.34}$$

and its eigenvalues are $(0, 0, \pm i/\sqrt{8})$.

Introducing the transformation

$$\boldsymbol{z} = \boldsymbol{0} + \boldsymbol{Q}\boldsymbol{w}, \tag{5.35}$$

where

$$\boldsymbol{Q} = \begin{bmatrix} -3 & -3 & \sqrt{8} & 0 \\ 0 & -3 & 0 & -1 \\ 1 & 1 & 0 & 0 \\ 0 & 1 & 0 & 0 \end{bmatrix} \quad \text{and} \quad \boldsymbol{Q}^{-1} = \begin{bmatrix} 0 & 0 & 1 & -1 \\ 0 & 0 & 0 & 1 \\ 1/\sqrt{8} & 0 & 3/\sqrt{8} & 0 \\ 0 & -1 & 0 & -3 \end{bmatrix},$$

into the system of eqns (5.33), yields the system

$$\frac{\mathrm{d}\boldsymbol{w}}{\mathrm{d}t} = \boldsymbol{W}(w^i, \eta^1, \eta^2, \varepsilon) \tag{5.36}$$

which is given by eqn (3.332) in the theory. The Jacobian of this system, evaluated at c, is now in the canonical form

$$\begin{bmatrix} 0 & 1 & 0 & 0 \\ 0 & 0 & 0 & 0 \\ 0 & 0 & 0 & -1/\sqrt{8} \\ 0 & 0 & 1/\sqrt{8} & 0 \end{bmatrix}, \tag{5.37}$$

as given by eqn (3.333), which shows clearly that the critical point is associated with a *Jordan block of order 2.* It is noted once more that, although the second diagonal block ($\boldsymbol{K}_3$) is associated with pure imaginary eigenvalues, this does not indicate a flutter point here.

The equilibrium surface in the vicinity of the critical point $\rho_c = \sqrt{2}$ may be constructed on the basis of eqn (3.335), after computing the necessary derivatives. If this procedure is carried out with $h_1 = 1$, $h_2 = \frac{1}{16}$, and $h_3 = -\frac{1}{2}$, one obtains

$$\hat{\varepsilon} = (w^1)^2 + w^1\mu, \quad \text{and} \quad w^3 = 2.12(w^1)^2, \tag{5.38}$$

where $\rho = \rho_c + \mu$ and $\hat{\varepsilon} = \varepsilon/5$.

Clearly, the critical point is an *anticlastic* general point of order 2 which is associated with the *fold* catastrophe. The imperfection sensitivity is described by

$$\hat{\varepsilon} = -\tfrac{1}{4}\overset{*}{\mu}{}^2 \tag{5.39}$$

which is a parabolic relationship (Fig. 5.6*a*).

Now let $h_1 = 1$, $h_2 = \frac{1}{4}$, $h_3 = -0.7$, $k_1 = \frac{1}{8}$, $k_2 = \frac{1}{16}$, and $k_3 = 2.73$; effectively modifying the structural system under consideration. It can

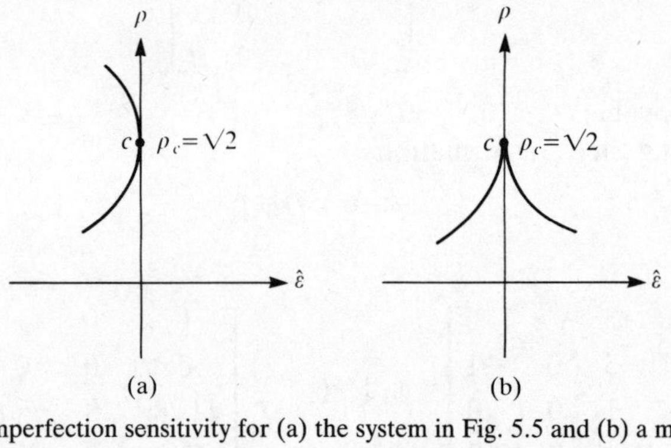

Fig. 5.6 The imperfection sensitivity for (a) the system in Fig. 5.5 and (b) a modified system (see text).

then be verified that $W_{211}=0$ and c is a *singular general point.* The equilibrium surface may be constructed with the aid of eqn (3.336), and takes the form

$$\hat{\varepsilon} = (w^1)^3 + w^1\mu,$$
$$w^3 = -0.71(w^1)^2, \tag{5.40}$$

which is associated with the *cusp* catastrophe. The imperfection sensitivity is described by

$$\hat{\varepsilon}^2 = -\tfrac{4}{27}\overset{*}{\mu}{}^3, \tag{5.41}$$

which is shown in Fig. 5.6*b*.

As emphasized in the general theory, Section 3.2.2, Case 2, unlike the generic case in which a real eigenvalue crosses the origin, both the fold and cusp catastrophes associated with a Jordan block of order 2 may not describe the behaviour of the system fully, and a family of limit cycles may exist in the vicinity of the critical point. This phenomenon can be examined by following the analysis presented in Section 5.2.2, but it will not be pursued here. Also note that this system may serve as a simple mechanical model for a qualitative study of flow-induced oscillations of pipes.

5.2.2 *Static instability of a non-linear network*

In this section an electrical network, consisting of a d.c. source E, a resistor R, an inductor L, a capacitor C, and a tunnel-diode (Fig. 5.7), is considered. The current through the inductor and the voltage across the capacitor are chosen as the state variables z^1 and z^2, respectively. The state equations of the network are then derived as

$$z^1_t = \frac{1}{L}(E - Rz^1 - z^2)$$
$$z^2_t = \frac{1}{C}(z^1 - f(z^2)), \tag{5.42}$$

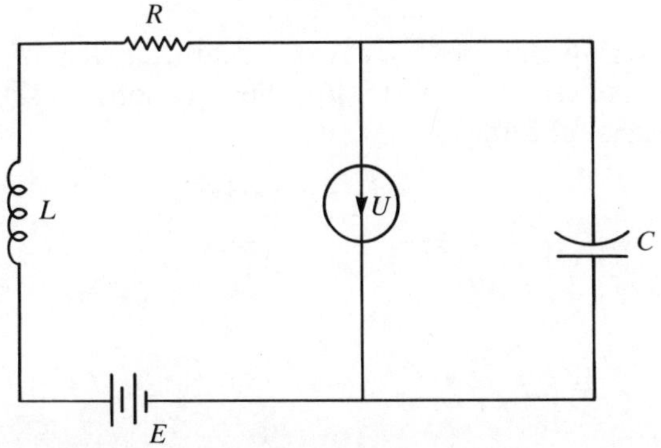

Fig. 5.7 A non-linear electrical network.

where $f(z^2)$ is a polynomial approximation for the tunnel-diode characteristic given (Chua 1969) by

$$f(z^2) = 0.01776z^2 - 0.10379(z^2)^2 + 0.22962(z^2)^3 - 0.22631(z^2)^4 + 0.08372(z^2)^5. \tag{5.43}$$

The equilibrium states of the network are governed by

$$Rz^1 + z^2 = E \quad \text{and} \quad z^1 = f(z^2). \tag{5.44}$$

The Jacobian matrix evaluated at an arbitrary equilibrium state is of the form

$$[Z_{ij}] = \begin{bmatrix} -\frac{R}{L} & -\frac{1}{L} \\ \frac{1}{C} & -\frac{a}{C} \end{bmatrix}, \tag{5.45}$$

where 'a' is the value of the derivative $\mathrm{d}f/\mathrm{d}z^2$ at the equilibrium state of interest.

Let $L = 1.115\ \mu\mathrm{H}$, $C = 1\ \mathrm{pF}$, and $E = E_c = 1.2\ \mathrm{V}$ (held constant).

The equilibrium equations (5.44) and $\det|Z_{ij}| = 0$ can then be solved simultaneously to obtain the primary critical point

$$z_c^1 = 0.9920\ \mathrm{mA}, \quad z_c^2 = 0.15414\ \mathrm{V}, \quad R_c = 1.0543\ \mathrm{k}\Omega, \tag{5.46}$$

on the non-linear equilibrium path. Here, eqn (5.46) has been obtained approximately by solving three equations directly; however, in more complicated problems, one may have to use a perturbation procedure (Huseyin 1972*c*, 1975) to generate successive approximations for the critical point.

Before exploring the local behaviour characteristics of the system in the vicinity of the critical point (5.46), the equations (5.42) are written in the non-dimensional form

$$\bar{z}_\tau^1 = \bar{E} - \eta\bar{z}^1 - \tfrac{5}{6}\bar{z}^2, \quad \bar{z}_\tau^2 = \tfrac{6}{5}\bar{z}^1 - kf(\bar{z}^2) \tag{5.47}$$

by defining

$$z^1 = E_c R_c \frac{C}{L}\bar{z}^1, \qquad z^2 = \tfrac{5}{6}E_c\bar{z}^2, \qquad R = R_c\eta,$$

$$E = E_c\bar{E}, \qquad t = CR_c\tau, \quad \text{and} \quad k = R_c.$$

Introducing now the transformation

$$\begin{bmatrix} \bar{z}^1 \\ \bar{z}^2 \end{bmatrix} = \begin{bmatrix} \bar{z}_c^1 \\ \bar{z}_c^2 \end{bmatrix} + \begin{bmatrix} -1 & 1 \\ \frac{6}{5} & 0 \end{bmatrix} \begin{bmatrix} y^1 \\ y^2 \end{bmatrix} \tag{5.48}$$

into eqns (5.47) yields a system of the form

$$\boldsymbol{y}_\tau = \boldsymbol{Y}(\boldsymbol{y}, \mu), \qquad (\eta = \eta_c + \mu), \tag{5.49}$$

with the Jacobian

$$[Y_{ij}]_c = \begin{bmatrix} 0 & 1 \\ 0 & 0 \end{bmatrix}. \tag{5.50}$$

It follows that the system under consideration falls within the scope of the theory presented in Section 3.1.1, Case 2, provided $Y_2' \neq 0$. Indeed, this derivative may be determined as

$$Y_2' = \partial Y_2 / \partial \eta|_c = 0.872 \neq 0.$$

Using (3.51) and (3.52) yields the first-order equations of the equilibrium path in the vicinity of c as

$$\mu = 38.88(y^1)^2$$

and

$$y^2 = -33.91(y^1)^2, \tag{5.51}$$

which represents a limit point or fold catastrophe (Mandadi and Huseyin 1979). In other words, a steady-state equilibrium solution of the network reaches an extremum at $R_c = 1.0543\,\text{k}\Omega$ where the system loses its stability and a jump to another state occurs.

5.2.3 *Hopf bifurcation associated with a non-linear network*

Consider the electrical circuit discussed by Hirsch and Smale (1974, *p.* 211) which is shown in Fig. 5.8*a*. It is assumed that the capacitor C and

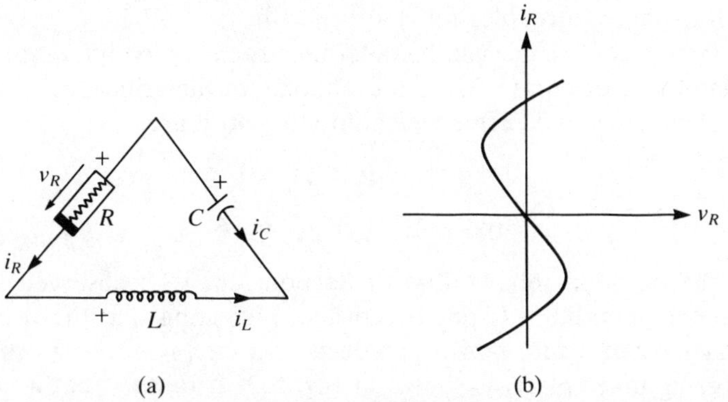

Fig. 5.8 (a) The electrical network discussed by Hirsch and Smale (1974) and (b) the non-linear current-voltage relationship for R for a given $\eta \neq 0$.

the inductor L are linear elements (of unit value) whereas the resistor R is characterized by the non-linear relationship

$$v_R(i_R, \eta) = (i_R)^3 - \eta i_R, \tag{5.52}$$

where v_R, i_R, and η represent the voltage, the current, and a variable parameter (i.e. temperature) of the resistor, respectively. A typical relationship associated with eqn (5.52) is illustrated in Fig. 5.8*b* for a given $\eta \neq 0$.

The state equations of this system are in the form

$$\begin{aligned} z_t^1 &= z^2 + \eta z^1 - (z^1)^3 \\ z_t^2 &= -z^1 \end{aligned} \tag{5.53}$$

where $i_L = z^1$ and $v_C = z^2$. This is recognized as a special case of Lienard's equation. Since $z^1 = z^2 = 0$ yields $z_t^1 = z_t^2 = 0$ for all η, and $\eta = 0$ results in a canonical form of the Jacobian on this equilibrium path, system (5.53) is already in the form of system (4.64) defined in Section 4.4. Introducing the time scaling $\tau = \omega t$, system (5.53) may then be expressed as

$$\omega \begin{bmatrix} w_\tau^1 \\ w_\tau^2 \end{bmatrix} = \begin{bmatrix} \eta & 1 \\ -1 & 0 \end{bmatrix} \begin{bmatrix} w^1 \\ w^2 \end{bmatrix} - \begin{bmatrix} (w^1)^3 \\ 0 \end{bmatrix}, \tag{5.54}$$

where ω is the frequency.

Clearly, $\eta = 0$ represents a critical point on the fundamental path $w^i = 0$, and the eigenvalues of the Jacobian along this path are given by

$$\lambda_{1,2} = \{\eta \pm (\eta^2 - 4)^{1/2}\}/2 \tag{5.55}$$

where it is assumed that $|\eta| < 2$. It follows that $\lambda_{1,2}(\eta)$ has a negative (positive) real part for $\eta < 0$ ($\eta > 0$), and the equilibrium path $w^i = 0$ is, therefore, stable (unstable) for $\eta < 0$ ($\eta > 0$).

The system coefficients can be determined readily by differentiating the right-hand side of eqn (5.54) and evaluating at the critical point c where $\eta_c = 0$. Thus, the coefficients which do not vanish are

$$\begin{aligned} W_{12} = -W_{21} = \omega_c = 1, \\ W'_{11} = 1, \qquad W_{1111} = -6, \end{aligned} \tag{5.56}$$

where the notation follows that of Section 4.4. It is observed that the *transversality* condition (4.68) is satisfied, indicating that the roots cross the imaginary axis with non-zero velocity. In fact, it is easy to show that the crossing takes place as shown in Fig. 5.9. It follows that a family of periodic solutions of eqn (5.54) bifurcates off from the fundamental equilibrium path at the critical point c. Note that for a two-dimensional

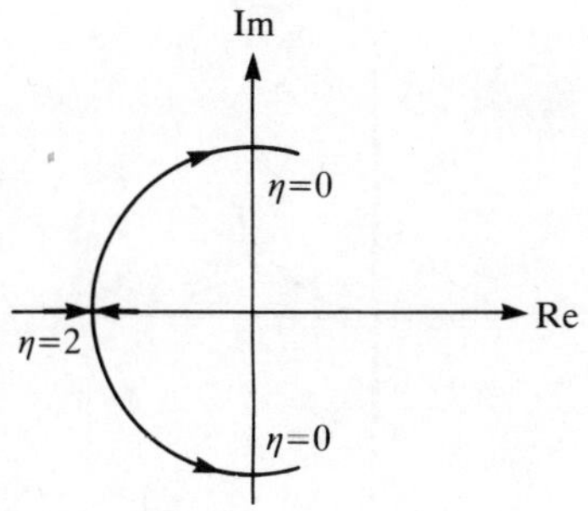

Fig. 5.9 Path of the eigenvalues for the system shown in Fig. 5.8*a*.

system, eqn (4.97) takes the form

$$\gamma_{22}-\gamma_{11}=-\frac{3}{4}\left\{W_{1111}+W_{1122}+W_{2112}+W_{2222}+\frac{1}{\omega_c}(W_{111}W_{211}+W_{211}W_{212}\right.$$
$$\left.+W_{212}W_{222}-W_{111}W_{112}-W_{112}W_{122}-W_{122}W_{222})\right\}.$$

The formula (4.95) then yields

$$\ddot{\eta}=\tfrac{3}{2},$$

which results in the bifurcation path

$$\mu=\tfrac{3}{4}(p_{11})^2, \qquad (\eta=0+\mu), \tag{5.57}$$

where p_{11} is the amplitude of the first harmonic.

The periodic solutions are obtained from eqn (4.103) as

$$\begin{bmatrix} w^1 \\ w^2 \end{bmatrix}=p_{11}\begin{bmatrix} \cos\tau-\frac{3}{32}(p_{11})^2\sin 3\tau \\ -\sin\tau-\frac{1}{32}(p_{11})^2\cos 3\tau \end{bmatrix}. \tag{5.58}$$

On the other hand, the second derivative of the frequency follows from eqn (4.96) as

$$\ddot{\omega}=0,$$

and the frequency expression is, therefore, simply

$$\omega=\omega_c=1. \tag{5.59}$$

It is observed that the bifurcation path (5.37) is *supercritical,* and it represents a stable family of limit cycles (Fig. 5.10). In physical terms, the circuit's steady states for $\eta<0$ may be described as 'dead' so that all the currents and voltages are zero. As η passes through zero, this dead equilibrium path becomes unstable, and *self-excited oscillations* develop. In this vicinity, there exists a limit cycle corresponding to a given η such that its amplitude and frequency are determined by eqns (5.57) and (5.59), respectively.

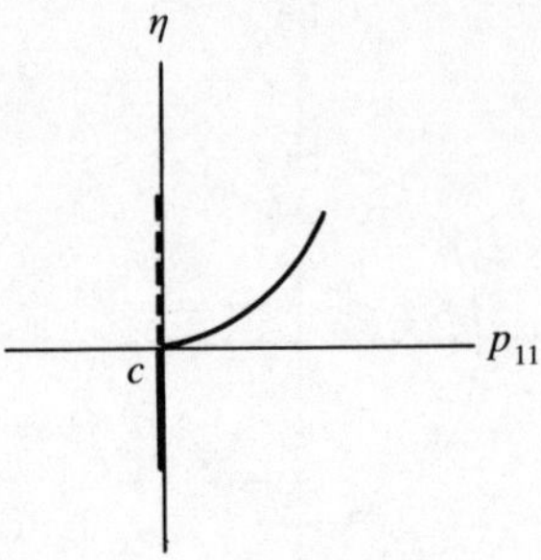

Fig. 5.10 The bifurcation path for the system (Fig. 5.8*a*) is supercritical and represents a stable family of limit cycles.

Problem 5.3. Consider the electrical circuit treated by Desoer and Kuh (1969, *p.* 212) which is shown in Fig. 5.11. Let the inductor L and the capacitance C be linear elements (of unit values), and let the non-linear characteristic of the conductance be given by the relationship

$$i_G(v_G, \eta) = v_G^3 - \eta v_G,$$

where η is a parameter. Treating the capacitor voltage v_C and inductor current i_L as state variables, and observing the duality between this circuit and the one described by eqn (5.53), show that this system exhibits a family of bifurcating limit cycles identical to that of system (5.53) (see Atadan 1983).

5.2.4 A mechanical system and its electrical analogue

Consider the mechanical system and its electrical analogue shown in Fig. 5.12 (Atadan and Huseyin 1984*b*). Let the spring of the mechanical system be linear with a spring constant $k = 1$, and assume a mass of $m = 1$. Similarly, let the capacitor and the inductor in the electrical circuit be linear elements of capacitance $C = 1$ and inductance $L = 1$, respectively. On the other hand, assume that the dampers and resistors have non-linear characteristics which may be specified in an appropriate way. Thus, consider first a one-parameter system in which the force P_1 exerted by the first damper and the velocity v_2 associated with the second

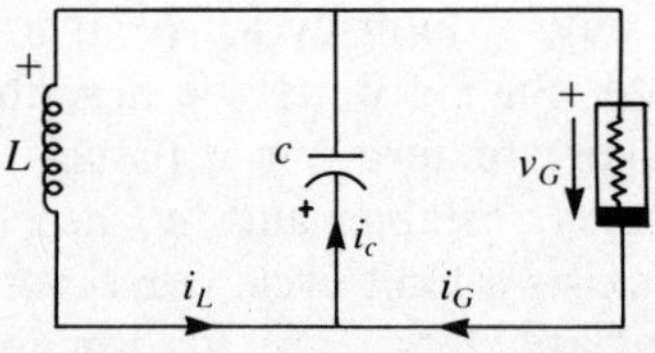

Fig. 5.11 System considered in Problem 5.3.

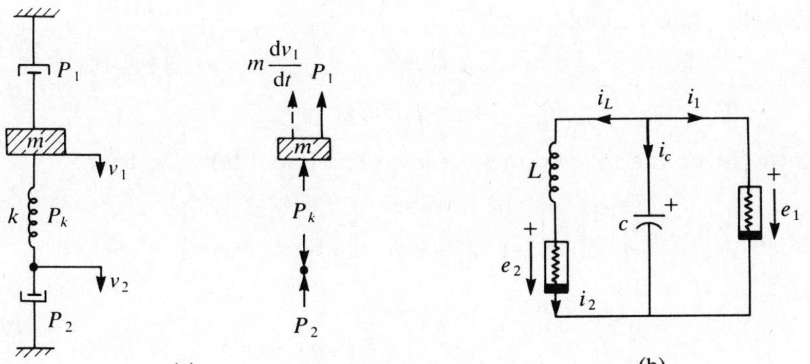

(a) (b)

Fig. 5.12 (a) A mechanical system and (b) its electrical analogue.

damper are specified by the non-linear functions

$$\begin{aligned} P_1 &= f_1(v_1, P_k, \eta), \\ v_2 &= f_2(P_2, v_1, \eta), \end{aligned} \tag{5.60}$$

where η is a parameter, P_2 is the force exerted by the second damper and P_k is the force in the spring. For the electrical analogue, one specifies

$$\begin{aligned} i_1 &= f_1(e_1, i_L, \eta) \\ e_2 &= f_2(i_2, e_c, \eta), \end{aligned} \tag{5.61}$$

where the notation e is used, instead of v, to denote voltages in order to avoid confusion with the velocities. From here on, discussion will be based on the mechanical system only, for convenience.

The equations of motion follow from Fig. 5.12*a* as

$$\begin{aligned} \frac{\mathrm{d}v_1}{\mathrm{d}t} + P_1 + P_k &= 0, \\ P_2 - P_k &= 0, \end{aligned} \tag{5.62}$$

since $m = 1$. Differentiating the second equation with respect to time, and recalling that the spring constant $k = 1$, yields

$$\frac{\mathrm{d}P_2}{\mathrm{d}t} = v_1 - v_2. \tag{5.63}$$

The state equations of the system may then be expressed as

$$\begin{aligned} z_t^1 &= z^2 + f_1(z^1, z^2, \eta), \\ z_t^2 &= -z^1 - f_2(z^1, z^2, \eta), \end{aligned} \tag{5.64}$$

where the state variables are

$$\begin{aligned} z^1 &= -v_1, \\ z^2 &= P_k = P_2. \end{aligned} \tag{5.65}$$

Introducing the time scaling $\tau = \omega t$, eqns (5.64) take the form

$$\omega \begin{bmatrix} z^1_\tau \\ z^2_\tau \end{bmatrix} = \begin{bmatrix} 0 & 1 \\ -1 & 0 \end{bmatrix} \begin{bmatrix} z^1 \\ z^2 \end{bmatrix} + \begin{bmatrix} f_1(z^1, z^2, \eta) \\ -f_2(z^1, z^2, \eta) \end{bmatrix}. \tag{5.66}$$

Now let

$$\begin{aligned} f_1 &= az^1\eta - 3(z^1)^2 + bz^1(\eta)^2 + z^2\eta, \\ f_2 &= cz^2\eta - d(z^2)^3/6, \end{aligned} \tag{5.67}$$

where a, b, c, and d are certain non-zero constants.

It is first observed that $z^i = 0$ defines an equilibrium path, and the Jacobian evaluated on this path is in the form

$$[Z_{ij}]_{z=0} = \begin{bmatrix} a\eta + b(\eta)^2 & 1+\eta \\ -1 & -c\eta \end{bmatrix}. \tag{5.68}$$

The eigenvalues $\lambda_{1,2}(\eta) = \alpha(\eta) \pm i\omega(\eta)$ are defined by

$$\begin{aligned} \alpha(\eta) &= \{a\eta - c\eta + b(\eta)^2\}/2 \\ \omega(\eta) &= \{4 + 4\eta - (a+c)^2(\eta)^2 - 2(a+c)b(\eta)^3 - b^2(\eta)^4\}^{1/2}/2. \end{aligned} \tag{5.69}$$

Clearly, $\eta = 0$ is a critical point and the Jacobian (5.68) at this point is in a canonical form. It then follows that system (5.66) is, in fact, in the form of system (4.156) defined in Section 4.5. It is understood that the W-system and the X-system are now equivalent since one has a (2×2) system here.

Shifting from eqn (5.66) to $\omega \boldsymbol{x}_\tau = \boldsymbol{X}(\boldsymbol{x}, \eta)$ in notation, the non-zero system coefficients are evaluated as

$$\begin{aligned} &X_{12} = -X_{21} = \omega_c = 1, \qquad X'_{11} = a, \qquad X'_{12} = 1, \\ &X'_{22} = -c, \qquad X_{111} = -6, \qquad X''_{11} = 2b, \quad \text{and} \quad X_{2222} = d, \end{aligned} \tag{5.70}$$

which are used to obtain

$$\begin{aligned} \bar{\gamma}_{11} - \bar{\gamma}_{22} &= 3d/4, \\ \alpha' &= (a-c)/2, \quad \text{and} \\ \alpha'' &= b. \end{aligned} \tag{5.71}$$

Depending on the constants a, b, c, and d several different phenomena may be exhibited as η passes through $\eta_c = 0$. It is readily verified that if $a \neq c$ and $d \neq 0$, the system exhibits *Hopf bifurcation*—since $\alpha' \neq 0$. On the other hand, if

$$a = c, \qquad b \neq 0, \qquad d \neq 0, \tag{5.72}$$

then

$$\alpha' = 0 \quad \text{and} \quad \alpha'' \neq 0,$$

and the conditions (4.167) are satisfied, indicating that a *symmetric bifurcation* (*tri-furcation*) phenomenon may be exhibited. Indeed, the *existence condition* (4.181) is satisfied if

$$bd < 0, \tag{5.73}$$

and under this condition a family of limit cycles bifurcates from the equilibrium path at $\eta_c = 0$. Assuming that $a = c = 1$ and $d = -32b$, the bifurcation paths (4.180) take the form

$$\mu = \pm 2p_{11}, \qquad (\eta = 0 + \mu), \tag{5.74}$$

and the frequencies are given by

$$\begin{aligned}\Omega &= \pm\omega'\{(\bar{\gamma}_{22} - \bar{\gamma}_{11})/6\alpha''\}^{1/2}p_{11}\\ &= \pm p_{11},\end{aligned} \tag{5.75}$$

where ω' can be determined either directly from eqns (5.69) or by means of eqns (4.161). The equations of the corresponding limit cycles are constructed on the basis of eqns (4.147) and (4.148) which can be easily adapted to the X-system to yield

$$\begin{bmatrix} x^1 \\ x^2 \end{bmatrix} = \begin{bmatrix} \cos\tau \\ -\sin\tau \end{bmatrix} p_{11} + \frac{1}{2}\begin{bmatrix} -2\sin 2\tau \\ 3 \mp 4\cos\tau \pm 2\sin\tau - \cos 2\tau \end{bmatrix}(p_{11})^2 + \dots \tag{5.76}$$

It is also observed that the sign of 'd' now plays an important role in determining the stability of solutions. Indeed, the equilibrium path through $\eta_c = 0$ is *stable* or *unstable* according to whether $d > 0$ or $d < 0$, respectively. This can be seen from the behaviour of $\alpha(\eta)$ which is given by eqn (5.69). Clearly, $\alpha(\eta) < 0$ in the vicinity of $\eta_c = 0$ if $d = -32b > 0$, and $\alpha(\eta) > 0$ if $d < 0$ (Fig. 5.13*a*). On the other hand, according to Section 4.6.2, the bifurcating family of limit cycles is unstable (*asymptotically orbitally stable*) if $d > 0$ ($d < 0$), because in this case $\bar{\gamma}_{11} - \bar{\gamma}_{22} = 3d/4$ is positive (negative). This phenomenon is illustrated in Fig. 5.13*b*.

Finally, it is recalled that if the conditions

$$a \neq c \quad \text{and} \quad d \neq 0 \tag{5.77}$$

hold, the system exhibits the phenomenon of Hopf bifurcation, and an initially stable equilibrium path loses its stability upon passing through the critical point while self-excited oscillations emerge. The symmetric bifurcation, on the other hand, occurs when $a = c$, and if in addition $bd < 0$, the equilibrium path maintains its stability through the critical

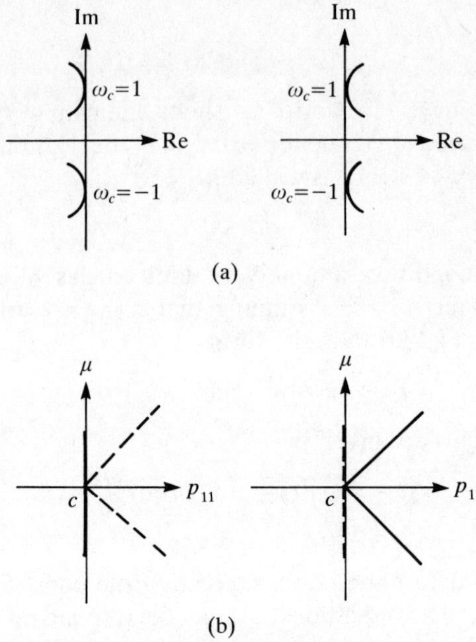

Fig. 5.13 (a) Behaviour of $\alpha(\eta)$ and (b) the bifurcating family of limit cycles.

point. This may give the impression that the unstable limit cycles bifurcating from the critical point are harmless as far as the stability of the system is concerned. However, the critical bifurcation point is obviously imperfection sensitive and a small perturbation in the structural properties of the system may render $a \neq c$, resulting in an instability via Hopf bifurcation. For $d<0$, the equilibrium path is entirely unstable and, in the vicinity of the critical point, the operational regime is dominated by stable periodic motions.

Problem 5.4. Examine the Hopf bifurcation associated with system (5.66) under the conditions (5.77). Derive the equations of the bifurcating path, the frequency-amplitude relation, and the corresponding family of limit cycles. Discuss the stability properties of the system in the vicinity of the critical point.

A two-parameter system. Consider once more the system described by eqn (5.66); but replace the non-linear characteristics (5.67) by

$$f_1 = z^2\eta^1 - az^1(\eta^1)^2 + (a/3)z^1(\eta^1)^3 - 3(z^1)^2 \quad \text{and}$$
$$f_2 = 3z^1\eta^2 - z^2\eta^2 + (\tfrac{1}{6})(z^2)^3, \tag{5.78}$$

where $a \neq 0$, and η^1 and η^2 are two independent parameters (Atadan and Huseyin 1985*a*).

The initial equilibrium surface associated with system (5.66) is then given by $z^i(\eta^\alpha) \equiv 0$, since $z^i = 0$ for all η^α $(\alpha = 1, 2)$. The Jacobian evaluated on this surface is in the form

$$[Z_{ij}]_{z=0} = \begin{bmatrix} -a(\eta^1)^2 + \dfrac{a}{3}(\eta^1)^3 & 1 + \eta^1 \\ -1 + 3\eta^2 & -\eta^2 \end{bmatrix}. \tag{5.79}$$

The real part of the eigenvalues can be determined as

$$\alpha(\eta^1, \eta^2) = -\tfrac{1}{6}\{3\eta^2 + 3a(\eta^1)^2 - a(\eta^1)^3\},$$

and

$$3\eta^2 + 3a(\eta^1)^2 - a(\eta^1)^3 = 0$$

may be regarded as a *stability boundary*.

In particular, at the critical point $\eta_c^1 = \eta_c^2 = 0$ on this boundary, the real part of the eigenvalues of the Jacobian vanishes in such a way that eqn (5.79) takes the canonical form (4.256). The behaviour of the system will now be analysed in the vicinity of this critical point where $\alpha_c = 0$ and $\omega_c = 1$. Clearly, one may shift to the W-system of Section 4.7 directly by introducing $z^i = w^i$ into (5.66), which may be expressed as

$$\omega \begin{bmatrix} w_\tau^1 \\ w_\tau^2 \end{bmatrix} = \begin{bmatrix} 0 & 1 \\ -1 & 0 \end{bmatrix} \begin{bmatrix} w^1 \\ w^2 \end{bmatrix} + \begin{bmatrix} w^2\eta^1 - aw^1(\eta^1)^2 + \dfrac{a}{3}w^1(\eta^1)^3 - 3(w^1)^2 \\ 3w^1\eta^2 - w^2\eta^2 + \frac{1}{6}(w^2)^3 \end{bmatrix}. \tag{5.80}$$

The system coefficients which do not vanish are then determined as

$$\begin{aligned} &W_{12} = -W_{21} = \omega_c = 1, \\ &W_{12}^1 = 1, \qquad W_{21}^2 = 3, \qquad W_{22}^2 = -1, \\ &W_{11}^{11} = -2a, \qquad W_{11}^{111} = 2a, \qquad W_{111} = -6, \qquad W_{2222} = 1, \end{aligned} \tag{5.81}$$

which are substituted into eqns (4.260) to (4.262) to obtain

$$\begin{aligned} &\alpha^1 = \alpha^{12} = \alpha^{22} = 0, \\ &\alpha^2 = -\tfrac{1}{2}, \qquad \alpha^{11} = -a, \\ &\omega^1 = \tfrac{1}{2}, \qquad \omega^2 = -\tfrac{3}{2}. \end{aligned} \tag{5.82}$$

These derivatives indicate clearly that the conditions (4.285) are satisfied, and the system under consideration falls, therefore, within the scope of Section 4.7.2. Following the theory presented in Section 4.7.2, one evaluates the derivatives

$$\begin{aligned} &\dot{p}_{21}^1 = \ddot{p}_{10} = \ddot{p}_{12} = \ddot{r}_{22} = 0, \\ &\dot{r}_{21}^1 = \tfrac{1}{2}, \qquad \ddot{p}_{20} = 3, \qquad \ddot{p}_{22} = -1, \qquad \ddot{r}_{12} = -2 \\ &\gamma_{22} - \gamma_{11} = -\tfrac{3}{4} \quad \text{and} \quad \gamma_{12} + \gamma_{21} = 18. \end{aligned} \tag{5.83}$$

Using these derivatives leads to

$$\ddot{\eta}^2 = \tfrac{1}{4}, \quad \eta^{2,11} = -2a,$$
$$\omega^{,1} = \tfrac{1}{2}, \qquad \ddot{\omega} = -\tfrac{27}{8}. \tag{5.84}$$

The *behaviour surface* (4.289) follows readily as

$$\mu^2 = \tfrac{1}{8}(p_{11})^2 - a(\mu^1)^2, \tag{5.85}$$

which is an *anticlastic saddle surface* if $a > 0$ (Fig. 4.16), and *elliptic paraboloid* if $a < 0$ (Fig. 4.15). The discussion presented in Section 4.7.2 with regard to these figures will not be repeated here.

The frequency-amplitude relationship (4.290) yields

$$\Omega = \tfrac{1}{2}\mu^1 - \tfrac{27}{16}(p_{11})^2. \tag{5.86}$$

Similarly, the family of bifurcating limit cycles (4.284) takes the form

$$\begin{bmatrix} w^1 \\ w^2 \end{bmatrix} = \begin{bmatrix} \cos\tau \\ -\sin\tau \end{bmatrix} p_{11} + \frac{1}{2}\begin{bmatrix} -2\sin 2\tau \\ 3 - \cos 2\tau \end{bmatrix}(p_{11})^2 + \frac{1}{2}\begin{bmatrix} 0 \\ \sin\tau \end{bmatrix} p_{11}\mu^1 + \dots \tag{5.87}$$

Consider now another critical point ($\eta^1_c = 3$, $\eta^2_c = 0$) on the stability boundary at which $\alpha_c = 0$ and ω can be shown to be $\omega_c = 2$. One may introduce

$$\eta^1 = 3 + \mu^1$$
$$\eta^2 = \mu^2 \tag{5.88}$$

at the outset, together with the transformation

$$\begin{bmatrix} z^1 \\ z^2 \end{bmatrix} = \begin{bmatrix} 2 & 0 \\ 0 & 1 \end{bmatrix}\begin{bmatrix} w^1 \\ w^2 \end{bmatrix} \tag{5.89}$$

to obtain

$$\omega\begin{bmatrix} w^1_\tau \\ w^2_\tau \end{bmatrix} = \begin{bmatrix} 0 & 2 \\ -2 & 0 \end{bmatrix}\begin{bmatrix} w^1 \\ w^2 \end{bmatrix} +$$
$$+ \begin{bmatrix} 3aw^1\mu^1 + 2aw^1(\mu^1)^2 + \dfrac{a}{3}w^1(\mu^1)^3 + \tfrac{1}{2}w^2\mu^1 - 6(w^1)^2 \\ 6w^1\mu^2 - w^2\mu^2 + \tfrac{1}{6}(x^2)^3 \end{bmatrix} \tag{5.90}$$

from which the non-zero system coefficients follow as

$$W_{12} = -W_{21} = \omega_c = 2,$$
$$W^1_{11} = 3a, \quad W^1_{12} = \tfrac{1}{2}, \quad W^2_{21} = 6, \quad W^2_{22} = -1, \tag{5.91}$$
$$W^{11}_{11} = 4a, \quad W^{111}_{11} = 2a, \quad W_{111} = -12, \quad W_{2222} = 1.$$

These derivatives lead to

$$\alpha^1 = \frac{3a}{2} \quad \text{and} \quad \alpha^2 = -\tfrac{1}{2}, \tag{5.92}$$

which indicate that the system under consideration is now concerned with the generalized Hopf bifurcation No. 1 treated in Section 4.7.1. One also obtains the following derivatives (Atadan and Huseyin 1985*a*):

$$\begin{aligned}
&\omega^1 = \tfrac{1}{4}, \qquad \omega^2 = -3,\\
&\dot{p}^{,1}_{21} = -\tfrac{3}{2}a, \qquad \dot{r}^{,1}_{21} = (1+36a)/8,\\
&\ddot{p}_{10} = 0, \qquad \ddot{p}_{20} = 3, \qquad \ddot{p}_{12} = 0, \qquad \ddot{p}_{22} = -1,\\
&\ddot{r}_{12} = -2, \qquad \ddot{r}_{22} = 0,\\
&\gamma_{22} - \gamma_{11} = -\tfrac{3}{4} \quad \text{and} \quad \gamma_{12} + \gamma_{21} = 36.
\end{aligned} \tag{5.93}$$

These derivatives, in turn, lead to

$$\begin{aligned}
&\mu^{2,1} = 3a, \qquad \ddot{\mu}^2 = \tfrac{1}{4},\\
&\omega^{,1} = (1-36a)/4, \qquad \ddot{\omega} = -\tfrac{27}{4},
\end{aligned} \tag{5.94}$$

and the *behaviour surface* (4.281) takes the form

$$\mu^2 = 3a\mu^1 + \tfrac{1}{8}(p_{11})^2, \tag{5.95}$$

as depicted in Fig. 4.14.

The frequency-amplitude relationship is given by

$$\Omega = \tfrac{1}{4}(1-36a)\mu^1 - \tfrac{27}{8}(p_{11})^2 \tag{5.96}$$

and the family of limit cycles (4.284) takes the form

$$\begin{aligned}
\begin{bmatrix} w^1 \\ w^2 \end{bmatrix} = \begin{bmatrix} \cos\tau \\ -\sin\tau \end{bmatrix} p_{11} + \frac{1}{2}\begin{bmatrix} -2\sin 2\tau \\ 3-\cos 2\tau \end{bmatrix}(p_{11})^2\\
-\frac{1}{8}\begin{bmatrix} 0 \\ 12a\cos\tau - (1+36a)\sin\tau \end{bmatrix} p_{11}\mu^1 + \ldots .
\end{aligned} \tag{5.97}$$

Problem 5.5. Consider a one-parameter system described by

$$\omega \begin{bmatrix} z^1_\tau \\ z^2_\tau \\ z^3_\tau \end{bmatrix} = \begin{bmatrix} (\mu)^3/3 & 1+\mu & 0 \\ -1 & 0 & 0 \\ 0 & 0 & -2 \end{bmatrix} \begin{bmatrix} z^1 \\ z^2 \\ z^3 \end{bmatrix} + \begin{bmatrix} -3(z^1)^2 \\ -6(z^2)^3 - (z^3)^2(\mu)^2 \\ z^1z^2z^3\mu \end{bmatrix}.$$

Show that the critical point $\mu = 0$ represents a cusp bifurcation treated in Section 4.5.2 and that the bifurcation path is given by

$$\mu = 3(p_{11})^{2/3}.$$

Determine the frequency-amplitude, frequency-parameter relationships, and the family of limit cycles (Huseyin and Atadan 1984*b*).

5.3 Thermodynamics: phase transitions

Consider the well-known Van der Waals equation

$$\left(P+\frac{a}{V^2}\right)(V-b)=cT \tag{5.98}$$

where P, V, and T denote the pressure, volume, and temperature, respectively, and a, b, and c are certain constants. This equation models the phenomena of phase transitions from liquid to gas to liquid. From the stability theory point of view, such transitions occur at critical points where one form loses its stability and the system seeks another stable form.

Treating eqn (5.98) as an equilibrium equation, relating the *parameters* P and T to the *behaviour variable* V, the critical points on this *equilibrium surface* may be located by the standard procedures of Chapters 2 and 3. Thus, expressing eqn (5.98) as

$$PV^3-(bP+cT)V^2+aV-ab=0, \tag{5.99}$$

and differentiating with respect to V twice, yields

$$3PV^2-2(bP+cT)V+a=0 \tag{5.100}$$

and

$$6PV-2(bP+cT)=0. \tag{5.101}$$

Solving eqns (5.99), (5.100), and (5.101) simultaneously results in a *singular general* point c of order 3 (Sections 2.4.2 and 3.2.3) in the *configuration* space spanned by V, P, and T as

$$V_c=3b, \qquad P_c=\frac{a}{27b^2}, \qquad T_c=\frac{8a}{27bc}, \tag{5.102}$$

since the first, second, and third-order system derivatives all vanish at this point.

One expects that eqn (5.102) is the *cusp* point, and shifts the coordinates to this point by means of the transformation (Huseyin 1977)

$$\begin{bmatrix}\dfrac{P}{P_c}\\[2mm] \dfrac{T}{T_c}\end{bmatrix}=\begin{bmatrix}1\\1\end{bmatrix}+\frac{1}{3}\begin{bmatrix}-1 & 1\\ \frac{1}{8} & \frac{1}{4}\end{bmatrix}\begin{bmatrix}\varphi^1\\ \varphi^2\end{bmatrix}, \qquad \frac{v_c}{V}=1+u, \tag{5.103}$$

which also converts the parameter vector to a canonical form. Thus, in the vicinity of c, the Van der Waals equation takes the simple form

$$\varphi^1+3(u)^3+u\varphi^2=0, \tag{5.104}$$

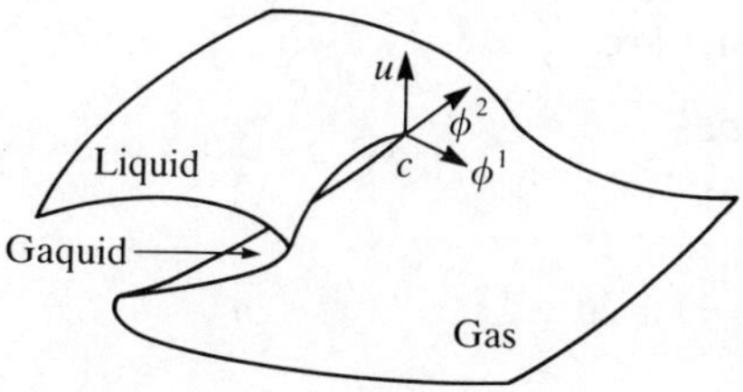

Fig. 5.14 Thermodynamic phase transitions.

where the density u takes on the role of a behaviour variable (Fowler 1972). This equation is associated with the *singular general point of order 3* or the *cusp catastrophe.* Phase transitions from liquid to gas and/or from gas to liquid occur on the critical fold line given by

$$9(u)^2 + \varphi^2 = 0, \tag{5.105}$$

and the stability boundary is obtained by eliminating u between eqns (5.104) and (5.105) as

$$(\overset{*}{\varphi}{}^1)^2 = -\tfrac{4}{81}(\overset{*}{\varphi}{}^2)^3. \tag{5.106}$$

Physically, the jumps occurring on the critical line (5.105) represent *boiling* or *condensation.* However, in changing from liquid to gas, for example, the same state can be reached by a smooth path rather than a jump, if the pressure and temperature parameters are controlled appropriately so that an equilibrium path around the critical point c is followed, thus avoiding the fold lines (Fig. 5.14). The interior sheet of the surface represents unstable states referred to as '*gaquid*'.

5.4 Bio-chemical processes (static instability)

In chemical and biological processes, the instability of the *thermodynamic branch,* which corresponds to an equilibrium path, has received considerable attention. It has been shown, for instance, that *symmetry breaking bifurcations* (static instabilities) and limit cycles branching off from the thermodynamic branch (Hopf bifurcation) play a significant role in the evolution processes (e.g. Nicolis and Prigogine 1977).

As an illustrative example, consider Edelstein's model described by Glansdorff and Prigogine (1971). The reaction scheme

$$\begin{aligned} \eta^1 + z^1 &\rightleftarrows 2z^1 \\ z^1 + z^2 &\rightleftarrows \vartheta \\ \vartheta &\rightleftarrows z^2 + \eta^2 \end{aligned} \tag{5.107}$$

is concerned with a biological process. The overall reaction is $\eta^1 \rightleftarrows \eta^2$, and it is assumed that $z^2 + \vartheta = a$, where a is a constant. The dynamical equations of the model are

$$\begin{aligned}\frac{dz^1}{dt} &= a - z^2 - (z^1)^2 - z^1z^2 + \eta^1 z^1\\ \frac{dz^2}{dt} &= 2a - 2z^2 - z^1z^2 - \eta^2 z^2\end{aligned} \tag{5.108}$$

where the z^i and η^α are concentrations, and the η^α are treated as parameters so that the behaviour of the system is effectively redefined and examined for each fixed set of η^α.

The equilibrium surface defined by $dz^i/dt = 0$ can be shown to be associated with a *singular general point* (*cusp*). In this section, however, the local behaviour of the system around a particular critical point—which is located on one of the folds of the surface—will be analysed quantitatively.

The Jacobian associated with eqns (5.108) is given by

$$[Z_{ij}(\eta^1, \eta^2)] = \begin{bmatrix}\eta^1 - 2z^1 - z^2 & -(1+z^1)\\ -z^2 & -(\eta^2 + z^1 + 2)\end{bmatrix}. \tag{5.109}$$

Let $a = 30.0018$; it can then be shown that $\eta^1_c = 8.5548$ and $\eta^2_c = 0.2$ represent a critical point on the surface $dz^i/dt = 0$ and the corresponding values of the state variables are $z^1 = 1.2465$ and $z^2 = 17.41$. The Jacobian evaluated at this point is

$$[Z_{ij}]_c = \begin{bmatrix}-11.3482 & -2.2465\\ -17.4100 & -3.4465\end{bmatrix}, \tag{5.110}$$

and its eigenvalues are $\lambda_1 = 0$, $\lambda_2 = -14.7947$.

Introducing the transformations (3.208),

$$\boldsymbol{z} = \boldsymbol{z}_c + \boldsymbol{Q}\boldsymbol{y} \quad \text{and} \quad \boldsymbol{\eta} = \boldsymbol{\eta}_c + \boldsymbol{P}\boldsymbol{\varphi}, \tag{5.111}$$

where

$$\boldsymbol{Q} = \begin{bmatrix}1 & 1\\ -5.0515 & 1.5342\end{bmatrix} \quad \text{and} \quad \boldsymbol{P} = \begin{bmatrix}-0.11 & 1\\ -1 & -0.11\end{bmatrix},$$

results in the system

$$\frac{d\boldsymbol{y}}{dt} = \boldsymbol{Y}(\boldsymbol{y}, \boldsymbol{\varphi}) \tag{5.112}$$

which is system (3.209) of Section 3.2.

The Jacobian matrix of the new system is in the canonical form

$$[\boldsymbol{Y}_{ij}]_c = \begin{bmatrix}0 & 0\\ 0 & -14.7947\end{bmatrix} \tag{5.113}$$

which is *similar* to matrix (5.110) through Q. The parameter vector is

$$[Y_{1\alpha}]_c = [-2.6719 \quad 0], \tag{5.114}$$

which indicates that the critical point under consideration is *general* and the problem falls within the scope of Section 3.2.2, Case 1. Note that the transformation matrix $\boldsymbol{P}$ could have been chosen such that the first element of the vector (5.114) would be (-1) as in the theory. It is, however, easy to adapt the theory to this case by simply keeping $Y_{1\alpha_1} = -2.6719$. Indeed, using this value and eqns (3.230) and (3.231), yields

$$\varphi^1 = 0.066(y^1)^2 + 0.37y^1\varphi^2 + 0.0056(\varphi^2)^2$$

and (5.115)

$$y^2 = 0.08\varphi^2 + 0.2706(y^1)^2,$$

which indicate that the critical point c is a *general* point of order 2, and the phenomenon in this vicinity may be described as a *fold* catastrophe.

The critical line on the surface (5.115) is given by (Huseyin 1982)

$$y^1 = -2.8030\varphi^2, \qquad y^2 = 0.08\varphi^2, \tag{5.116}$$

and the stability boundary by

$$\overset{*}{\varphi}{}^1 = -0.5130(\overset{*}{\varphi}{}^2)^2, \tag{5.117}$$

which shows that the surface (5.115) is *anticlastic*, and the stability boundary is convex towards the region of stability. As remarked earlier, a more complete picture in this problem may be obtained by first locating the *singular* general point and then establishing the equations of the equilibrium surface in the vicinity of this point.

5.5 The Brusselator

The aim of this section is to illustrate the application of the dynamic instability theory to a chemical process which is capable of exhibiting dynamic bifurcations.

Unimolecular and bimolecular chemical reactions do not lead to dynamic instability of the thermodynamic branch, because the eigenvalues of the Jacobian evaluated on this branch cannot have positive real parts. At least a cubic non-linearity is required for this phenomenon to take place. An interesting tri-molecular model was first introduced by Prigogine and Lefever (1968) which is known as the Brusselator.

Consider the chemical reaction scheme

$$\begin{aligned} A &\to z^1 \\ B + z^1 &\to z^2 + D \\ 2z^1 + z^2 &\to 3z^1 \\ z^1 &\to E \end{aligned} \tag{5.118}$$

where the tri-molecular reaction is represented by the third step which produces the required non-linearity. Here A, B, D, and E are the initial and final products whose concentrations are assumed to be controlled throughout the reaction process, and the z^i are the concentrations of two variable intermediates. After scaling (Nicolis and Prigogine 1977), the rate equations associated with the scheme (5.118) are given by the partial differential equations

$$\begin{aligned} \frac{\partial z^1}{\partial t} &= A - (B+1)z^1 + (z^1)^2 z^2 + D_1\,\Delta z^1 \\ \frac{\partial z^2}{\partial t} &= Bz^1 - (z^1)^2 z^2 + D_2\,\Delta z^2 \end{aligned} \tag{5.119}$$

where Δ is the Laplace's operator, and D_1 and D_2 are the diffusion coefficients associated with z^1 and z^2, respectively. This system may exhibit static as well as dynamic instabilities, and a non-linear perturbation analysis due to Auchmuty and Nicolis (1975) demonstrates the emergence of sinusoidally varying spatial patterns. Here it will be assumed, however, that the z^i are space independent, so that eqns (5.119) lead to a set of ordinary differential equations,

$$\begin{aligned} \frac{dz^1}{dt} &= A - (\eta+1)z^1 + (z^1)^2 z^2, \\ \frac{dz^2}{dt} &= \eta z^1 - (z^1)^2 z^2, \end{aligned} \tag{5.120}$$

where B is replaced by η. This is to indicate that B is to be treated as a parameter while A is assumed to be a non-zero constant.

The fundamental equilibrium path defined by eqns (5.120) is in the form

$$\begin{aligned} z^1 &= A \\ z^2 &= \frac{\eta}{A}. \end{aligned} \tag{5.121}$$

The Jacobian evaluated on this path is given by

$$[\mathbf{Z}_{ij}(\eta)] = \begin{bmatrix} -1+\eta & A^2 \\ -\eta & -A^2 \end{bmatrix}, \tag{5.122}$$

and the eigenvalues $\lambda_{1,2}(\eta) = \alpha(\eta) \pm i\omega(\eta)$, where

$$\alpha(\eta) = \tfrac{1}{2}(\eta - 1 - A^2)$$

and

$$\omega(\eta) = \tfrac{1}{2}\{-\eta^2 + 2(1 + A^2)\eta - (1 - A^2)^2\}^{1/2},$$

are complex conjugate in a Region G defined by

$$1 - 2A + A^2 < \eta < 1 + 2A + A^2.$$

It then follows that

$$\eta_c = 1 + A^2 \tag{5.123}$$

is a critical point at which

$$\alpha_c = 0 \quad \text{and} \quad \omega_c = A. \tag{5.124}$$

Introducing the transformations

$$\begin{bmatrix} z^1 \\ z^2 \end{bmatrix} = \begin{bmatrix} A \\ \eta/A \end{bmatrix} + \begin{bmatrix} 1 & 0 \\ -1 & 1/A \end{bmatrix} \begin{bmatrix} w^1 \\ w^2 \end{bmatrix} \tag{5.125}$$

and

$$\eta = (1 + A^2) + \mu, \qquad \tau = \omega t \tag{5.126}$$

into (5.120) yields the canonical form

$$\omega \begin{bmatrix} w^1_\tau \\ w^2_\tau \end{bmatrix} = \begin{bmatrix} \mu & A \\ -A & 0 \end{bmatrix} \begin{bmatrix} w^1 \\ w^2 \end{bmatrix} + \begin{bmatrix} f(w^i, \mu) \\ 0 \end{bmatrix}, \tag{5.127}$$

where

$$f(w^i, \mu) = \left(\frac{1}{A} - A\right)(w^1)^2 + 2w^1w^2 - (w^1)^3 + \frac{1}{A}(w^1)^2w^2 + \frac{1}{A}(w^1)^2\mu.$$

Obviously, system (5.127) falls within the scope of Section 4.4.1 since

$$\alpha' = \tfrac{1}{2} \neq 0$$

This transversality condition is confirmed by determining

$$W'_{11} = 1 \quad \text{and} \quad W'_{22} = 0, \tag{5.128}$$

as in the theory. By following the theory, one also obtains the derivatives

$$\begin{aligned} W_{111} &= 2(1 - A)/A, \qquad W_{112} = 2, \qquad W_{1111} = -6, \\ W'_{12} &= W'_{21} = 0, \qquad W_{1112} = 2/A, \end{aligned} \tag{5.129}$$

and

$$\gamma_{22} - \gamma_{11} = 3(A^2 + 2)/2A^2 > 0 \tag{5.130}$$

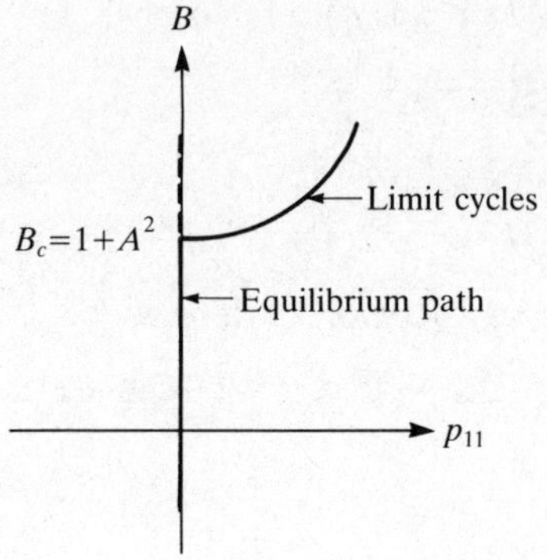

Fig. 5.15 System behaviour.

which yields (Atadan and Huseyin 1984*c*) the bifurcation path

$$\mu=\{(A^2+2)/4A^2\}(p_{11})^2, \tag{5.131}$$

upon substituting eqn (5.130) into formula (4.99).

It is immediately recognized that the bifurcating family of limit cycles is stable, and the system under consideration will develop *self-excited oscillations* upon passing through the critical point (5.123) where the equilibrium path becomes unstable (Fig. 5.15).

The limit cycles are obtained from eqn (4.103) as

$$\begin{bmatrix} w^1 \\ w^2 \end{bmatrix}=\begin{bmatrix} \cos\tau \\ -\sin\tau \end{bmatrix}p_{11}+\frac{1}{2}\left\{\frac{1}{A^2}\begin{bmatrix} 0 \\ (A^2-1) \end{bmatrix}+\frac{1}{3A^2}\begin{bmatrix} 4A \\ (1-A^2) \end{bmatrix}\cos 2\tau \right.$$
$$\left.+\frac{2}{3A^2}\begin{bmatrix} -A \\ (1-A^2) \end{bmatrix}\sin 2\tau\right\}(p_{11})^2+\dots. \tag{5.132}$$

The amplitude-frequency relationship (4.100) takes the form

$$\omega=A-\{4(1-A^2)^2+A^2\}(p_{11})^2/24A^3. \tag{5.133}$$

Similar results were obtained by Hassard *et al.* (1981) who derived an expression for the period rather than the frequency. If one expresses the period T as

$$T=\frac{2\pi}{\omega}=\frac{2\pi}{\omega_c}\{1+\tau_2(p_{11})^2+\dots\}, \tag{5.134}$$

then, a first-order approximation for the frequency is in the form of

$$\omega=\omega_c\{1-\tau_2(p_{11})^2\}. \tag{5.135}$$

Here, $\omega_c=A$ and comparing eqn (5.133) with (5.135) yields

$$\tau_2=\{4(1-A^2)^2+A^2\}/24A^4 \tag{5.136}$$

which may be substituted into eqn (5.134) to obtain a first-order approximation for the period of oscillations.

5.6 Aircraft at high angles of attack

Dynamic behaviour of an aircraft at high angles of attack can be highly non-linear and very complex. The angle of attack is the angle formed by a plane's steady flight path and the chord line through the wing. By increasing the angle of attack the plane may be given added lift; however, with increasing angles of attack stability problems become more complicated, involving non-linear motions. Linearized stability analysis becomes inadequate. Recently, Mehra and Carroll (1980) presented a method of non-linear analysis based on the bifurcation theory and catastrophe theory. By presenting the equilibrium surface and the stability boundary in a nine-dimensional space, these authors demonstrated how different types of instabilities and families of limit cycles arise as the parameters vary. On the other hand, Hui and Tobak (1982) focused their attention on the phenomenon of oscillatory instability and showed analytically that the steady flight of an aircraft at high angles of attack becomes unstable at a critical value of this angle and Hopf bifurcation sets in. Here, the approach developed in Chapter 4 will be illustrated on the problem formulated by Hui and Tobak.

Consider an aircraft in steady flight at an angle of attack η. Certain disturbances may result in an unsteady pitching motion of the aircraft which is coupled to an unsteady flow field. Assuming that aero-elastic effects are negligible, the rigid body pitching motion of the aircraft about a pivot axis is described by

$$\begin{aligned}\frac{\mathrm{d}z^1}{\mathrm{d}t} &= z^2,\\ I\frac{\mathrm{d}z^2}{\mathrm{d}t} &= M(t),\end{aligned} \tag{5.137}$$

where z^1 is the angular displacement of the motion from η, $M(t)$ is the pitching moment of aerodynamic forces about the pivot axis, and I is the moment of inertia about the same axis. The coupling between this motion and the unsteady flow field occurs through $M(t)$. Hui and Tobak (1982) decouple the equations of pitching motion from the unsteady flow equations in the vicinity of a critical value of η, where the 'damping-in-pitch' derivative vanishes. Thus, the eqns (5.137) take the form

$$\begin{aligned}\frac{\mathrm{d}z^1}{\mathrm{d}t} &= z^2,\\ \frac{\mathrm{d}z^2}{\mathrm{d}t} &= F(z^1, z^2, \eta),\end{aligned} \tag{5.138}$$

where
$$F(z^1, z^2, \eta) = k\{f(\eta + z^1) - f(\eta) + g(\eta + z^1)z^2\} \tag{5.139}$$

Here, k is a constant, $g(\eta) = -D(\eta)$, $\mathrm{d}f(\eta)/\mathrm{d}\eta = -S(\eta)$, and $D(\eta)$ and $S(\eta)$ are the damping-in-pitch derivative and stiffness derivative, respectively.

By virtue of eqn (5.139), the equations of motion (5.138) may be expressed as

$$\begin{bmatrix} z_t^1 \\ z_t^2 \end{bmatrix} = \begin{bmatrix} 0 & 1 \\ -kS(\eta) & -kD(\eta) \end{bmatrix} \begin{bmatrix} z^1 \\ z^2 \end{bmatrix} + \begin{bmatrix} 0 \\ B_{2jk}(\eta) z^j z^k + C_{2jkl}(\eta) z^j z^k z^l + \dots \end{bmatrix}, \tag{5.140}$$

where the Taylor's expansion of $F(z^1, z^2, \eta)$ yields

$$B_{2jk} = \frac{1}{2!} \frac{\partial^2 F}{\partial z^j \, \partial z^k}\bigg|_{z=0}, \qquad C_{2jkl} = \frac{1}{3!} \frac{\partial^3 F}{\partial z^j \, \partial z^k \, \partial z^l}\bigg|_{z=0}.$$

The eigenvalues of the Jacobian, $\lambda_{1,2}(\eta) = \alpha(\eta) \pm \mathrm{i}\omega(\eta)$, are defined by

$$\begin{aligned} \alpha(\eta) &= -\tfrac{1}{2} k D(\eta), \\ \omega(\eta) &= \tfrac{1}{2}\{4kS(\eta) - k^2 D^2(\eta)\}^{1/2}. \end{aligned} \tag{5.141}$$

A critical flutter state is exhibited if, for a certain $\eta = \eta_c$, $D(\eta_c) = 0$ while $S(\eta) > 0$. If, furthermore

$$\mathrm{d}D(\eta)/\mathrm{d}\eta|_{\eta=\eta_c} \neq 0,$$

then, the transversality condition is satisfied and Hopf bifurcation occurs.

In order to be able to apply the theory, introduce the transformations

$$z = Q(\eta)x \quad \text{and} \quad \tau = \hat{\omega} t, \tag{5.142}$$

where

$$Q(\eta) = \begin{bmatrix} 1 & 0 \\ \alpha(\eta) & \omega(\eta) \end{bmatrix} \quad \text{and} \quad Q^{-1}(\eta) = \frac{1}{\omega(\eta)} \begin{bmatrix} \omega(\eta) & 0 \\ -\alpha(\eta) & 1 \end{bmatrix},$$

to obtain

$$\hat{\omega} \begin{bmatrix} x_\tau^1 \\ x_\tau^2 \end{bmatrix} = \begin{bmatrix} \alpha(\eta) & \omega(\eta) \\ -\omega(\eta) & \alpha(\eta) \end{bmatrix} \begin{bmatrix} x^1 \\ x^2 \end{bmatrix} + \begin{bmatrix} 0 \\ \text{H.O.T.} \end{bmatrix}$$

(H.O.T. = higher order terms).

It is observed that the Q-transformation places system (5.142) within the scope of Section 4.5.

At the critical point $\alpha(\eta_c) = 0$, one has the following derivatives:

$$\begin{aligned} &X_{211} = 2B_{211}/\omega_c, \qquad X_{212} = 2B_{212}, \qquad X_{222} = 2\omega_c B_{222}, \\ &X_{2111} = 6C_{2111}/\omega_c, \qquad X_{2112} = 6C_{2112}, \\ &X_{2122} = 6\omega_c C_{2122}, \qquad X_{2222} = 6\omega_c^2 C_{2222}. \end{aligned}$$

The key coefficient $(\bar{\gamma}_{22} - \bar{\gamma}_{11})$ is given in the Appendix and takes the form

$$\bar{\gamma}_{22} - \bar{\gamma}_{11} = -\frac{3}{4}\Big\{X_{1111} + X_{1122} + X_{2112} + X_{2222} + \frac{1}{\omega_c}(X_{111}X_{211} + X_{211}X_{212} + X_{212}X_{222} - X_{111}X_{112} - X_{112}X_{122} - X_{122}X_{222})\Big\}, \tag{5.143}$$

which is then evaluated as

$$\bar{\gamma}_{22} - \bar{\gamma}_{11} = -\frac{9}{2}\Big\{C_{2112} + \omega_c^2 C_{2222} + \frac{2}{3\omega_c^2}(B_{211}B_{212} + \omega_c^2 B_{212}B_{222})\Big\}. \tag{5.144}$$

On the other hand,

$$\alpha' = \left.\frac{\mathrm{d}\alpha(\eta)}{\mathrm{d}\eta}\right|_c = -\tfrac{1}{2}kD'(\eta_c), \tag{5.145}$$

and the bifurcation path, therefore, is

$$\eta = \eta_c + \{(\bar{\gamma}_{22} - \bar{\gamma}_{11})/12\alpha'\}(p_{11})^2. \tag{5.146}$$

The family of limit cycles and the frequency-amplitude relationship can also be obtained from the formulae given in the theory. Depending on the basic properties of the system, eqn (5.144) may be positive or negative, thus resulting in a stable or unstable family of limit cycles. It has been demonstrated by Hui and Tobak (1982), who obtained a similar result through a different approach, that for a flat plate airfoil in supersonic/hypersonic flow, for example, the bifurcating family of limit cycles is unstable.

5.7 Urban systems

Quite often the dynamics of a city population involves rapid changes—such as the many-fold growth of an already large city in a relatively short period of time or sudden depopulation of a city's core after a long period of intense activity. Many authors have proposed abstract models to explain this behaviour. Amson (1974), for example, constructed a number of interesting catastrophic models which will be discussed here briefly. His work is based on various *urbanitic laws* concerning the behaviour of an abstract city. Basically, these laws relate density z to rental p and opulence t. The laws are interpreted by treating the density as a behaviour variable which changes under the influence of two parameters, the rental and opulence.

The modelling process is based on the primary assumption that the product of average space occupied and rental is proportional to opulence;

that is

$$\frac{1}{z}p = kt \tag{5.147}$$

where k is a constant. This relationship can be modified in a number of ways. Suppose, for example, that the rental is reduced by an amount proportional to the density so that kt is replaced by $(kt - az)$, resulting in

$$p = (kt - az)z$$

or

$$p + az^2 = ktz, \tag{5.148}$$

where a is an appropriate constant. Here, eqn (5.148) describes the equilibrium states of the system.

A second modification of eqn (5.147) is achieved by removing essential personal space (b) from $1/z$, together with the modification reflected in eqn (5.148). Thus, one has

$$p + az^2 = \frac{ktz}{1 - bz}. \tag{5.149}$$

It is readily recognized that the equilibrium surface (5.148) is associated with a *general* critical point of order 2, and the phenomenon is described by the *fold catastrophe* as described in Chapters 2 and 3. It is further noted that the surface (5.148) is *anticlastic* as depicted in Figs. 2.1*b* and 3.19*b*. The city becomes '*evanescent*' at the critical fold line

$$z = kt/2a \tag{5.150}$$

and the stability boundary is given by

$$\overset{*}{p} = \frac{k^2}{4a}\overset{*}{t}{}^2. \tag{5.151}$$

A more interesting behaviour is exhibited by model (5.149) which may be expressed as

$$(p + az^2)\left(\frac{1}{z} - b\right) = kt. \tag{5.152}$$

Obviously, eqn (5.152) resembles Van der Waals equation (5.98). In fact, if one lets $1/z = V$, eqn (5.152) becomes identical to eqn (5.98) in mathematical terms, and it is immediately concluded that this model is also associated with the *cusp catastrophe* (Huseyin 1977). Following the analysis presented in Section (5.3), the *singular* critical point can be determined as

$$z_c = \frac{1}{3b}, \qquad p_c = \frac{a}{27b^2}, \qquad t_c = \frac{8a}{27bk}. \tag{5.153}$$

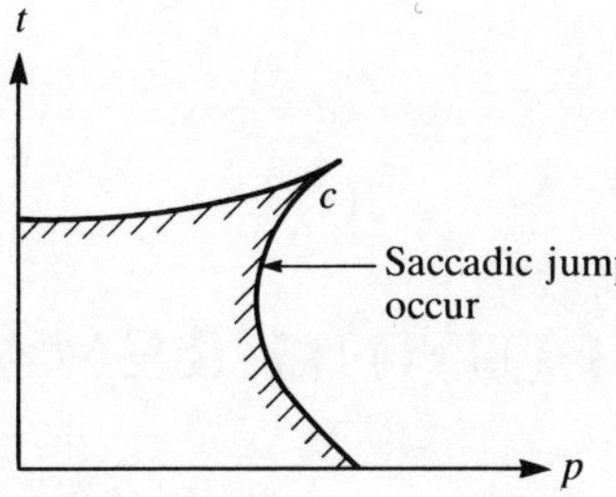

Fig. 5.16 The stability boundary for an urban system (eqn 5.149).

Introducing the transformations

$$\begin{bmatrix} p/p_c \\ t/t_c \end{bmatrix} = \begin{bmatrix} 1 \\ 1 \end{bmatrix} + \frac{1}{3}\begin{bmatrix} -1 & 1 \\ \frac{1}{8} & \frac{1}{4} \end{bmatrix}\begin{bmatrix} \varphi^1 \\ \varphi^2 \end{bmatrix}, \qquad \frac{z}{z_c} = 1 + u \tag{5.154}$$

leads to

$$\varphi^1 + 3(u)^3 + u\varphi^2 = 0 \tag{5.155}$$

which is, indeed, associated with the singular general point (cusp catastrophe).

The critical fold line and the stability boundary follow from eqn (5.155) as

$$9(u)^2 + \varphi^2 = 0 \tag{5.156}$$

and

$$(\overset{*}{\varphi}{}^1)^2 = -\tfrac{4}{81}(\varphi^2)^3, \tag{5.157}$$

respectively. The latter is depicted in Fig. 5.16. The density z undergoes a '*saccadic*' leap on this boundary (Amson 1974).

6

CONCLUDING REMARKS

Recent experimental and theoretical developments together with earlier observations provide sufficient evidence that the processes in nature and the behaviour of systems can be seemingly disorderly (complex) as well as orderly (simple).

This book is concerned with the latter. In fact, the entire treatment is confined to autonomous models of real systems. It is seen that as a parameter (or set of parameters) is varied, bifurcations from a time-independent stationary state to other time-independent states may occur. Similarly, bifurcations from a stationary state to time-periodic oscillatory states may take place. In other words, at the range of parameters considered, the behaviour of the system is *simple* or *orderly*. In fact, it was implicitly assumed for a long time that such steady-states (i.e. the equilibria and periodic motions) were the only alternatives for the asymptotic behaviour of solutions associated with a deterministic system. A set of three first-order equations studied by Lorenz (1963) changed this situation, and brought back a phenomenon so puzzling that Poincaré had to abandon it after he discovered it. It has now become clear, however, that even simple-looking autonomous systems (of three or higher dimensions) can exhibit a complicated *chaotic* motion which appears to be random or unpredictable. In other words, it is possible that, as the parameters are varied, the behaviour of a system changes from *simple* to *erratic* rather than from one simple state to another. Indeed, observed phenomena in physics, chemistry, biology, and other disciplines support the existence of seemingly chaotic time evolutions. A common example is concerned with the smoke from a cigarette, rising initially in a smooth stream which suddenly develops complicated oscillations. The onset of turbulence in fluid flows is also described as *chaos*. Ruelle and Takens (1971) discussed a new type of attractor, a '*strange attractor*', with regard to turbulence and chaos. Strange attractors are abstract mathematical objects, but computers generate pictures of them. In simple terms, a strange attractor consists of an infinity of points which correspond to the states of a chaotic system. It occupies a certain region of the state space, and all trajectories within the region remain in this region. In analogy with the familiar attractors, such as *asymptotically stable equilibrium points* and *asymptotically orbitally stable limit cycles,* the neighbouring trajectories are attracted to a *strange attractor.* A variety of strange attractors may exist, and one way of classifying these attractors may be by means of *Lyapunov exponents* according to Ruelle (see Haken 1983). A characteristic feature of the strange attractors is that within the attractor trajectories wander in a seemingly erratic manner and are extremely sensitive to initial conditions. Lorenz reports that when he wanted to examine one of his results in greater detail he fed an intermediate solution, which was computed to six decimal places, back to his computer after rounding it off to three decimal places. To his amazement, this small perturbation resulted in a totally new solution. The system of equations studied by Lorenz is concerned with a problem of atmospheric convection, and gives an approximate description of a horizontal fluid layer heated from below. The celebrated autonomous system of equations is

$$\frac{dz^1}{dt} = -10z^1 + 10z^2$$

$$\frac{dz^2}{dt} = -z^1z^3 + 28z^1 - z^2$$

$$\frac{dz^3}{dt} = z^1z^2 - \tfrac{8}{3}z^3.$$

If the intensity of heat increases to a critical level, convection occurs in a chaotic, turbulent manner, and the sensitive dependence on initial conditions explains inaccurate weather predictions! The dynamic behaviour of the solutions of the above system of equations has been the focus of considerable attention, and computer-generated pictures of the *Lorenz attractor* have appeared (e.g. Ruelle 1980). Nevertheless, according to Hirsch (1984), 'almost nothing has been proved about this particular system'.

There appears to be a number of routes to chaos, and the *tori* play a significant role in these *scenarios.* Basically, a dynamical system may pass through several instabilities, involving equilibrium states, limit cycles, quasiperiodic motions and chaotic motions. Two and higher-dimensional tori may describe quasiperiodic motions which take place at several basic frequencies. A two-dimensional torus, for example, may be visualized as a doughnut whose surface accommodates the trajectories of the system which involve two basic frequencies (Fig. 6.1). According to the Landau–Hopf scenario, many systems follow a route to chaos which initially involves a static bifurcation phenomenon. Then, the post-bifurcation equilibrium path exhibits a dynamic bifurcation to limit cycles (Hopf bifurcation) followed by a bifurcation of limit cycles into a two-dimensional torus at two basic frequencies. Then, a sequence of bifurcations from a torus to a higher-dimensional torus takes place, an

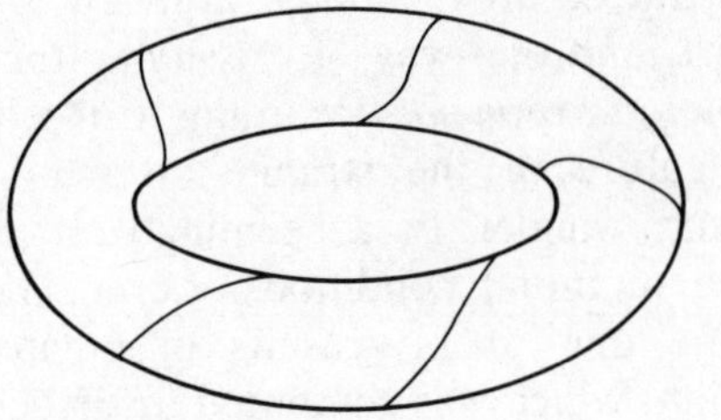

Fig. 6.1 A two-dimensional torus.

n-dimensional torus involving quasiperiodic motions with n basic frequencies. The Landau–Hopf scenario then suggests that chaos is described by motions on a torus of infinite dimensions (Landau and Lifschitz 1959; Hopf 1948).

As opposed to such an infinite sequence of bifurcations, Ruelle and Takens suggest that the bifurcation to a two-dimensional torus is followed by a bifurcation to a *strange attractor* rather than to a three-dimensional torus (Ruelle and Takens 1971; Ruelle 1980).

Another route to chaos is described as period-doubling (Feigenbaum 1979, 1980). According to this scenario, at successive critical values of a system parameter, a sequence of bifurcations from one periodic motion to another takes place such that the period of oscillations doubles at each step of the process. As the phenomenon of period-doubling continues, the range of parameter values between the steps becomes successively smaller. Eventually, a limit is reached where the behaviour of the system is no longer periodic, and chaos sets in. A remarkable aspect of all this is that the rate at which a system eventually becomes chaotic, as the parameter increases, is the same. In other words, if η_n denotes the value of the parameter at which the period doubles for the nth time, the rate

$$\delta_n = \frac{\eta_{n+1} - \eta_n}{\eta_{n+2} - \eta_{n+1}}$$

tends to a constant value δ quite rapidly. What is more amazing is that δ has the universal value

$$\delta = 4.6692016\ldots$$

which is valid for all systems exhibiting the phenomenon of successive period-doubling. This value of δ is one of the Feigenbaum numbers.

In view of the fact that one is seldom able to carry out the analysis beyond the first or second bifurcation, as may readily be inferred from the preceding chapters, Feigenbaum's theory has remarkable implications with regard to applications. There seem to be questions, however, about whether the complete Feigenbaum sequence actually takes place or not, since it is extremely difficult to observe experimentally the nature of higher order bifurcations due to the presence of noise (Haken 1983).

Experimental difficulties represent a serious obstacle in the way of rapid progress in this exciting field; nevertheless, some helpful inferences are drawn from certain carefully performed experiments concerning turbulence. For example, experiments carried out by Swinney and Gollub (1978) and others indicate that the *frequency spectrum* of a turbulent fluid—that is, the relationship between the frequency and the square of the amplitude—becomes *continuous* very rapidly as the parameter of the system (in this case the Reynolds number) is increased. Ruelle argues that this experimental evidence favours the strange attractor scenario for the onset of turbulence over the Landau–Hopf route involving quasiperiodic attractors. It is observed that the frequency spectrum associated with a quasiperiodic function exhibits discrete peaks at the basic frequencies while the spectrum is continuous in the case of strange attractors. An additional argument in this regard is concerned with the sensitivity to initial conditions. Although the n-dimensional torus of the Landau–Hopf scenario has a non-periodic and irregular feature, it is not sensitive to initial conditions. On the other hand, the tori with $n \geq 3$ are structurally unstable, and a small perturbation of the governing equations transforms it to a strange attractor (Ruelle and Takens 1971).

It should also be noted that many systems cannot be associated with strange attractors, while sensitivity to initial conditions in such systems is not uncommon. Hamiltonian systems, for example, cannot exhibit strange attractors. In fact, in the absence of asymptotic stability there can be no attractors of any kind. However, the discovery of 'chaos' goes back to Poincaré, and the grandfather of the dynamical systems theory was working with a Hamiltonian system when he came across certain 'bizarre' behaviour patterns. His discovery is associated with *homoclinic bifurcations* where a periodic orbit grows and transforms into a *homoclinic orbit* which passes through a *saddle* point.

It seems that Poincaré did spend some time trying to understand the seemingly chaotic behaviour characteristics that his results were indicating, before he eventually gave up. Despite the emergence of a number of significant advances in the past two decades, the subject remains stunningly complex and far from being well understood. Progress is hindered by experimental difficulties and lack of a satisfactory mathematical theory. Computers and various simulation techniques may be helpful in directing the forces of intuition but real advancement can only be achieved through the development of a satisfactory quantitative theory. Expectations with regard to applications are high. It is hoped that developments in chaos will ultimately lead to predictions concerning turbulence in water and the atmosphere, earthquakes, erratic oscillations in ecological systems and stock markets, ventricular fibrillations of the heart and many other disorderly phenomena. Consider, for instance, the familiar heart beat. A number of mathematical models have been developed over the years to explain the periodic behaviour of a normal heart, and Zeeman's model (1972), for example, is based on the catastrophe and stability theories described in this book (Huseyin 1977). Recently, efforts have been made to model the pathological behaviour of ventricular fibrillation of the heart in the light of chaos. This chaotic behaviour of the heart is deadly, and it is hoped that the current investigations, guided by new discoveries concerning chaos and computer simulations, will ultimately lead to some form of an early warning system.

The study of 'chaos' is still at an early stage, and it remains to be seen where the research will eventually lead. One thing seems to be certain: the subject is bound to receive much future attention from researchers in many fields.

APPENDIX

The derivatives of the amplitudes p_{im}, r_{im}, and other key coefficients referred to in Section 4.5 and subsequent sections with regard to the X-system are listed here in a compact format.

These formulas represent an alternative formulation of the derivatives and coefficients given by eqns (4.93), (4.97), and all the third-order derivatives appearing in eqn (4.107):

$$\left.\begin{aligned} p_{1m}^{(k)} &= -(mN_{12km} + N_{21km})/(m^2-1)\omega_c, \\ p_{2m}^{(k)} &= (N_{11km} - mN_{22km})/(m^2-1)\omega_c, \end{aligned}\right\} \quad (m \neq 1)$$

$$\begin{aligned} p_{qm}^{(k)} &= (AM_{2km} - A_m M_{1km})/(A^2 + A_m^2) \\ p_{(q+1)m}^{(k)} &= -(A_m M_{2km} + AM_{1km})/(A^2 + A_m^2), \\ p_{sm}^{(k)} &= -(\alpha_s N_{s1km} + m\omega_c N_{s2km})/(\alpha_s^2 + m^2\omega_c^2) \end{aligned} \qquad (A_1)$$

$$\left.\begin{aligned} r_{1m}^{(k)} &= (mN_{11km} - N_{22km})/(m^2-1)\omega_c, \\ r_{2m}^{(k)} &= (N_{12km} + mN_{21km})/(m^2-1)\omega_c, \end{aligned}\right\} \quad (m \neq 1)$$

$$\begin{aligned} r_{qm}^{(k)} &= -(AM_{4km} - A_m M_{3km})/(A^2 + A_m^2), \\ r_{(q+1)m}^{(k)} &= (A_m M_{4km} + AM_{3km})/(A^2 + A_m^2), \\ r_{sm}^{(k)} &= -(\alpha_s N_{s2km} - m\omega_c N_{s1km})/(\alpha_s^2 + m^2\omega_c^2), \end{aligned}$$

where (k) can take values from 1 to 4, $q = 3, 5, \ldots, Q$; $s = Q+2, Q+3, \ldots, n$; and

$$\begin{aligned} A &= 2\alpha_q \omega_q, \\ A_m &= \alpha_q^2 - \omega_q^2 + m^2\omega_c^2, \end{aligned}$$

$$\left.\begin{aligned} M_{1km} &= m\omega_c N_{q2km} + \alpha_q N_{q1km} + \omega_q N_{(q+1)1km}, \\ M_{2km} &= m\omega_c N_{(q+1)2km} - \omega_q N_{q1km} + \alpha_q N_{(q+1)1km}, \\ M_{3km} &= m\omega_c N_{q1km} - \alpha_q N_{q2km} - \omega_q N_{(q+1)2km}, \\ M_{4km} &= m\omega_c N_{(q+1)1km} + \omega_q N_{q2km} - \alpha_q N_{(q+1)2km}, \end{aligned}\right\} \qquad (A_2)$$

$$\begin{aligned} N_{i120} &= (X_{i11} + X_{i22})/2, \\ N_{i122} &= (X_{i11} - X_{i22})/2, \qquad N_{i222} = -X_{i12}, \end{aligned}$$

$$\begin{cases} \bar{\gamma}_{i1} \triangleq N_{i131} = \frac{3}{4}(X_{i111} + X_{i122}) + \frac{3}{2}\{X_{ij1}(2p_{j0}^{(2)} + p_{j2}^{(2)}) - X_{ij2}r_{j2}^{(2)}\}, \\ \bar{\gamma}_{i2} \triangleq N_{i231} = -\frac{3}{4}(X_{i112} + X_{i222}) + \frac{3}{2}\{X_{ij1}r_{j2}^{(2)} - X_{ij2}(2p_{j0}^{(2)} - p_{j2}^{(2)})\}, \end{cases}$$

$$N_{i133} = \frac{1}{4}(X_{i111} - 3X_{i122}) + \frac{3}{2}(X_{ij1}p_{j2}^{(2)} + X_{ij2}r_{j2}^{(2)}),$$

$$N_{i233} = -\frac{1}{4}(3X_{i112} - X_{i222}) + \frac{3}{2}(X_{ij1}r_{j2}^{(2)} - X_{ij2}p_{j2}^{(2)}),$$

$$\begin{aligned} N_{i140} = \tfrac{3}{8}(X_{i1111} + 2X_{i1122} + X_{i2222}) &+ \tfrac{3}{2}\{X_{ij11}(2p_{j0}^{(2)} + p_{j2}^{(2)}) \\ &- 2X_{ij12}r_{j2}^{(2)} + X_{ij22}(2p_{j0}^{(2)} - p_{j2}^{(2)})\} + 2(X_{ij1}p_{j1}^{(3)} - X_{ij2}r_{j1}^{(3)}) \\ &+ \tfrac{3}{2}X_{ijk}(2p_{j0}^{(2)}p_{k0}^{(2)} + p_{j2}^{(2)}p_{k2}^{(2)} + r_{j2}^{(2)}r_{k2}^{(2)}) + \ddot{\eta}\tilde{N}_{i140}, \end{aligned}$$

where

$$\tilde{N}_{i140} = 3(2X'_{ij}p_{j0}^{(2)} + X'_{i11} + X'_{i22}).$$

$$\begin{aligned} N_{i142} = \tfrac{1}{2}(X_{i1111} - X_{i2222}) &+ 3\{X_{ij11}(p_{j0}^{(2)} + p_{j2}^{(2)}) - X_{ij22}(p_{j0}^{(2)} - p_{j2}^{(2)})\} \\ &+ 2\{X_{ij1}(p_{j1}^{(2)} + p_{j3}^{(3)}) + X_{ij2}(r_{j1}^{(3)} - r_{j3}^{(3)})\} \\ &+ 2(N_{1231} + N_{2131})r_{i2}^{(2)} + \ddot{\eta}\tilde{N}_{i142}, \end{aligned}$$

where

$$\tilde{N}_{i142} = 3(2X'_{ij}p_{j0}^{(2)} + X'_{i11} - X'_{i22} - 4\omega').$$

$$\begin{aligned} N_{i242} = -(X_{i1112} + X_{i1222}) &+ 3\{(X_{ij11} + X_{ij22})r_{j2}^{(2)} - 2X_{ij12}p_{j0}^{(2)}\} \\ &+ 2\{X_{ij1}(r_{j1}^{(3)} + r_{j3}^{(3)}) - X_{ij2}(p_{j2}^{(3)} - p_{j3}^{(3)})\} \\ &- 2(N_{1231} + N_{2131})p_{j2}^{(2)} + \ddot{\eta}\tilde{N}_{i242}, \end{aligned}$$

where

$$\tilde{N}_{i242} = 6(X'_{ij}r_{j2}^{(2)} - X'_{i12} + 2\omega').$$

The constants $\bar{\delta}_{11}$ and $\bar{\delta}_{22}$ are given by

$$\begin{aligned} \bar{\delta}_{11} = \tfrac{5}{8}(X_{111111} &+ 2X_{111122} + X_{112222}) \\ &+ \tfrac{5}{2}\{X_{1j111}(3p_{j0}^{(2)} + 2p_{j2}^{(2)}) - 3X_{1j112}r_{j2}^{(2)} + 3X_{1j122}p_{j0}^{(2)} - X_{1j222}r_{j2}^{(2)}\} \\ &+ \tfrac{5}{2}\{X_{1j11}(3p_{j1}^{(3)} + p_{j3}^{(3)}) - 2X_{1j12}(r_{j1}^{(3)} + r_{j3}^{(3)}) + X_{1j22}(p_{j1}^{(3)} - p_{j3}^{(3)})\} \\ &+ \tfrac{15}{2}\{X_{1jk1}(2p_{j0}^{(2)}p_{k0}^{(2)} + p_{j2}^{(2)}p_{k2}^{(2)} + r_{j2}^{(2)}r_{k2}^{(2)} + 2p_{j0}^{(2)}p_{k2}^{(2)} - 2X_{1jk2}p_{j0}^{(2)}p_{k2}^{(2)}\} \\ &+ \tfrac{5}{2}\{X_{1j1}(2p_{j0}^{(4)} + p_{j2}^{(4)}) - X_{ij2}r_{j2}^{(4)}\} \\ &+ 5X_{1jk}(2p_{j0}^{(2)}p_{k1}^{(2)} + p_{j2}^{(2)}p_{k1}^{(3)} + p_{j2}^{(2)}p_{k3}^{(3)} + r_{j2}^{(2)}r_{k1}^{(3)} + r_{j2}^{(2)}r_{k3}^{(3)}), \end{aligned}$$

$$\bar{\delta}_{22} = -\tfrac{5}{8}(X_{211112} + 2X_{211222} + X_{222222})$$
$$\tfrac{5}{2}\{X_{2j111}r^{(2)}_{j2} - 3X_{2j112}p^{(2)}_{j2} + 3X_{2j122}r^{(2)}_{j2} - X_{2j222}(3p^{(2)}_{j0} - 2p^{(2)}_{j2})\}$$
$$+\tfrac{5}{2}\{X_{2j11}(3r^{(3)}_{j1} + r^{(3)}_{j3}) - 2X_{2j12}(p^{(3)}_{j1} - p^{(3)}_{j3}) + X_{2j22}(3r^{(3)}_{j1} - r^{(3)}_{j3})\}$$
$$-\tfrac{15}{2}\{X_{2jk2}(2p^{(2)}_{j0}p^{(2)}_{k0} + p^{(2)}_{j2}p^{(2)}_{k2} + r^{(2)}_{j2}r^{(2)}_{k2} - 2p^{(2)}_{j0}p^{(2)}_{k2} - 2X_{2jk1}p^{(2)}_{j0}r^{(2)}_{k2}\}$$
$$+\tfrac{5}{2}\{X_{2j1}r^{(4)}_{j2} - X_{2j2}(2p^{(4)}_{j0} - p^{(4)}_{j2})\}$$
$$+5X_{2jk}(2p^{(2)}_{j0}r^{(2)}_{k1} - p^{(2)}_{j2}r^{(3)}_{k1} + p^{(2)}_{j2}r^{(3)}_{k3} + r^{(2)}_{j2}p^{(3)}_{k1} - r^{(2)}_{k1}p^{(3)}_{k3})$$
$$-\tfrac{5}{3}(\bar{\gamma}_{12} + \bar{\gamma}_{21})p^{(3)}_{21}.$$

The constants ζ_{11} and ζ_{22} are given by

$$\zeta_{11} = \frac{15}{2\omega_c}\{(3X_{1112} + X_{1222})X'_{11} + X_{1122}(X'_{12} + X'_{21})\}$$
$$+\tfrac{5}{2}\{X_{1j1}(2\tilde{p}^{(4)}_{j0} + \tilde{p}^{(4)}_{j2}) - X_{1j2}\tilde{r}^{(4)}_{j2}\}$$
$$-\frac{15}{\omega_c}X_{1j2}\{2(2p^{(2)}_{j0} + p^{(2)}_{j2})X'_{11} + r^{(2)}_{j2}(X'_{12} + X'_{21})\}$$
$$+\tfrac{15}{2}(X'_{1111} + X'_{1122}) + 15\{X'_{1j1}(2p^{(2)}_{j0} + p^{(2)}_{j2}) - X_{1j20}r^{(2)}_{j2}\}$$
$$-\frac{10}{\omega_c}X'_{12}\bar{\gamma}_{11},$$

$$\zeta_{22} = \frac{15}{4\omega_c}\{(X_{2112} + 3X_{2222})(X'_{12} + X'_{21}) + 4X_{2122}X'_{11}\}$$
$$+\tfrac{5}{2}\{X_{2j1}\tilde{r}^{(4)}_{j2} - X_{2j2}(2\tilde{p}^{(4)}_{j0} - \tilde{p}^{(4)}_{j2})\}$$
$$+\frac{15}{\omega_c}X_{2j2}\{(2p^{(2)}_{j0} - p^{(2)}_{j2})(X'_{12} + X'_{21}) - 2r^{(2)}_{j2}X'_{11}\}$$
$$-\tfrac{15}{2}(X'_{2112} + X'_{2222}) + 15\{X'_{2j1}r^{(2)}_{j2} - X'_{2j2}(2p^{(2)}_{j0} - p^{(2)}_{j2})\}$$
$$+\frac{5}{\omega_c}X'_{11}(\bar{\gamma}_{12} - \bar{\gamma}_{21})$$

where $\tilde{p}^{(4)}_{jm}$ and $\tilde{r}^{(4)}_{jm}$ are obtained by replacing N_{ijkm} in eqns (A_1) and (A_2) by $\tilde{N}_{ijkm}$.

REFERENCES

Abraham, R. and Marsden, J. E. (1978). *Foundations of mechanics,* 2nd ed. Benjamin, Reading.

Abraham, R. and Shaw, C. D. (1982). *Dynamics: the geometry of behaviour,* vol. 1. Aerial Press, Santa Cruz, Ca.

Alexander, J. C. and Yorke, J. A. (1978). Global bifurcations of periodic orbits. *Amer. J. Math.* **100,** 263–92.

Allwright, D. J. (1977). Harmonic balance and the Hopf bifurcation. *Math. Proc. Camb. Phil. Soc.* **82,** 453–67.

Amson, J. C. (1974). Equilibrium and catastrophe models of urban growth. *Space-time concepts in urban and regional models.* (ed. E. L. Crisp) 108–28, Pion, London.

Andronov, A. A. and Pontryagin, L. S. (1937). Coarse Systems. *Dokl. Akad. Nauk., SSSR,* **14,** 247–51.

—— and Witt, A. (1930). Sur la théorie mathematiques des autooscillations. *C.R. Acad. Sci.* **190,** 256–8, Paris.

Arnold, V. I. (1974). Critical points of smooth functions. *Proc. Intern. Cong. Math.,* 19–39, Vancouver.

—— (1983). Geometrical methods in the theory of ordinary differential equations. *Comp. Studies in Math.,* Springer-Verlag, New York.

Atadan, A. S. (1983). On generalized Hopf bifurcations. *Ph.D. thesis,* Dept of Systems Design Eng., Univ. of Waterloo, Ont., Canada.

—— and Huseyin, K. (1982*a*). A note on a uniformly valid asymptotic solution for $(d^2y/dt^2) + y = a + \varepsilon y^2$. *J. Sound Vib.* **85,** 129–31.

—— —— (1982*b*). A perturbation method for the analysis of limit cycles associated with Hopf bifurcation. *Hadronic J.* **5,** 2125–45.

—— —— (1983*a*). Cusp and tangential bifurcations associated with the limit cycles of autonomous systems. *Proc. American Cont. Conf.,* 319–23, San Francisco, Ca.

—— —— (1983*b*). Stability of limit cycles associated with a symmetric bifurcation phenomenon. *Proc. IEEE Intl. Symp. Cir. Syst.,* 681–4, Newport Beach, Ca.

—— —— (1983*c*). Post-critical oscillatory behaviour of two-parameter autonomous systems. *Proc of 26th Midwest Symp. Cir. Syst.,* 292–4, Pueblo, Mexico.

—— —— (1984*a*). An intrinsic method of harmonic analysis for non-linear oscillations. *J. Sound Vib.* **95,** 525–30.

—— —— (1984*b*). Symmetric and flat bifurcations: an oscillatory phenomenon. *Acta Mech.* **53,** 213–32.

—— —— (1984*c*). Bifurcation analysis of aircraft pitching motions and the brusselator via an intrinsic perturbation procedure. *ICTAM,* August 19–24, Abstracts 607, Lyngby, Denmark.

—— —— (1985*a*). On the oscillatory instability of multiple-parameter systems. *Int. J. Engrg. Sci.* **23,** 857–73.

—— —— (1985*b*). Analysis of dynamical bifurcations in the case of a non-simple eigenvalue. *Intl. Symp. Cir. Syst.*, Tokyo, Japan, 871–2.

Auchmuty, J. F. G. and Nicolis, G. (1975). Bifurcation analysis of non-linear reaction-diffusion equations—I. Evaluation equations and the steady state solutions. *Bull. Math. Bio.* **37,** 323–65.

Bajaj, A. K. and Sethna, P. R. (1982). Bifurcations in three-dimensional motions of articulated tubes, I and II. *J. Appl. Mech.* **49,** 606–11 and 612–18.

—— ——, and Lundgren, T. S. (1980). Hopf bifurcation phenomena in tubes carrying a fluid. *SIAM J. Appl. Math.* **39,** 213–230.

Bolotin, V. V. (1963). *Non-conservative problems of the theory of elastic stability.* Pergamon Press, Oxford.

Britvec, S. J. and Chilver, A. H. (1963). Elastic buckling of rigidly-jointed braced frames. *J. Engng. Mech. Div. ASCE* **89,** 217–55.

Budiansky, B. and Hutchinson, J. W. (1966). A survey of some buckling problems. *AIAA J.* **4,** 1505–10.

Burgess, I. W. and Levinson, M. (1972). The post-flutter oscillations of discrete symmetric structural systems with circulatory loads. *Int. J. Mech. Sci.* **14,** 471–88.

Chafee, N. (1968). The bifurcation of one or more closed orbits from an equilibrium point of an autonomous differential system. *J. Diff. Eqs.* **4,** 661–79.

Chillingworth, D. (1975). The catastrophe of a buckling beam. Dynamical systems, Warwick (ed. A. Manning). *Lecture Notes in Math* **468,** 88–91. Springer-Verlag, Berlin.

Chilver, A. H. (1967). Coupled modes of elastic buckling. *J. Mech. Phys. Solids* **15,** 15–28.

Chow, S. N. and Hale, J. (1982). *Methods of bifurcation theory.* Springer-Verlag, New York.

——, Paret, J. M., and Yorke, J. A. (1978). Global Hopf bifurcation from a multiple eigenvalue. *Non-linear Anal. Theory Meth. Appl.* **2,** 753–63.

Chua, L. O. (1969). *Introduction to non-linear networks.* McGraw-Hill, New York.

Coddington, E. L. and Levinson, N. (1955). *Theory of ordinary differential equations.* McGraw-Hill, New York.

Cohen, D. S., Coutsias, E., and Neu, J. C. (1979). Stable oscillations in single species growth model with hereditary effects. *Math. Biosci.* **44,** 255–68.

Croll, J. G. A. and Walker, A. C. (1972). *Elements of structural stability.* Macmillan, London.

Desoer, C. A. and Kuh, E. S. (1969). *Basic circuit theory.* McGraw-Hill, New York.

Euler, L. (1744). *Methodus inveniendi lineas curvas maximi minimive propriatate gaundentes* (appendix, De curvis elasticis), Marcum Michaelem Bousquet, Lausanne and Geneva.

Feigenbaum, M. J. (1979). The onset spectrum of turbulence. *Phys. Lett.* **74A,** 375–8.

—— (1980). Universal behaviour in non-linear systems. *Los Alamos Sci.* **1,** 4–27.

Flockerzi, D. (1981). Bifurcation formulas for O.D.E.'s. *Non-Lin. Anal.* **5,** 249–63.

Fowler, D. H. (1972). The Riemann–Hugoniot catastrophe and Van der Waals' equation. *Towards a theoretical biology* (ed. C. H. Waddington) vol. 4, 1–7. Edinburgh University Press, Edinburgh.

Gantmacher, F. R. (1959). *The theory of matrices,* vols. 1 and 2. Chelsea, New York.

Gaspar, Z., Huseyin, K., and Mandadi, V. (1978). Discussion related to the paper by K. Huseyin and V. Mandadi, 'On the imperfection sensitivity of compound branching. *Ing. Arch.* **47,** 315–18.

Gilmore, R. (1981). *Catastrophe theory for scientists and engineers.* Wiley, New York.

Glansdorff, P. and Prigogine, I. (1971). *Thermodynamic theory of structures, stability and fluctuations.* Wiley, New York.

Golubitsky, M. and Langford, W. F. (1981). Classification and unfoldings of degenerate Hopf bifurcations. *J. Diff. Eqns.* **41,** 375–415.

—— and Shaeffer, D. (1979). An analysis of imperfect bifurcation. *Ann. N.Y. Acad. Sci.* **316,** 127–33.

Guckenheimer, J. (1979). A brief introduction to dynamical systems. *Lectures in Appl. Math.*, vol. 17 (ed. F. C. Hopensteadt), Amer. Math. Soc., Providence, Rhode Island.

—— (1984). Multiple bifurcation problems of co-dimension two. *SIAM J. Math. Anal.* **15,** 1–49.

—— and Holmes, P. (1983). *Bifurcations and non-linear oscillations.* Springer-Verlag, New York.

Gurel, O. and Rössler, O. E. (eds.) (1979). *Bifurcation theory and applications in scientific disciplines.* New York Academy of Science, New York.

Hagedorn, P. (1981). *Non-linear oscillations.* (translated by W. Stadler). Oxford Univ. Press.

Haken, H. (1978). *Synergetics.* Springer-Verlag, Berlin, Heidelberg, New York.

—— (1983). *Advanced synergetics.* Springer-Verlag, Berlin, Heidelberg, New York.

Hale, J. K. (1969). *Ordinary differential equations.* Wiley, New York.

Hansen, J. S. (1977). Some two-mode buckling problems and their relation to catastrophe theory. *AIAA J.* **15,** 1638–44.

Hartman, P. (1973). *Ordinary differential equations.* (Revised edition), Philip Hartman, Baltimore.

Hassard, B. D., Kazaninoff, N. D., and Wan, Y. H. (1981). Theory and applications of Hopf bifurcations. *London Math. Soc. Lec. Notes Series 41.* Cambridge Univ. Press.

Hayashi, C. (1964). *Non-linear oscillations in physical systems.* McGraw-Hill, New York.

Herrmann, G. (1967). Stability of equilibrium of elastic systems subjected to non-conservative forces. *Appl. Mech. Rev.* **20,** 103.

Herschkowitz-Kaufman, M. (1975). Bifurcation analysis of non-linear reaction diffusion equations—II. Steady state solutions and comparisons with numerical solutions. *Bull. Math. Bio.* **37,** 589–636.

Hirsch, M. W. (1984). The dynamical systems approach to differential equations. *Bull. Amer. Math. Soc.* **11,** 1–41.

—— and Smale, S. (1974). *Differential equations, dynamical systems and linear algebra.* Academic Press, New York.

Ho, D. (1972). The influence of imperfections on systems with coincident buckling loads. *Int. J. Non-linear Mech.* **7,** 311–21.

—— (1974). Buckling load of non-linear systems with multiple eigenvalues. *Int. J. Solids Struct.* **10,** 1315–30.

Holmes, P. J. (1977). Bifurcations to divergence and flutter in flow induced oscillations: a finite dimensional analysis. *J. Sound Vib.* **53,** 471–503.

—— and Marsden, J. E. (1978). Bifurcations to divergence and flutter in flow induced oscillations: an infinite dimensional analysis. *Automatica* **14,** 367–84.

Hopf, E. (1942). Abzweigung einer periodischen Lösung von einer stationären Lösung eines Differentialsystems. *Ber. Math-Phys. Sachsische Akademie der Wissenschaften Leipzig* **94,** 1–22.

—— (1948). A mathematical example displaying features of turbulence. *Comm. Pure Appl. Math.* **1,** 303–22.

Hoppensteadt, F. C. (ed., 1979). Non-linear oscillations in biology. *Lectures in Appl. Math.* vol. 17. AMS, Providence, Rhode Island.

Howard, L. N. (1979). Non-linear oscillations in biology. *Lectures in Appl. Math.* vol. 17 (ed. F. C. Hoppensteadt). AMS, Providence, Rhode Island.

Hui, W. H. and Tobak, M. (1982). Bifurcation analysis of non-linear stability of aircraft at high angles of attack. AIAA-82-0244, 20th Aerospace Sci. meeting, 1–10.

Hunt, G. W. (1977). Imperfection-sensitivity of semi-symmetric branching. *Proc. Roy. Soc. Lond. Ser. A* **357,** 193–211.

—— (1981). An algorithm for the non-linear analysis of compound bifurcations. *Phil. Trans. Roy. Soc. Ser. A* **300** (1455), 443–71.

Huseyin, K. (1976). Elastic stability of structures under combined loading. *Ph.D. Thesis,* University of London.

—— (1969). The convexity of the stability boundary of symmetric structural systems. *Acta Mech.* **8,** 205–11.

—— (1970*a*). Fundamental principles in the buckling of structures under combined loading. *Int. J. Solids Struct.* **6,** 479–87.

—— (1970*b*). The elastic stability of structural systems with independent loading parameters. *Int. J. Solids Struct.* **6,** 677–91.

—— (1970*c*). The stability boundary of systems with one-degree-of-freedom, Part I. *Meccanica* **5,** 308–11.

—— (1970*d*). The stability boundary of systems with one-degree-of-freedom, Part II. *Meccanica* **5,** 312–16.

—— (1971*a*). The post-buckling behaviour of structures under combined loading, *ZAMM* **51,** 177–82.

—— (1971*b*). Instability of symmetric structural systems with independent loading parameters. *Quart. Appl. Math.* **28,** 571–86.

—— (1972*a*). On the stability of critical equilibrium states and post-buckling, part I. Study No. 6 *Stability* (ed. H. H. E. Leipholz). 15–40, SMD, Univ. of Waterloo, Canada.

—— (1972*b*). Singular critical points in the general theory of elastic stability. *Meccanica* **7,** 58–68.

—— (1972*c*). On the estimation of the stability boundary of symmetric structural systems. *Int. J. Non-linear Mech.* **7,** 31–50.

—— (1973). The multiple-paramter perturbation technique for the analysis of non-linear systems. Int. J. Non-linear Mech. **8,** 431–43.

—— (1975). *Non-linear theory of elastic stability.* Noordhoff, Leyden, The Netherlands.

—— (1976). Vibrations and stability of mechanical systems. *Shock Vib. Dig.* **8,** 56–66.

—— (1977). The multiple-parameter stability theory and its relation to catastrophe theory. *Proc. of the Interdisciplinary Conf. on "Prob. Anal. in Sci. and Eng."* (eds F. Branin and K. Huseyin), 229–55. Academic Press, New York.

—— (1978*a*). Stability of gradient systems. *Stability of elastic structures* (ed. H. H. E. Leipholz). CISM courses No. 238, 100–80. Springer-Verlag, Vien, New York.

—— (1978*b*). *Vibrations and stability of multiple-parameter systems.* Sijthoff and Noordhoff, Alphen aan den Rijn, The Netherlands.

—— (1981*a*). On the stability of equilibrium paths associated with autonomous systems. *J. Appl. Mech.* **48,** 183–7.

—— (1981*b*). Vibrations and stability of mechanical systems II. *Shock Vib. Dig.* **13,** 21–9.

—— (1982). Bifurcations, instabilities, and catastrophes associated with non-potential systems. *Hadronic J.* **5,** 931–74.

—— (1984*a*). On degenerate bifurcations concerning non-conservative systems. *ZAMM* **64,** T53–4.

—— (1984*b*). Vibrations and stability of mechanical systems III. *Shock Vib. Dig.* **16,** 15–22.

—— and Atadan, A. S. (1981). On the Hopf bifurcation. *Proc. 8th CANCAM* (ed. N. K. Srivastava), 115–16. Univ. of Moncton.

—— —— (1983). On the analysis of Hopf bifurcations. *Int. J. Engng. Sci.* **21,** 247–62.

—— —— (1984*a*). On a double Hopf bifurcation. *Proc. IEEE Intl. Symp. Cir. Syst.,* 1348–52. Montreal.

—— —— (1984*b*). On generalized Hopf bifurcations. ASME Trans., *J. Dynamic Systems, Meas. and Control* **106,** 327–34.

—— and Mandadi, V. (1977*a*). Classification of critical conditions in the general theory of stability. *Mech. Res. Comm.* **4,** 11–15.

—— —— (1977*b*). On the imperfection sensitivity of compound branching. *Ing. Arch.* **46,** 213–22.

—— —— (1980). On the instability of multiple-parameter systems. Sectional Lecture, *Proc. IUTAM Congress,* Toronto (eds F. P. J. Rimrott and B. Tabarrok) 281–94. North-Holland Pub. Co., Amsterdam.

—— and Plaut, R. H. (1973). The elastic stability of two-parameter non-conservative systems. *J. Appl. Mech.* **40,** 175–80.

Hutchinson, J. W. and Budiansky, B. (1966). Dynamic buckling estimates. *AIAA J.* **4,** 525–30.

Iooss, G. and Joseph, D. (1980). *Elementary stability and bifurcation theory.* Springer-Verlag, New York.

Johns, K. C. (1974). Imperfection sensitivity of coincident buckling systems. *Int. J. Non-linear Mech.* **9,** 1–21.

Jost, R. and Zehnder, E. (1972). A generalization of the Hopf bifurcation theorem. *Helv. Phys. Acta* **45,** 258–76.

Keener, J. P. (1976). Secondary bifurcation in non-linear diffusion-reaction equations. *Studies in Appl. Math.* **55,** 187–211.

Keller, J. B. and Antman, S. (1969). *Bifurcation theory and non-linear eigenvalue problems.* Benjamin, New York.

Kielhöfer, H. (1979). Hopf bifurcation at multiple eigenvalues. *Arch. Rat. Mech. Anal.* **69,** 53–83.

—— (1980). Degenerate bifurcation at simple eigenvalues and stability of bifurcating solutions. *J. Func. Anal.* **38,** 416–41.

Knops, R. J. and Wilkes, E. (1973). Theory of elastic stability, *Handbuch der Physik,* VIa/3 (ed. C. Truesdell). Springer-Verlag, Berlin.

Koiter, W. T. (1945). On the stability of elastic equilibrium. *Dissertation.* University of Delft.

—— (1963). Elastic stability and post-buckling behaviour. *Proc. Symp. Non-linear Problems* (ed. R. E. Langer). Univ. Wisconsin Press, Madison.

—— (1965). The energy criterion of stability for continuous elastic bodies. *Proc. R. Neth. Acad. Sci. Ser* [*B*] **68,** 178–202.

Kolkka, R. W. (1984). Singular perturbations of bifurcations with multiple independent bifurcation parameters. *SIAM J. Appl. Math.* **44,** 257–69.

Kopell, N. and Howard, L. N. (1974). Pattern formation in the Belousov reaction. *Lectures on Mathematics in the Life Series* vol. 7, 201–16, Amer. Math. Soc., Providence, Rhode Island.

Kuo, C. C., Morion, L. and Dugundji, J. (1972). Perturbation and harmonic balance methods for non-linear panel flutter. *AIAA J.* **10,** 1479–84.

Lagrange, J. L. (1788). *Mécanique analytique.* Courcier, Paris.

Landau, L. D. and Lifschitz, E. M. (1959). *Course of theoretical physics,* vol. 6, Fluid Mechanics. Pergamon, London, New York.

Langford, W. F. (1979). Periodic and steady-state mode interactions lead to tori. *SIAM J. Appl. Math.* **37,** 22–48.

La Salle, J. and Lefschetz, S. (1961). *Stability by Lyapunov's direct method.* Academic Press, New York.

Leipholz, H. H. E. (1970). *Stability theory.* Academic Press, New York.

—— (1977). *Direct variational methods and eigenvalue problems in engineering.* Noordhoff, Leyden, The Netherlands.

—— (1980). *Stability of elastic systems.* Sijthoff and Noordhoff, The Netherlands.

Lorenz, E. N. (1963). Deterministic non-periodic flow. *J. Atmos. Sci.* **20,** 130–41.

Lyapunov, A. (1892). *Problème général de la stabilité du mouvement.* Kharkov. Reproduced in *Ann. Math. Studies* vol. 17, Princeton University Press, Princeton (1949).

Mahaffey, R. A. (1976). A harmonic oscillator description of plasma oscillations. *Phys. Fluids* **19,** 1327–91.

Mandadi, V. and Huseyin, K. (1977). A classification of critical conditions in the stability theory of non-gradient systems. *Mech. Res. Comm.* **4,** 179–83.

—— —— (1978). The effect of symmetry on the imperfection-sensitivity of coincident critical points. *Ing. Arch.* **47,** 35–45.

—— —— (1979). Non-linear instability behaviour of non-gradient systems. *Hadronic J.* **2,** 657–81.

—— —— (1980). Non-linear bifurcation analysis of non-gradient systems. *Int. J. Non-linear Mechanics* **15,** 159–72.

Marsden, J. E. and Hughes, J. R. (1983). *Mathematical foundations of elasticity.* Prentice-Hall, New Jersey.

—— and McCracken, M. (eds., 1976). *The Hopf bifurcation and its applications.* Springer-Verlag, New York.

Mees, A. I. (1981). *Dynamics of feedback systems.* Wiley, Chichester.

—— and Chua, L. O. (1979). The Hopf bifurcation theorem and its applications to non-linear oscillations in circuits and systems. *IEEE,* **CAS-26,** 235–54.

Mehra, R. K. and Carroll, J. V. (1980). Bifurcation analysis of aircraft high angle-of-attack flight dynamics. *AIAA paper No. 80-1599.*

Merrill, S. J. (1978). A model of the stimulation of B-cells by replicating antigen—I and II. *Math. Biosci.* **41,** 125–41 and 143–55.

Mickens, R. E. (1981). A uniformly valid asymptotic solution for $\mathrm{d}^2y/\mathrm{d}t^2 + y = a + \varepsilon y^2$. *J. Sound Vib.* **76,** 150–152.

Minorsky, N. (1962). *Non-linear oscillations.* Princeton Univ. Press, Princeton, New Jersey.

Morino, L. (1969). A perturbation method for treating non-linear panel flutter problems. *AIAA J.* **7,** 405–11.

Nayfeh, A. H. and Mook. D. T. (1979). *Non-linear oscillations.* Wiley-Interscience, New York.

Nicolis, G. and Prigogine, I. (1977). *Self-organization in non-equilibrium systems.* Wiley, New York.

Papkovich, P. H. (1934). Works on the structural mechanics of ships. *Proc. 4th Int. Cong. of Appl. Mech.* Cambridge, England.

Plaut, R. H. (1976). Post-buckling analysis of non-conservative elastic systems. *J. Struct. Mech.* **4,** 395–416.

—— (1977). Branching analysis at coincident buckling loads of non-conservative elastic systems. *J. Appl. Mech.* **44,** 317–21.

—— (1978). Post-buckling behaviour of continuous non-conservative elastic systems. *Acta Mech.* **30,** 51–64.

Poincaré, H. (1892). *Les Méthodes Nouvelles de la Mécanique Céleste.* vol. 1. Ganthier-Villars, Paris.

—— (1885). Sur l'equilibre d'une masse fluide animée d'un mouvement de rotation. *Acta Math.* **7,** 259–380.

Poore, A. B. (1976). On the theory and application of the Hopf–Friedrichs bifurcation theory. *Arch. Rat. Mech. Anal.* **60,** 371–93.

Poston, T. and Stewart, I. N. (1976). Taylor expansions and catastrophes. *Research notes in mathematics* **7**. Pitman, London, San Francisco, Melbourne.

Potier-Ferry, M. (1979). Perturbed bifurcation theory. *J. Diff. Eqns.* **33,** 112–46.

—— (1981*a*). Imperfection sensitivity of a nearly double bifurcation point. *Proc. of stability in the mechanics of continua* (ed. F. H. Schroeder) 201–14. Springer-Verlag, Berlin.

—— (1981*b*). Multiple bifurcation, symmetry and secondary bifurcation. *Research notes in mathematics* **46,** 158–67. Pitman, London, San Francisco, Melbourne.

Prigogine, I. and Lefever, R. (1968). Symmetry breaking instabilities in dissipative sytems—II. *J. Chem. Phys.* **48,** 1695–1700.

Reis, A. J. and Roorda, J. (1979). Post-buckling behaviour under mode interaction. *J. Eng. Mech. Div.* **105,** 609–21.

Renton, J. D. (1967). *Thin walled structures* (ed. A. H. Chilver) 1–59. Chatto and Windus, London.

Roorda, J. (1965). Stability of structures with small imperfections. *J. Eng. Mech. Div. ASCE* **91,** 87–96.

—— (1972). Concepts in elastic structural stability. *Mechanics today* (ed. N. Nasser) vol. 1, 322–72. Pergamon Press, New York.

Rosenblat, S. (1977). Global aspects of bifurcation and stability. *Arch. Rat. Mech. Anal.* **66,** 119–34.

Ruelle, D. (1980). Strange attractors. *The Mathematical Intelligencer* **2,** 126–37.

Ruelle, D. and Takens, F. (1971). On the nature of turbulence. *Comm. Math. Phys.* **20,** 167–92.

Samuels, P. (1979). The relationship between post-buckling behaviour at coincident branching points and the geometry of an umbilic point of the energy surface. *J. Struct. Mech.* **7,** 297–324.

—— (1982). Instability of dual eigenvalue fourth order systems. *J. Struct. Mech.* **10,** 209–25.

Santilli, R. M. (1983). *Foundations of theoretical mechanics II.* Springer-Verlag, New York.

Sattinger, D. H. (1973). Topics in stability and bifurcation theory. *Lecture Notes in Math.* **309**. Springer-Verlag, New York.

Scheidl, R., Troger, H. and Zeman, K. (1984). Coupled flutter and divergence bifurcation of a double pendulum. *Int. J. Non-linear Mech.* **19,** 163–76.

Sethna, P. R. and Schapiro, S. M. (1977). Non-linear behaviour of flutter unstable dynamic systems with gyroscopic and circulatory forces. *J. Appl. Mech.* **44,** 755–62.

Sewell, M. J. (1965). The static perturbation technique in buckling problems. *J. Mech. Phys. Solids* **13,** 247–65.

Sewell, M. (1977). Elementary catastrophe theory. *Problem analysis in science and engineering* (eds F. H. Branin and K. Hyseyin) 391–26. Academic Press, New York.

Smith, L. L. and Morino, L. (1976). Stability analysis of non-linear differential autonomous systems with applications to flutter. *AIAA J.* **14,** 333–341.

Stewart I. (1981). Applications of catastrophe theory to the physical sciences. *Physica D* **2,** 245–305.

Sprig, F. (1983). Sequence of bifurcations in a three-dimensional system near a critical point. *ZAMP* **34,** 259–76.

Supple, W. I. (1967). Coupled branching configurations in the elastic buckling of symmetric structural systems. *Int. J. Mech. Sci.* **9,** 97–112.

—— (ed. 1973). *Structural instability.* IPC Science and Technology Press Ltd., Guildford.

Swinney, H. L. and Gollub, J. P. (1978). The transition to turbulence. *Physics Today* **31,** 41–9.

Takens, F. (1973). Unfoldings of certain singularities of vector fields: generalized Hopf bifurcations. *J. Diff. Eqns.* **14,** 476–93.

—— (1974). Singularities of vector fields. *Publ. Math. IHES* **43,** 47–100.

Thom, R. (1975). *Structural stability and morphogenesis* (translated by D. K. Fowler). Benjamin, Reading.

Thompson, J. M. T. (1963). Basic principles in the general theory of elastic stability. *J. Mech. Phys. Solids* **11,** 13–20.

—— (1982*a*). *Instabilities and catastrophes in science and engineering.* Wiley, New York.

—— (1982*b*). Catastrophe theory in mechanics: progress or digression. *J. Struct. Mech.* **10,** 167–75.

—— and Gaspar, Z. (1977). A buckling model for the set of umbilic catastrophes. *Math. Proc. Cambridge Phil. Soc.* **82,** 497–507.

—— and Hunt, G. W. (1973). *A general theory of elastic stability.* Wiley, New York.

Troger, H. (1976). Bemerkungen zum Begriff 'Struturelle Stabilität'. *ZAMM* **56,** T78–9.

—— (1981). Zur korrekten Modellbildung in der dynamik diskreter System. *Ing. Arch.* **51,** 31–43.

Willems, J. L. (1970). *Stability theory of dynamical systems.* Nelson, London.

Wilson, A. G. (1981). *Catastrophe theory and bifurcation.* Univ. of California Press, Berkeley and Los Angeles.

Yu, P. (1984). Sensitivity of static and dynamic bifurcations. *M.Sc. Thesis,* Dept. of Systems Design Eng. Univ. of Waterloo, Ont., Canada.

—— and Huseyin, K. (1985). On a problem concerning bifurcations into tori (to be published).

Zeeman, E. C. (1972). Differential equations for the heart beat and nerve impulse. *Towards a theoreticl biology* (ed. C. H. Waddington) **4,** 8–67.

—— (1976). The classification of elementary catastrophes of codimension ≤5. *Lecture notes in mathematics* **525,** 263–327. Springer-Verlag, Berlin, Heidelberg, New York.

Ziegler, H. (1968). *Particles of structural stability*. Blaisdell Publ. Co., Waltham, Massachusetts.

INDEX